Introduction to Microwave Remote Sensing

Introduction to Microwave Remote Sensing

Iain H. Woodhouse

The University of Edinburgh
Scotland

Taylor & Francis
Taylor & Francis Group
Boca Raton   London   New York

A CRC title, part of the Taylor & Francis imprint, a member of the
Taylor & Francis Group, the academic division of T&F Informa plc.

Published in 2006 by
CRC Press
Taylor & Francis Group
6000 Broken Sound Parkway NW, Suite 300
Boca Raton, FL 33487-2742

© 2006 by Iain H. Woodhouse
CRC Press is an imprint of Taylor & Francis Group

No claim to original U.S. Government works
Printed in the United States of America on acid-free paper
10 9 8 7 6 5 4 3 2

International Standard Book Number-10: 0-415-27123-1 (Hardcover)
International Standard Book Number-13: 978-0-415-27123-3 (Hardcover)
Library of Congress Card Number 2005050635

Library of Congress Cataloging-in-Publication Data

Woodhouse, Iain H.
    Introduction to microwave remote sensing / Iain H. Woodhouse.
        p. cm.
    Includes bibliographical references and index.
    ISBN 0-415-27123-1
    1. Microwave remote sensing. I. Title.

G70.4.W66 2005
621.36'78--dc22   2005050635

Taylor & Francis Group
is the Academic Division of Informa plc.

Visit the Taylor & Francis Web site at
http://www.taylorandfrancis.com

and the CRC Press Web site at
http://www.crcpress.com

To my father, Hector Woodhouse, for bringing me up to know that science is fun.

foreword

"Self Induction's in the air,
Everywhere, everywhere
Waves are running to and fro
Here they are, there they go
Try to stop 'em if you can
You British Engineering man!"
"Maxwell was only half a Maxwellian".
— Oliver Heaviside, 1895.

New ideas take time to be accepted. Generations of scientists and engineers first need to re-examine basic assumptions, test theories and probe the limits of technology before maturity and clarity is obtained. Microwave Remote Sensing (MRS), the subject of this book, has arguably achieved this status and Iain Woodhouse presents use with a comprehensive and timely introduction to the topic.

The path to MRS leads all the way back to the fierce debate over the existence of electromagnetic waves in the late nineteenth century. In this case, time has demonstrated that the electromagnetic spectrum is a wide and wonderful natural resource and that our experience of it with our human senses is limited to a small band of optical frequencies. Our interpretation and observation of the world around us, both in qualitative and quantitative ways, tends to be governed by our reliance on our biological 'passive sensors', requiring the narrow filter of sunlight to see the world in colour and texture. Technology has changed all that. We can now devise our own sources of radiation with infinite subtlety, to view our environment and select the part of the spectrum best suited to our needs. In this way we can 'see through microwave eyes', or perhaps, as Iain Woodhouse suggests in this book, 'listen through microwave ears'.

This is becoming increasingly important to our world. Issues like climate change, ozone depletion and ocean salinity are widely recognised as big problems for the future, whose solution requires international collaboration and a new global politics on a scale never before experienced by mankind. Microwaves provide an important window on these issues and, by exploiting their properties and developing new sensors and technologies, we can provide the sort of scientific data required by future generations to make the right decisions at the right time.

There is however one key problem, we cannot 'see' these elusive microwaves and the technology to 'stop and collect them' is both complicated and expensive. To add insult to injury, as Iain Woodhouse suggests, 'microwave images' also require a new and different kind of interpretation to that used in our familiar optical spectrum. All of this tends to make MRS traditionally a rather arcane subject, often requiring years of advanced study and hence remaining solidly in the realm of the specialist engineer, physicist or more recently, radar applications scientist. All of this will change in the future and this book is a bold attempt to provide a comprehensive survey of the field at a level accessible to many. It is built on a solid understanding of the physics behind microwave interactions but is presented in a readable and descriptive manner. It provides a unique blend of historical context, basic physical principles and applications science. It covers the full range of applications and details numerous examples of the successful application of microwaves in remote sensing. It ranges in coverage from the earliest attempts at microwave radiometry through to the very latest techniques for producing high-resolution global scale microwave radar images. Importantly it is one of the first texts to provide a clear exposition of the development of radar polarimetry and interferometry, two new technologies whose fusion promises much for the future. As alluded to above, technology forecasting is a dangerous process but this book provides a timely and informative review of an important subject in the 'toolbox' of remote sensing. The reader will not only gain valuable insight into the principles behind microwave sensors but also achieve a deeper understanding and appreciation of this important technology.

Shane R Cloude
University of Adelaide, Australia

preface

There's good news and there's bad news.

The bad news is that microwave remote sensing is a difficult subject. It requires a fair bit of mathematics and physics to be able to understand many of the topics and there are many novel concepts that will seem rather difficult to grasp first time around. Coherence. Polarimetry. Interference. If these are new to you, then don't panic. They are difficult concepts and will take a while to get used to. But remember: "perseverance furthers".

As my contribution, I've tried to make sure that the more challenging material can be skimmed over without loss of consistency in the rest of the text. There are more words than equations in this text. I've done that to put the emphasis on ideas and concepts, rather than the detailed physics behind them. But sooner or later, if you want to truly get to grips with microwave remote sensing, you will have to tackle some mathematics and basic physics.

But there's good news...

The good news is that microwave remote sensing is fascinating. It's only really now, as we enter the 21st century, that microwave techniques in Earth observation have matured out of infancy and are starting to make a real impact on the our understanding and management of planet Earth. Perhaps more than any other remote sensing technique (and I write with a clear bias) microwaves are unique in their scope, their range of applications, their versatility and their quite staggering ability to make quantitative measurements. From a distance of 800km, microwave instruments can make centimetre-scale measurements of the ground surface, or determine atmospheric concentrations to within parts per billion.

It is no longer the case that remote sensing professionals can live without understanding the basics of microwave techniques. But because of its complexity of ideas and the intensity of some of the mathematics and physics behind it, microwave remote sensing risks always remaining elitist and apparently inaccessible. My intentions for this text were, if nothing else, to offer some kind of peek over the fence into the realms of microwave remote sensing, and at least gain enough knowledge to adopt respect and wonder.

To this end, I've included a few things to help make the book easier to use. The margin notes allow easy reference without disturbing the flow of the text. I have tried to explain the steps in mathematical derivations, and there is an appendix summarising some useful mathematics, including logarithms, trigonometry, matrices and complex numbers. Equations are usually accompanied by the relevant units in square brackets. And I have also tried to emphasise the shift in conceptualisation that is required for those readers with some prior knowledge of optical remote sensing. In particular, following the brief summary of microwave remote sensing in Chapter 1, Chapter 2 begins the book proper with a summary of the history of the subject. This is included not simply for background, but to dispel any thoughts you might have of linking the history of microwave remote sensing to that of optical remote sensing. The two are barely even distant cousins when it comes to their heritage, and giving this historical context is an important first step in establishing the special nature of microwaves and how they are exploited.

This is followed by the basic physics of the subject in Chapters 3 (the physical properties of microwaves themselves) and 4 (the polarimetric properties of electromagnetic waves). The first of these chapters includes some extremely important material on the fundamental principles of interference and coherence.

Chapter 5 then explores the physics of how microwaves interact with matter. The underlying rationale for any remote sensing technique is that you can measure some useful physical property of the Earth's surface or atmosphere. This chapter summarises the key bio- and geo-physical variables that influence the emission and scattering of microwaves.

My other approach to help readers conceptualise this subject differently, is to make comparisons between microwave systems and audio perception. This is most apparent in Chapter 6 where the workings of the ear, rather than the eye, are explained as a metaphor for microwave instrumentation.

Two chapters then deal with passive atmospheric measurements and passive surface measurements. The rest of the chapters relate to specific techniques or methods of microwave radar. And I apologise in advance for having a clear bias to these techniques – this is my research area, and so it is always easy to find more to say about these topics. Chapter 9 deals with general radar principles and describes the operation of altimeters and scatterometers. Chapter 10 then expands on these ideas to describe how high resolution images can be created using radars and finally, Chapter 11 describes the techniques of interferometry for both passive and active sensors.

Enjoy!

Acknowledgements   The first words in this text were written as part of my PhD thesis, 10 years ago. That is a long time to get feedback and comments from many people over the years. I would particularly like to thank Chris Varekamp — it was Chris's encouragement while I was in Wageningen that convinced me to take what was then only a set of lecture notes for students, and adapt it into a publishable book. As a consequence, I developed large parts of this text at Wageningen and so was greatly guided by the approach of Dirk Hoekman who put me on track to understanding many of the topics in this book, as did Peter van Oevelen and Joost van der Sanden. And while it would be difficult to list everyone, I would also like to thank all the friends, colleagues and students who have commented upon or proof read various sections of the text at various stages of its production: Dean Turner, Ed Wallington, Fran Taylor, Laura Stewart, Karin Viergever, Iain Cameron, Avril Behan. There are no doubt others who have given me some feedback here and there, and apologies for not mentioning you all, but you have to stop somewhere.

A special thank you goes to Shane Cloude: for writing the Foreword, but also for introducing me to the subject of synthetic aperture radar way back in 1990 while I studied for my MSc. Who knows what I might have ended up doing had it not been for that early introduction to the scattering matrix.

Finally, I should very much like to thank my wife and children: Karen, Sandy and Jamie. Karen has persevered through me saying, "it's almost finished!" for about 5 years. Sandy and Jamie didn't exist when I started writing this text, and I may well have finished it long before now had they not arrived. And while they wouldn't have known otherwise, I thank them for those occasional times that I did decide to work late on the manuscript rather than spend more time with them.

Iain Hector Woodhouse
University of Edinburgh, Scotland

about the author

Iain H. Woodhouse earned his B.Sc. degree in Physics and Natural Philosophy from the University of Edinburgh, Scotland, in 1989, and M.Sc. degree in Remote Sensing from the University of Dundee in 1990. Following a period at the Marconi Research Centre working on radar system design, he completed a Ph.D. degree at Heriot-Watt University, Edinburgh, in atmospheric remote sensing. From 1995–1998 he held a fellowship at Wageningen Agricultural University, The Netherlands, where he was privileged to work in Dirk Hoekman's research group.

Since 1999, he has been employed as a lecturer at the University of Edinburgh's School of GeoSciences and became the Head of the Edinburgh Earth Observatory in 2004. His main research interests are in active remote sensing of vegetation, particularly forests, with some sideline interests in remote sensing of deserts.

CONTENTS

1
WHY MICROWAVES?

Opportunities multiply as they are seized.
— Sun Tzu, *The Art of War*.

The fundamental reason for using microwaves for remote sensing is that they are different. That sounds a rather trivial statement, but is true nonetheless. By using the microwave region of the electromagnetic (EM) spectrum we gain capabilities that complement remote sensing methods used in other spectral regions — microwave interactions, in general, are governed by different physical parameters to those that effect other forms of EM radiation. For example, the amount of microwave energy of a particular wavelength scattered off a green leaf is proportional to its size, shape and water content, rather than the amount of chlorophyll (or "greenness").

Microwaves have further advantages: some can penetrate clouds and can even get through the top layer of dry soils or sand (by as much as a few metres in some circumstances). And since thermal emission is observed in passive imagers, and you provide your own illumination with active imagers, the measurements can be made at any time without relying on background sources such as the Sun. A further advantage for atmospheric remote sensing over, say, infrared techniques is that microwave wavelengths can be chosen such that ice clouds and other particulates (such as aerosols) have negligible effects on the signal.

Of course, there are also some disadvantages. The long wavelengths mean that large (about a metre or more) antennas are required even to achieve the kinds of spatial resolutions appropriate for regional-scale studies (many kilometres). And active microwave systems, such as Synthetic Aperture Radar (SAR) instruments, tend to be the heaviest, largest, most power consuming, and most data prolific instruments that are ever likely to fly on Earth observing satellites (and are therefore the most unpopular for all non-SAR devotees!). In addition, as will become apparent later in these

1

notes, the interpretation of data from such instruments is rarely simple.

The term "microwave" is used throughout this text as a generic term to include centimetre, millimetre and submillimetre wavelength regions of the electromagnetic spectrum. Within the context of imaging radar this definition will also be stretched a little into the radio wave region when we additionally consider wavelengths up to a metre or longer.

1.1   Overview of Microwave Systems

On a rather general level, we may distinguish two types of microwave remote sensing instrument: active and passive.

Passive sensors, or *radiometers*, measure the microwave energy that is radiated (by thermal emission) or reflected (from the sun or other radiating objects) by the Earth's surface or atmosphere. A careful selection of observing frequency will allow measurements of the atmosphere, the ground, ice or rain. Figure 1.1 shows the atmospheric attenuation over the range of the electromagnetic spectrum. Those radiometers that are used to measure emission from the atmosphere are termed microwave *sounders*.

Active sensors, such as radar systems, generate their own illumination by transmitting pulses of microwave radiation and then using a specialised receiver system to measure the reflected (or more precisely, scattered) signal from the area of interest. They primarily (but not exclusively) use wavelengths greater than 3cm (<10GHz) where the atmosphere becomes virtually transparent.

1.1.1   Information from Passive Microwave Imagers

The microwave emission from objects are primarily dependent upon the objects physical temperature and its dielectric properties (i.e. their frequency dependent radiative/absorptive properties). We will see later that it is practical to define the measured microwave intensities in terms of temperature-like properties.

The dielectric properties are related to the physical make-up of the object: i.e. the materials it is made from. For most practical applications of passive imagers (for land observation and sea ice) the most significant factors effecting the measured intensities are temperature, salinity and liquid water content. The surface roughness also has an influence on the directivity of the emission.

1.1.2   Information from Passive Microwave Sounders

A unique feature of microwave spectroscopy for radiometry is the ability to design instrumentation which simultaneously gives the best possible spectral resolution *and* sensitivity of thermal emission measurements at long wavelengths. Measurements can be made of a number of physical parameters and molecular concentrations required for monitoring the global

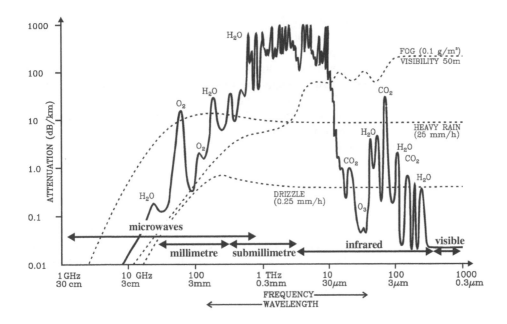

FIGURE 1.1 The attenuation of the atmosphere for a nominal clear atmosphere, with additional attenuation shown for fog, heavy rain and drizzle. Notice that the regions used most often for terrestrial remote sensing are in the "windows" (visible and microwave) while atmospheric sounding is carried out in the "walls" (millimetre to infrared). (After Preissner 1978, cited in Peckham 1991).

atmospheric system and for testing and developing atmospheric models; e.g. temperature, pressure, liquid water, rainfall, and a collection of molecular concentrations, including water vapour and ozone.

1.1.3   Information from Active Microwave Instruments

The optimism that still surrounds SAR imaging systems stems from both its unique properties (which are quite distinct from conventional optical and IR imagers) and its staggering potential to measure physical properties of ground features at high resolution. The properties of microwaves also allow imaging through clouds (ideal for observing regions perpetually covered in cloud, such as the surface of Venus and the humid tropics on Earth), and because SAR is an active system, both day and night imaging is possible, which offers the opportunity to obtain continuous coverage by Earth orbiting satellites.

In addition, the singular nature in which microwaves interact with surface features means that information obtained can be indicative of moisture content, salinity and physical characteristics (shape, size, orientation) — not just reflectivity. While it is generally convenient to consider a radar image as a representation of the reflecting properties (backscatter) of the terrain at radar wavelengths at each position within the image, we shall see in later chapters that this is perhaps a rather too simplistic way of thinking about such images.

Of course, every silver lining has a cloud: the digital processing and interpretation of such images are rarely straightforward. In terms of visual interpretation the most obvious problems are *speckle* and geometry. The effect of speckle (which is dealt with in Chapter 10) can be all but removed at the expense of spatial resolution, but the unnatural geometry is an inherent effect resulting from the imaging process of radar which maps objects as a function of their *slant-range* from the instrument rather than *look-angle* (as in optical systems). This can also be partially compensated for by *geocoding* or *interferometry*, or by other rectification methods, but areas of ambiguity always remain.

A further problem is our limited knowledge of how microwaves interact with surfaces and although there are various theoretical models at this time there is by no means a definitive way to characterise the interactive processes.

There are also headaches for the engineers and system managers. A typical SAR system weighs in at about 400kg, requires kilo-Watts of power (it is an active system remember), uses an antenna about 10×1 m and transmits data at a rate of many hundreds (if not thousands) of Mbps. So, for Earth observation it is the largest, heaviest, most power consuming and most data prolific instrument that could be put on a satellite. By way of illustration, the Microwave Limb Sounder on the Upper Atmosphere Research Satellite was only half as massive as the Active Microwave Instrument on the ERS satellites, used a tenth of the power, had an antenna one fifth of the size and transmitted less data per second by a factor of 100,000.

Despite these drawbacks, synthetic aperture radar (SAR) has developed into a highly successful imaging tool since its origins in the 1950s. The first SARs were carried on aircraft, but in 1978 NASA launched a SAR onboard the Seasat satellite. Although it failed after only 6 weeks the data was still being studied over ten years later.

Synthetic aperture radar has arguably been the most exciting and progressive field of remote sensing over the last decade, and is likely to continue to be so well into the new century[1]. Already ESA (the European Space Agency) has had successful missions with ERS-1 & 2, Japan's space agency (JAXA) with JERS-1, and Canada with Radarsat, all of them carrying SAR instruments. These have been followed by ESA's Envisat in 2002, and (hopefully by the time you read this) JAXA's ALOS and Canada's Radarsat 2. Additionally, the first commercially-oriented (rather than research oriented) radar satellite, TerraSAR-X is also shortly due for launch.

NASA meanwhile has continued its program of short term SAR missions on board the Space Shuttle, including the impressive 11-day Shuttle

---

[1]   OK, I concede that this statement may only refer to land or sea observations by spaceborne instruments. Other exciting developments include further atmospheric sounding and passive sensing of ocean salinity.

Radar Topography Mission in February 2000, as well as many short missions using their airborne AIRSAR instrument. Canada, The Netherlands, Denmark, France and Germany also continue with specific airborne SAR campaigns, and Intermap Technologies now fly the first truly operational commercial airborne SAR mapping system.

Elsewhere in the solar system, following the success of the SAR carried by the Magellan probe to Venus in 1990, the Marsis (Mars Advanced Radar for Subsurface and Ionosphere Sounding) instrument has been sent to Mars on the Mars Express Mission, and the Cassini Radar has been sent to Saturn and its largest moon, Titan. At the time of writing the Cassini radar is days away from its closest fly-by of Titan, and Marsis has been given the green light to deploy its antenna after a long delay due to unanticipated problems with the deployment mechanism.

1.1.4   How Can This Information be Used?

There are countless applications for which microwave remote sensing has proved capable of providing information comparable with, and sometimes beyond, that obtained with optical/IR sensors. Beyond these successes, there are also a few applications that are unique to microwave systems (e.g. those involving interferometry).

Here is a selection of some of the applications that have utilised microwave remote sensing: Planning shipping routes through the ice fields of the Arctic; observing deforestation in the tropical forests of Amazonia and S.E. Asia; measuring soil moisture variations in the semi-arid areas of the Sahel in Niger; monitoring ship movements and oil spills in seas and oceans; measuring atmospheric temperature and concentrations of ozone and water vapour in the upper atmosphere; making estimates of biomass in temperate forest regions; determination of regions of surface frost and thaw in Siberia; daily measurements of global wind speed and direction over the oceans; long term monitoring of the Antarctic ice shelf; making estimates of snow melt and run-off; inventory mapping of crop types in agricultural areas; mapping of ground movements (after earthquakes or subsidence) to within a few centimetres; mapping of glacial movements and estimates of velocity vectors; detection of ancient roads and river beds beneath metres of Saharan sand; acquisition of near-global topography; estimation of forest canopy heights.

The list continues to grow...

2
A BRIEF HISTORY OF MICROWAVES

We all *know* what light is; but it is not easy to *tell* what it is.

            — Dr. Samuel Johnson (1776)

The labours of men of genius, however erroneously directed, scarcely ever fail in ultimately turning to the solid advantage of mankind.

            — Mary Shelley, *Frankenstein, Vol.I.*

The development and exploitation of microwaves helped to shape the world as we know it today. Radio and television, mobile communication, microwave ovens and fanciful death rays all share a common heritage with microwave radiometry and radar systems. Experiments with microwaves led to one of the greatest advances in theoretical physics and laid the foundation of special relativity. The development of radar was probably the single most important contribution to the Allies winning World War II. And half a century of passive microwave measurements has increased our understanding of the solar system and changed our concepts of the size and history of the entire universe.

I begin this history of microwaves over 200 years ago, to a time when electricity, magnetism and light were three distinct phenomena, each wondrous in their own right, but entirely unconnected in the minds of the scientist of the day. Of course, this is a biased history written to set the scene for later chapters and form a basis for explaining the science, rather than intended as a scholarly historical study.

2.1   In the beginning...

There is a persistent mythology within the remote sensing community that the history of remote sensing is rooted in the development of photography in the 19th century. While it may be true that the great, great grandfather of Spot panchromatic imagery may be Nadar's balloon-borne photograph of Paris, and this in turn relied on the development of photography by fellows such as Niepce and Fox Talbot, it is not true for all forms of remote sensing in operation today. The history of microwave remote sensing shares so little with the history of optical remote sensing, that had photography never been invented, microwave remote sensing would have progressed in much the same way as it did[2]. This lack of recognition of the distinct history of microwave remote sensing has had a habit of leading many students and researchers alike to misconstrue the particular nature of microwave systems — particularly imaging radar.   This is not helped by the majority of introductory remote sensing text books that represent the use of microwaves as a consequence and progression from optical systems, rather than as a parallel, and virtually unrelated, development.

As with optical sensing, the foundations of microwave remote sensing lie in the 19th century, not in the workings of the pinhole camera and photosensitive chemicals, but in the study of electricity and magnetism.

2.2   Out of the Darkness: Maxwell and Hertz

During the half-dozen decades that spanned the turn of the 18th to 19th centuries, three apparently unrelated phenomena were being independently investigated by a wide range of scientists and philosophers, as well as a fair share of people whose intention, not to mention their scientific credibility, was a little suspect. The three phenomenon were electricity, magnetism, and light, and in their own way, each of them was influencing the culture of the developed world. The unification of these phenomena that eventually took place over the final decades of the 19th century spawned the beginning of modern physics, changing the face of western society and influencing almost every modern "invention" since.

Electricity   Electricity was not a new discovery by the 18th century — it had been a subject of recorded study as far back as the ancient Greeks and the Romans, primarily with the use of static electricity generated from leather and amber, or through the use of natural sources such as electric eels. But it was the Dutch development of the "Leyden jar", a form of simple battery, that allowed experimenters a degree of control that they had not had before. One such experimenter was Luigi Galvani (1738-98) an Italian physicist based in Bologna, whose name is immortalised in the word *galvani*sation, a process which describes the coating of a metal surface in zinc to protect against corrosion. Another experiment for which he is well-known,

---

[2]   Although I will admit that there is one exception to this extreme view, which is that early imaging radar required photographic film to record the vast amounts of data it collected.

is the application of an electric current to detached frogs' legs causing them to move in a lifelike fashion. He termed this effect "animal electricity", inspiring a generation of "galvanists" espousing Galvani's influential (though incorrect) interpretation of his results.   The French Revolution (1789) helped to fuel such study by conveniently providing a steady supply of raw materials — researchers did not have to rely on dead animals, but rather they could use human cadavers, or constituent parts thereof, and map in some detail the network of nerves throughout the body.

Electrophysiology fast became a macabre novelty rather than just a subject for scientific inquiry. Galvani's nephew, Giovanni Aldini, for instance, toured Europe with his gruesome public displays whereby, to the great delight of his audience, a demonstrator would induce convulsions and realistic motion in the lifeless flesh of, usually, some recently executed criminal. Most horrific of all, was the application of electric current to muscles on a human head, which would grimace and contort in response to the electric current. In 1818, a contemporary of Aldini, Andrew Ure (1778-1857) gave a similar public display in Glasgow and came close to suggesting that electricity may even be the key to the restoration of life. By interesting synchronicity, but undoubtedly not coincidence, 1818 also saw a young Mary Shelley write "Frankenstein, a Modern Prometheus"[3]. The reputation of such displays and the natural (but false) assumption by some people that electricity was the key to life, was undoubtedly an inspiration for this classic gothic novel.

The strange magic of natural magnetism was also in the public eye *Magnetism* at the end of the 18th century due to the popularity of Dr. Franz Anton Mesmer (1734–1815), a Viennese medical doctor. Mesmer believed that magnets could induce healing powers in those who held them and he displayed such a procedure during popular sessions that he held for French society at the end of the 18th century from about 1778. Eventually it became apparent that the magnets themselves were not required, but the influence remained, so, in recognition of Gavlani's work, he called this phenomenon "animal *magnetism*". Eventually Mesmer's activities came under scrutiny by the scientific establishment and in 1784 he became the subject of an investigation by a special commission of the French Academy of Sciences. The commission included the US Ambassador at the time, Benjamin Franklin (1706–1790), an "amateur" scientist famous for *Benjamin Franklin* his supposed escapade involving flying a kite with a key dangling from it during a thunderstorm in 1752, which led him to be the first person to explain the electrical origin of lightning and invent the lightning conductor[4]. The result of the commission was that *Mesmer-ism* was based on suggestion, not magnetism of any sort. We now know that this technique was an

---

[3]   Do not be put off by the numerous shockingly bad versions of Frankenstein in film and on TV — the original book is a masterpiece of gothic fiction and a must-read for all scientists because of the issues it raises.

[4]   Amongst his other achievements, it is also worth noting that he also helped draft the US constitution.

early form of what is now referred to as *hypnotic suggestion*.

Simply out of interest, it is worth mentioning here that in addition to Ben Franklin, the investigative committee included Antoine Lavoisier (discoverer of oxygen and considered the "father of modern chemistry"), Jean Bailly (astronomer and first mayor of Paris) and Dr. Joseph-Ignace Guillotin (after whom the guillotine is named, although he was not its inventor). A decade later Lavoisier lost his head to the guillotine in the French Revolution (the guillotine being the tool that so helpfully contributed to the work of the galvanists!).

Light

Scientific investigations into optics and light (which at this time only included infrared, visible and UV) were also undergoing some change. Thomas Young (1773–1829) and Augustin Jean Fresnel (1788–1827) were independently re-introducing the idea of light as a wave, to the great dismay of many scientists of the time who still held onto Newton's particle theory of light which had lasted almost 100 years. Young, an English physician and physicist, also described the principle of colour vision — that information from three colour sensitive cells could combine to give colour sensitivity. His model was that there are three different groups of cones in the retina, each of which is particularly sensitive to one of three colours: red, green and blue. Other colours (besides red, green, and blue) are seen when the cone cells are stimulated in different combinations. The sensation of white is produced by the combination of the three primary colours, and black results from the absence of stimulation. The theory was later further developed by Hermann von Helmholtz and so is now known as the Young-Helmholtz theory. Such ideas laid the groundwork for the development of colour photography and then colour displays (and the RGB colour composite, much loved by modern remote sensers).

It was a short step from physiological optics to considering the nature of light. Young's interest in this was reinforced by some work he had done in the mid-1790s on the transmission of sound, the nature of which he believed was analogous to light. In 1802 Young first demonstrated a simple proof of the wave theory of light. He directed light from a single source with a very distinct colour through a narrow slit and then directed that same light to through two more narrow slits placed within centimetres of each other. The light from the two slits then directed towards a screen where Young observed that instead of a smoothly illuminated region there was a pattern comprised of bright and dark bands. This is the so-called "Young's double-slit experiment" and was the first demonstration of the phenomenon of interference in light. With this experiment Young established the wave nature of light. He then used his new wave theory to explain the colours of thin films (such as soap bubbles), and, relating colour to wavelength, he calculated the approximate wavelengths of the seven colours described by Newton. In 1817 he proposed that light waves were transverse (vibrating at right angles to the direction of travel), rather than longitudinal (vibrating in the direction of travel as is the case for sound waves), which allowed an explanation of *polarisation*, which describes the direction of vibration of the wave.

In passing, it is worth noting that it was around this time, in 1827 to be exact, that Joseph Niépce (1765-1833) publicly demonstrated the first photograph.

While popular culture of the early 19th century were being entertained by the mysteries of magnetism or the wonders of electricity, or were simply grateful for the benefits of the new applications of light, there were also a number of scientists more concerned with studying the fundamental properties and the underlying nature of such phenomena. Perhaps most significantly, between 1845 and 1850 a number of key experiments took place that began to change our understanding of these three phenomena. At this time there was a growing realisation that electricity and magnetism were related — the Danish physicist Hans *Electricity and Magnetism*

Christian Oersted (1777-1851), for instance, had made electricity flow down a wire and cause a nearby compass needle to move. This was very intriguing since the wire and the compass were not in any physical contact. Michael Faraday (1791–1867), probably one of the greatest experimental physicists, did a similar experiment but this time he made a magnetic force turn on and off and thereby generated a current of electricity in a nearby wire. Time-varying electricity had somehow generated magnetism, and time-varying magnetism had somehow generated electricity. This was called *induction*, and at the time was considered deeply mysterious.

Faraday's explanation was that the magnet had an invisible "field" (his *Faraday* term) of force that extended into the surrounding space, stronger close to the magnet, weaker further away. It was proposed that there was some kind of electrical-magnetic disturbance that somehow had properties of both electricity *and* magnetism.

Then in 1845 Faraday observed a relationship between electricity, magnetism and *light* by noticing that a strong magnetic field could effect the nature of a light beam through a medium. And in 1849 Armand Fizeau (1819–1896) made the first determination of the speed of light without using astronomical observations, which was followed one year later by Jean Foucault[5] (1819–1868) who determined the remarkable fact that the speed of light in water was less than in air. This was the final blow for the last remaining devotees of Newtonian corpuscular theory (at least until the development of quantum physics, when the idea of photons, or particles, of light again became a useful model).

A remarkable turning point in the history of physics then took place during the following two decades — so remarkable, in fact, that many consider it as the beginning of "modern" physics. The key to this historic leap in our understanding of the physical world was James Clerk Maxwell (1831-79) a brilliant, although by all accounts rather dull, the-

---

[5] Foucault was a French physicist who invented the gyroscope (1852) and the Foucault pendulum (1851). A gyroscope is essentially a spinning wheel set in a movable frame and is worth a mention since gyroscopes form the basis of many navigation instruments still used in remote sensing platforms.

oretical physicist, originally from Edinburgh. Maxwell combined all the laboratory evidence accrued by the likes of Faraday and Oersted on electricity and magnetism and how they behaved in matter and he concisely summarised them as four fairly simple-looking equations:

$$\nabla \cdot \mathbf{M} = \frac{\rho}{\varepsilon}$$

$$\nabla \cdot \mathbf{B} = 0$$

$$\nabla \times \mathbf{E} = -\frac{\partial}{\partial t}\mathbf{B}$$

$$\nabla \times \mathbf{B} = \mu\mathbf{J} + \mu\varepsilon\frac{\partial}{\partial t}\mathbf{E}.$$

These are now known, appropriately, as "Maxwell's Equations" and they are given here merely to demonstrate the apparent simplicity of these equations (even if you do not understand the mathematical notation). Almost everything there is to know about electromagnetic radiation is contained within these four equations.

It is not necessary to remember them for the rest of this book, or even understand them, but I put them in for completeness (you might see them written in different forms, but the meanings are the same). The electric field vector is represented by $\mathbf{E}$, and the magnetic field vector by $\mathbf{B}$. The electric charge is represented by ρ and $\mathbf{J}$ is the electric current. The constants, ε and μ will be described in more detail in Chapter 3 and describe the electric and magnetic properties of a material.

If all he did was produce these equations, it would still have been a great leap forward, but he went one stage further — he asked some questions about what these equations would mean in different circumstances. Importantly, he wondered what these equations would look like in empty space, in a vacuum, in a place where there were no electrical charges, no electrical currents and no magnets? His answer was that electric and magnetic fields propagated through empty space as if they were *waves*. The oscillating fields were self-perpetuating: an oscillating electric field induced an oscillating magnetic field which in turn induced an electric field.

Electromagnetic Waves Maxwell called them "electro-magnetic waves".

He then did a further step: he calculated the theoretical speed of these waves, and found that:

$$v_0 = \frac{1}{\sqrt{\varepsilon\mu}}.$$

The values of ε and μ had been measured in the laboratory and when plugged into the equation Maxwell found that the electric and magnetic fields in a vacuum ought to propagate, astonishingly, at the same speed as had been determined by Fizeau for visible light. The agreement was far too close to be coincidence. Suddenly, electricity and magnetism were tied to the nature of light — more than that, Maxwell postulated, they were exactly the same phenomenon.

Maxwell published his theory of electromagnetic waves in 1868, ten years after the first known aerial photograph, taken from a captive bal-

loon from an altitude of 1,200 feet over Paris by Gaspard-Felix Tourna-
chon ("Nadar") (1820-1910). It is the balloon event that is often hailed
as the historical origin of what might be referred to today as remote sensing. However, I only note it here for historical context — in microwave
remote sensing, it is Maxwell's electromagnetic theory that is the turning point, even though it took almost twenty years for it to be confirmed
by experiment in 1886 (unfortunately 7 years after Maxwell's death) by   *Confirmation By Hertz*
Heinrich Hertz (1857-1894). After 8 years of working on Maxwell's the-
ory Hertz demonstrated experimentally that the electric force propagates
through space at the speed of light. His experiment consisted of a transmitter — essentially an oscillatory electric discharge across a spark gap
forming an oscillating electric dipole, and a receiving antenna — basically
an open loop of wire with a brass knob on one end and a fine copper point
on the other. A small spark across these indicated incident EM waves. He
did many of the same things with the invisible electromagnetic waves as
had been done in optics — he focused the radiation, determined its polarisation (orientation), reflected it, refracted it, caused it to interfere setting up
standing waves and measured the wavelength (which were a little less than
a metre, so close to the wavelengths used by low frequency microwave
radars today).

By the beginning of the 20th century it was clear that visible light, radiant heat and radio waves were all forms of the same phenomenon: electromagnetic radiation. The key variable of these waves was frequency (or
wavelength, the two being linearly related) and the entire range of possible
wave types was called the *electromagnetic spectrum.* The now conventional understanding of the electromagnetic spectrum — running in wavelength from gamma rays to X-rays to ultraviolet light to visible light to infrared to radio waves — is due to Maxwell. Note, however, that the modern convention of referring to different regions of the EM spectrum with
discrete names (X-ray, UV, visible, infrared, microwave, radio) is merely
an indication of how these different frequency EM waves are detected —
the EM spectrum itself is continuous.

The waves Hertz used are on the border of the two broad classes called
radio waves and *micro*-waves. Generally, microwaves are EM radiation
with wavelengths from 1mm to 30cm — about 10^9Hz (1 GHz) to about
3×10^{11} Hz (300 GHz) — but modern microwave radar, for instance, can
use EM waves with a wavelength of 1 m of more. The terminology at the
boundaries of these regions is ambiguous and a little arbitrary.

Generally, accelerating charges are the source of electromagnetic waves.
This applies for centripetal (from rotation) as well as linear acceleration.
Any artificial or natural source of EM radiation can therefore be related
back to an accelerating charge — usually an electron.

Maxwell made enormous contributions to astronomy and physics, from   *A Final Word on Maxwell*
the conclusive demonstration that the rings of Saturn are composed of
small particles to the elastic properties of solids, to the disciplines now
called the kinetic theory of gases and statistical mechanics. It was he who

first demonstrated that an enormous number of tiny molecules, moving on their own and repeatedly colliding with each other and bouncing elastically, leads not to chaos and confusion, but to precise statistical distributions. (The bell-shaped curve that describes the speeds of molecules in a gas is now called the Maxwell-Boltzman distribution.) He invented a mythical being, now "Maxwell's Demon" whose actions generated a paradox that took modern information theory and quantum mechanics to resolve. But Maxwell's greatest contribution will always be his discovery that electricity and magnetism, of all things, join together to become light.

As a footnote to the 19th century, one other important development took place, namely the invention of the telephone in 1876, the credit for which usually going to Alexander Graham Bell (1847–1922), the Scottish-Canadian inventor and founding member of the National Geographic Society. The significance of this invention was that it spawned a whole new era of electrical engineers devoted to communication technology, a legacy that persists to this day in the technology and terminology of microwave remote sensing.

2.3   Radios, Death Rays and Radar

Seven years after Hertz's small-scale experiments, scientists at Cambridge transmitted radio signals over a distance of a kilometre and by 1901, a young Italian by the name of Guglielmo Marconi (1874-1937) was successful in transmitting radio waves over the Atlantic. The century of global communication had begun and the ground work had been laid for the development of microwave remote sensing and radar.

The main technical progress on radar as we now know it was from the 1920s to the end of World War II. Much of the development was carried out in secret, so a number of nations developed radar technology independently: Britain, the USA and Germany were the main contributors to its successful implementation, although Italy and Japan also provided important advances to the subject. However, the idea of radar was not new: by 1920 a number of ideas were in the public domain that could be interpreted as early radar concepts. The idea of echolocation was certainly not new — acoustic echolocation had been used, for instance, by large ships in the Arctic to avoid icebergs. A whistle or horn could be used to transmit a sound pulse, and the time delay to the echo used to determine the distance. In 1900 Nikola Tesla (1856-1943), a Serbian-American inventor obviously aware of this technique, conceived of the possibility of employing radio waves not only to detect, but also to measure the movement of distant objects, although he never worked out the details of his idea. Christian Hulsmeyer, however, did develop a working system using radio waves and received a patent for an obstacle detector using radio waves in 1904.

It is also worth mentioning, if only out of interest, that Hugo Gernsback[6] (1884–1967), a well-known science fiction writer, published a rather

---

[6]   Science fiction fans should recognise the name — the Science Fiction Achievement

dull, but technically visionary, romantic story in 1911 entitled, "RALPH 124C 41+". In the story there is a detailed description of the detection apparatus used by the hero:

> "A pulsating polarized ether wave, if directed on a metal object can be reflected from a bright surface or from a mirror. The reflection factor, however, varies with different metals. Thus the reflection factor from silver is 1000 units, the reflection from iron is 645, alomagnesium 460, etc. If therefore, a polarized wave generator were directed toward space, the waves would take a direction shown in the diagram, provided the parabolic wave reflector was used as shown. By manipulating the entire apparatus like a searchlight, waves would be sent over a large area. Sooner or later these waves would strike the metal body of the flyer, and these waves would be reflected back to the sending apparatus. Here they would fall on the actinoscope, which records only reflected waves, not direct ones.
> "From the actinoscope the reflection factor is then determined, which shows the kind of metal from which the reflection comes. From the intensity and the elapsed time of the reflected impulses, the distance between the earth and the flyer can then be accurately and quickly calculated."

For its day this is a particularly good description of a ground-to-air radar system. In 1922 Marconi presented to the American Institute of Electrical Engineers a fuller description of a similar technique which would "immediately reveal the presence and bearing of the other ship in fog or thick weather". Many of the first actual experiments, however, where not undertaken by Marconi, but by researchers at the U.S. Naval Research Laboratory (NRL) using continuous-wave systems that transmitted a constant signal rather than a series of pulses.

It is clear, then, that by the 1930s the radio technology and early radar *Radar* concepts were well-established. However, it is the subsequent work by the Scottish scientist Robert Watson-Watt (1912–1973) that is widely seen as the birth of truly operational pulsed radar systems. Since 1919 Watson-Watt, the meteorologist-in-charge at the British Royal Aircraft Establishment, and Arnold Wilkins, had been working on locating distant thunderstorms by radio. They used radio direction finding techniques using directional aerials with cathode ray tube displays. In Dec 1924 Prof. E.V. Appleton, who was in contact with Watson-Watt, eventually discovered that radio emissions would also bounce back from an ionised layer high in the upper atmosphere (now known as the *Appleton layer*) and that by timing the interval between emission and the return of the signal, the distance between the antenna and the layer could be measured. Watson-Watt was able to refine Appleton's techniques for measuring the distance between

---

Awards are named "Hugos" in his honour.

transmitter and receiver and the storm he was trying to locate, to the extent of being able to give its direction and range. On 2 April 1935, Watson-Watt received a patent for his Radio Detection and Ranging (RADAR)[7] device which was capable of locating and ranging aircraft using pulses of microwaves rather than continuous waves. At around the same time, the NRL were also perfecting the application of pulsed radars specifically for detecting objects.

However, such was the UK's concern over developments in Germany and the Nazi regime's increasing aerial power in the late 1930s, that Watson-Watt was asked by the UK government if he could develop a microwave system capable of destroying an aircraft. Such ideas were not new. In 1924 Nikola Tesla claimed that he had invented a death ray capable of stopping an aeroplane in mid-flight. In the same year Grindell H. Mathews, a British scientist claimed to have invented a similar device, and T.F. Wall unsuccessfully applied for a patent on a death ray. For a while the trend in death rays subsided, as no definitive invention appeared, then in 1934 Tesla claimed a new one based on an entirely new principle of physics — he claimed it could destroy 10,000 planes at distance of 250 miles. Each ray would require the construction of a $2 million power plant, located at high strategic points. A network of 12 such plants would protect the US from aerial invasion. He talked about his idea for a few years after, but no-one showed an interest. After his death, nothing was found in his papers, though his idea has clear resonance with Ronald Reagan's "star wars" programme in the US at the end of the cold war.

Returning to the story, Watson-Watt was unconvinced by the death-ray notion, and his calculations indicated that at best it might be possible to cook the pilot, but not destroy the plane. As an alternative he proposed that it should be possible to develop a system to locate incoming enemy aircraft to provide an early warning, even at night and through cloud cover (since clouds are transparent to radio waves, and longer wavelength microwaves). Watson-Watt was given a team of scientists and engineers to develop the system and by the outbreak of war in Europe in 1939 a series of tall towers with radio transmitters and receivers, known as "Chain Home", had been constructed along the Eastern coastline of Britain[8].

World War II proved to be the turning point for microwave remote sensing. While the need for improved communications fuelled advances in electrical engineering radio technology, the potential of tracking the position of enemy aircraft day or night was too valuable a resource to ignore and so an enormous effort was put into developing radar technologies. During the war radar was used extensively to locate aircraft as well as storms. Chain Home was instrumental in the "Battle of Britain" — by

---

[7]   Note that the name 'RADAR' came along later and is attributed to the US Navy in 1940.

[8]   At the time of writing there is a campaign to restore the Bawdsey Transmitter Block and convert it into a radar museum (www.bawdseyradargroup.co.uk). Bawdsey was the first part of the Chain Home system and, while not an attractive looking building (it was designed to withstand a direct hit from aerial bombing after all), it does have special historical significance in being the first operational radar installation ever constructed.

providing the location, altitude and bearing of Luftwaffe bombing sorties, the radar operators were able to direct the small number of Allied fighters with great accuracy to intercept the incoming aircraft[9].   This focused use of limited resources offered sufficient advantage to convince Hitler to postpone his invasion of Britain and focus instead on his Eastern front.

Similarly, during the intervening years before D-Day in 1944, the eventual development of a radar that could be carried by aircraft allowed the Allies to hunt down U-boats by detecting their periscopes above the water, even in darkness, and gave the Allies a final edge in the "Battle of the Atlantic".

In the context of remote sensing, the development of airborne radar was an important step forward.   Before the development of the cavity magnetron in 1940, an instrument that allowed the generation of powerful "short wave" radio signals, the required antenna size was many metres in length.   The Chain Home system, for instance, used 12m high aerials. An antenna any longer than 1.5m simply could not be carried by an aircraft, and since the required antenna size is proportional to the wavelength, shorter wavelengths were required. The klystron had been the standard radio wave generator but could not generate short radio waves with sufficient power.   The cavity magnetron was therefore a key breakthrough because it could reliably generate 10cm waves (which would now be called microwaves) with high power.   This allowed small antennas to be mounted on aircraft that could then be used to locate ships and U-boats at sea and intercept enemy aircraft as an air-to-air radar.   The all-weather, day-night operation of the radar also meant that it could be used as a navigational aid to bombers.   The most sophisticated airborne radar of the war, known as H2S, was designed with the express purpose of looking at the surface of the Earth from an aircraft to aid navigation to and from bombing targets. Coastlines and major towns could be identified in the radar images due to the contrast in backscatter of the microwaves.

Ironically, it is a blunder, not a success, that perhaps provides the best illustration of the wartime value of radar.   In 1941, on top of a mountain in Hawaii, a group of American soldiers were testing out the new technology of radar with some equipment given to them by the British scientists. Suddenly a set of blips appeared on the monitor screen. The operators got very excited, as it seemed to be picking up some aircraft far off the coast, and out of visual range.   As the blips got stronger and nearer they began to appear like a signal you might expect from a large group of planes, and the operators started to wonder if they might be incorrectly operating the

---

9   It is said that radar was so successful that the Allied Intelligence decided they needed a cover story for why so many planes were being found in darkness. Their answer: that carrots helped people see in the dark. As with all good urban myths a link to real science made it all the more convincing. A deficiency of Vitamin A causes "night blindness" whereby one finds it difficult to see in low lighting conditions. Carrots are rich in beta-carotene (that's what makes them orange) and our livers convert this to vitamin A so that carrots can correct a vitamin A deficiency, although a surplus will not improve vision. That carrots "help you see in the dark" was a well-known "fact" that I was often reminded of as a child.

Standard Radar Nomenclature		ITU Nomenclature		
Band	Freq. range		Band	Freq. range
HF	3–30 MHz	Dekametric (HF)	7	3–30 MHz
VHF	30–300 MHz	Metric (VHF)	8	30–300 MHz
P (UHF)	0.3–1 GHz	Decimetric (UHF)	9	0.3–3 GHz
L	1–2 GHz	"		
S	2–4 GHz	"		
C	4–8 GHz	Centimetric	10	3–30 GHz
X	8–12 GHz	"		
Ku	12–18 GHz	"		
Ka	18–27 GHz	Millimetric	11	30–300 GHz
V	27–40 GHz	"		
W	75–110 GHz	"		
mm	110–300 GHz	"		

TABLE 2.1   Designation of radio and microwave bands. The columns on the left are the standard terminology for radar but these are also often used for passive designations. Ultra high frequency (UHF) is now more usually referred to as P-band. Note that the actual frequencies allocated for radar useage by the International Telecommunications Union (ITU) are smaller bands within these broad classifications. (After IEEE 1984).

system. Perhaps there was something wrong with the equipment. They radioed for assistance but got no answer, but were relieved that they had completed the experiment and the radar did seem to work. By the time they left to inform someone of the strange appearance on the screen, the massive airborne raid on Pearl Harbor had already begun.

Frequency Bands

Today we are left with one frustrating legacy from these early wartime developments — the nomenclature of the frequency bands.

Radar systems use frequencies from 1–90 GHz and this range can be conveniently grouped together into "bands" that (used to) relate to the different equipment required to generate and detect them. Since these were defined during wartime, they had to choose a naming system that was not immediately obvious — for security reasons they actively chose a set of letters with no logical system — i.e. designed to confuse rather than clarify. To this day we still use the P, L, S, C, X, etc, naming scheme (see Table 2.1).

2.4   The Venus Ruler and Little Green Men

After the end of the war, the applications of microwaves diversified. The military continued its development of radar as an early warning device (which became more important after the development of long range aircraft and missiles) and meteorologists further developed their techniques for ground based observations of the atmosphere, thunderstorms and rainfall.

However, despite activities in these disciplines, the end of the war left an excess of expertise and technology, and out of this two new disciplines emerged: radio astronomy and radar astronomy. Although microwave and radio wavelength observations had been made of astronomical bodies before the war, the rapid technological development during the 1940s and the associated leap forward in expertise, opened up a whole new branch of research.

Sir Bernard Lovell (1913–), for instance, had been one of the key scientists on the development of the H2S air-to-ground radar, but after the war set up a facility for measuring microwaves with extraterrestrial origin. This facility eventually evolved into Jodrell Bank, one of the world's most important radio telescopes that helped formulate the discipline of radio astronomy. Radio astronomy is the passive sensing of signals from outside the atmosphere of the Earth and is responsible for at least three 20th century discoveries of astronomical (if you pardon the pun) significance. The first is the discovery of quasars. In Moscow in the early 1960s, Soviet astronomers held a press conference in which they announced that the intense radio emission from a mysterious distant object called CTA-102 was varying regularly, like a sine wave, with a period of about 100 days. No periodic distant source had ever before been found, and the scientists thought they had detected evidence of an extraterrestrial civilisation. In fact, they had discovered what is now known as a quasi-stellar object — a *quasar*. *Radio Astronomy*

A similar event happened in 1967 when British scientist Jocelyn Bell found an intense radio source turning on and off with astonishing precision. What was it? *Jocelyn Bell*

Their first thought was that it was a message from another world, or maybe an interstellar navigation and timing beacon for extraterrestrial spacecraft. They even gave it a name, among themselves, of LGM-1 — the acronym standing for Little Green Men.

However, further tests showed it to be something quite natural — a massive star shrunk to the size of a city and spinning very fast. We now refer to such stars as *pulsars*. This was the first ever observation of a pulsar.

The third discovery was an accident. In 1965 two scientists, Arno Penzias and Robert Wilson of Bell Labs in the US, had been calibrating a small radio horn (a type of antenna) designed for satellite communication. They became deeply puzzled by the persistent presence of excess noise in their measurements, after they had eliminated the possibility that it could have been terrestrial, solar or galactic in origin. This noise seemed to be distributed throughout the sky in a completely uniform manner. After consulting some astrophysicists they found that they had discovered the cosmic background radiation which had been predicted by the Big Bang theory some years earlier. This discovery is almost unrivalled in its cosmological importance establishing, as it did, the Big Bang theory as the prevailing cosmological model of the time.

The relevance of radio astronomy as a driver for Earth observation is apparent in the fact that the first passive radiometry from spacecraft was not for observing Earth, but was the two-channel microwave radiometer

(16 and 22GHz) flown by the Mariner 2 probe to Venus in 1962. It took a further six years before a radiometer was flown on a satellite to observe the Earth.

Radar Astronomy   The use of radar to observe the planets and moons of the solar system also arose as a consequence of the end of the war. American military engineers researching microwave radars for detecting aircraft and missiles in the late 1950s were tempted, more out of adventure than any practical military application, to try to bounce radar signals off the surface of Venus. As well as being a technical challenge, the measurement would allow a very precise measurement of the distance from Earth to Venus, and subsequently allow an improved estimate of the Astronomical Unit, the distance between the Earth and the Sun, which is a fundamental "ruler" for measuring distances within the rest of our galaxy. In 1961 the Jet Propulsion Laboratory in Pasadena, California, succeeded in obtaining the first real-time detection of a radar signal from Venus. It was to prove the beginning of many observations by radar of our solar system.

2.5   Imaging Radar

The origins of imaging radar come from progressive development of airborne radars like the H2S and the growing realisation that such instruments could be used for reconnaissance as well as navigation. The ability to create images in darkness and through cloud being an obvious advantage, but the potential to identify targets through vegetation being a further incentive. A key development was the use of long antennas with high frequencies that produced narrow fan-beams that were projected sideways from the aircraft, perpendicular (or in fact, often squinted slightly forwards or backwards) to the direction of flight. This provided extensive coverage since the fan-beam could be scanned along a wide swath as the aircraft travelled along the flight path. Such instruments were the first Side-Looking Airborne Radar (SLAR) and they were developed for military reconnaissance purposes in the 1950s. It is at this stage that photography and microwaves have their only historical overlap, in so far as long rolls of photographic film were the only appropriate means of recording the vast amounts of data collected by airborne radar[10].

An important development in 1952 was the technique of "Doppler beam sharpening" developed by Carl Wiley of the Goodyear Aircraft Corporation as a means of improving the spatial resolution of long wavelength

---

[10]   In fact radar has more heritage in common with a domestic microwave oven than it does with the modern pocket camera. It was during a radar research project just after the war that Dr. Percy Spencer, a radar engineer with the Raytheon Corporation, noticed that while working next to his new magnetron, the bar of chocolate in his pocket melted. Rather than demonstrating immediate concern for his health, he experimented with some other articles of food — a popcorn kernel that popped when in the vicinity of the magnetron, and an egg that eventually exploded. Eventually he determined that the food was absorbing stray microwaves which was causing the food to heat up. This realisation eventually led him to develop the first microwave oven.

imaging radar. The longer wavelengths required unfeasibly larger antennas to achieve the same spatial resolution, so Wiley developed a means of using the Doppler shift in the echoes to achieve a much higher resolution. This technique is now referred to as aperture synthesis, as it allows a small antenna to achieve the effective resolution of a much larger, synthesised antenna (or aperture).   *Aperture Synthesis*

There was a large time lag between the development of these military systems and their declassification for civilian use. The first large scale civilian application of airborne imaging radar was in 1967 — a 20,000 km^2 survey was carried out in Darien province of Panama, a region that had never been photographed or mapped in its entirety before because of almost perpetual cloud cover and inaccessible tropical forest. The success of this project led to further mapping surveys. In 1971 a 500,000 km^2 survey of Venezuela was carried out which resulted in improved border definitions, and permitted a systematic inventory and mapping of the country's water resources, including the discovery of the previously unknown source of several major rivers[11]. In the same year, Project Randam (Radar of the Amazon) was begun. This proved to be a huge mapping project of 8,500,000 km^2 which was used for geologic analysis, timber inventory, transportation route location and mineral exploration, as well as a number of other applications.

These successful projects led the way in showing imaging radar as being an irreplaceable mapping tool.

2.6 Microwave Remote Sensing from Space

By the time that the first meteorological satellites were being launched at the beginning of the 1960s, the technology for detecting passive microwaves was becoming sufficiently advanced to warrant carrying an instrument into space to observe the atmosphere. Microwave emissions at different wavelengths were known to contain information on atmospheric chemistry (water vapour being of particular interest) and temperature. Making passive measurements of the atmosphere is known as atmospheric *sounding*.

The earliest satellite observations of the Earth were primarily to observe clouds and were made at visible and infrared wavelengths (*e.g.* the TIROS series in 1960–65). It was not until 1968 when microwave radiometers for remote sensing of the atmosphere was initiated when the Soviet satellite Cosmos 243 carried four nadir-viewing[12] radiometers into orbit. Four years later the Americans launched Nimbus-5 which carried a nadir-viewing radiometer called NEMS (Nimbus-E Microwave Spectrometer[13]) which was designed primarily for temperature sounding for mete-

---

[11]   For comparison, the entire area of the United Kingdom is only 244,022 km^2.

[12]   The *nadir* is the point directly beneath the satellite track.

[13]   U.S. satellites are named consecutively by letter before launch, which is changed to its corresponding number after launch, *e.g.* Nimbus-E becomes Nimbus-5, *etc.*

orological purposes. Since that time instruments have evolved from the original single-channel radiometers with low spatial resolution to higher-resolution sensors with many channels, and measurements have been used to determine rainfall, cloud liquid water (*hydrometeors*), water vapour and vertical profiles of temperature. The reader is referred to Peckham (1991) and Ulaby (1986) for a more comprehensive history of satellite remote sensing of the atmosphere.

Recent advances in technology and satellite payload capability have now made possible advanced sounders such as the EOS-MLS (Microwave Limb Sounder) on NASA's Aura satellite launched in 2004. With the new generation of satellites at the beginning of the 21st century global distributions of many chemical species will be added to the list of measurements made by satellites.

Spaceborne Radar   The first satellite-borne radars were altimeters and scatterometers. Microwave altimeters were used for navigation in the Apollo moon landings in 1969–70 and the S-193 scatterometer was built for the Skylab missions in 1973 and 74.

Getting an imaging radar into space took a little longer. It was not until the development of reliable civilian radar technology by NASA during the early 1970's that it was eventually made possible. The launch of Seasat in 1978 revolutionised active microwave remote sensing from space (in the civilian context, at least). Seasat flew a microwave altimeter and a multi-beam scatterometer (the Seasat-A Satellite Scatterometer, or SASS) primarily aimed at studying the oceans, as well as the first civilian space-borne imaging synthetic aperture radar. Despite lasting only a few months, the Seasatinstrument clearly defined a niche role for active microwave instruments in studying the ocean at a range of scales. This subsequently led to a number of missions flown on the Space Shuttle in the 1980's using Seasat technology, and successful synthetic aperture radar instruments on several other orbiting platforms, including the European Space Agency's ERS 1 and 2 satellites that provided a data set spanning more than a decade.

Microwave technology, both passive and active, is now moving to a level of maturity that affords it a unique place in the observation of the Earth and other bodies in the solar system.

2.7   Further Reading

For the history of electricity, magnetism and electromagnetic radiation I can recommend Hecht's "Optics" and sections of Carl Sagan's "The Demon-Haunted World" . A number of other microwave books offer different perspectives on the historical development of microwave remote sensing: Ulaby et al (1981), Curlander and McDonough (1991) and Skolnik (2002) to name a few.

3
PHYSICAL FUNDAMENTALS

...you must begin with the fundamental principles and learn
these well, as you would in learning any other art.
— J.Hugard and F. Braue, *The Royal Road to Card Magic*

Whatever your interest in microwave remote sensing — from instrument design to data analysis and application — it is important to have
a sound understanding of the physical properties of electromagnetic radiation. It is a fundamental requirement to know how to describe microwaves,
both qualitatively and quantitatively. Qualitatively it is necessary to have
some idea of how they are created, how they are measured and the way
they interact with other continuous media or discrete objects. Quantitatively, we need to have a mathematical description of microwaves, so we
can quantify these physical processes in terms of measurable parameters
— frequencies, amplitudes, intensities, directions, and so on.

This chapter summarises the general physical properties of electromagnetic radiation, whereas Chapter 4 deals with the specifics of polarised radiation. Chapter 5 then looks at how these waves interact with materials
and objects.

3.1   Physical Properties of EM Waves

A full description of current theories of electromagnetic (EM) radiation
is well beyond the scope of this text, and besides, its exact nature is currently beyond our comprehension. We may model it with some conceptual
idea, or characterise it with mathematical, quantifiable parameters, but at
the end of the day, it has so far defied any complete description that is com-
prehensible within our limited experience as human beings. At least this is
still the case at the time of writing.

There are two common conceptual ideas that are used to describe EM

23

radiation and they have been used interchangeably throughout its long history: that of a wave, and that of a particle. In some situations EM radiation is more usefully described in terms of a wave, and in others as a flow of small packets (particles) of energy, called *photons*. In the case of microwaves, most of the important phenomena are best described with wave theory: frequency and wavelength, refraction, diffraction, interference, polarisation and scattering. For the purposes of these notes, I will therefore consider almost exclusively the wave description. I say "almost" because when considering the emission and absorption by specific molecules in the atmosphere it will be more convenient to temporarily switch to an explanation based on photons.

3.1.1   Electromagnetic Radiation as Waves

As outlined in the previous chapter, the term *electromagnetic* stems from the apparent properties of the radiation: that it consists of time varying electric and magnetic fields, which can propagate through space from one region to another, even when there is no matter in the intervening region. Such a propagating oscillatory phenomenon has the properties of a wave.

There are a number of different mathematical ways of describing a wave, each of which has an oscillatory nature and satisfies the same general conditions that are required to describe a wave. Many of the properties of waves can be studied mathematically without actually defining the exact shape of the wave. For simplicity, however, we do describe the shape of EM waves, and to make the mathematics as simple as possible it make sense to choose the simplest wave form that we can. We therefore base the wave profile on a sine or cosine curve. These are variously known as

Harmonic Waves

sinusoidal waves, *simple harmonic* waves, or *harmonic* waves[14].

Wave function

Let us consider a wave travelling along the z-axis, which we will describe by the function $\psi(z)$ (where we use the Greek letter *psi* to represent the function). It is convention to have the z-axis along the direction of travel of the wave (because the x and y plane will be used later to describe the polarisation). The profile of this simple wave function is:

$$\psi(z) = A \sin kz \qquad (3.1)$$

Wavenumber

where k is some positive constant known as the *wavenumber*, and kz is in units of *radians*[15]. The sine function can only have values between -1 to $+1$ so that the maximum value of $\psi(z)$ is A. The maximum disturbance of the wave is known as the *amplitude* of the wave and is therefore equal to A. Note that this is the distance from the z-axis to the maximum height of the wave, not the distance from a maximum to a minimum.

---

[14]   Those of you familiar with Fourier analysis will know that since any wave shape can be synthesised by a superposition of harmonic waves they therefore have a special significance.
[15]   We could have equally used a cosine instead of a sine function, but it is convention to use the latter.

This equation only describes the shape of the wave as a function of distance along the z-axis, but we are interested in waves that change with time — i.e. travelling waves moving in the positive z-direction with speed v. $\psi(z)$ is therefore also a function of time, so we should really write the function as $\psi(z, t)$. In fact, in order for Equation (3.1) to describe such a wave we need to replace z by its time varying equivalent, $(z - vt)$, since after a time t, the wave will have moved a distance vt. The reason that you subtract vt rather than add it, is that if the wave is travelling in the positive z-direction, then it is that part of the function that is towards the negative direction that will be at point z sometime later. (If you are travelling on the leading edge of the wave, the wave "history" lies behind you, not in front of you.)

The wave function is then given by

$$\psi(z, t) = A \sin k(z - vt). \tag{3.2}$$

This is now a complete description of the wave, at any location along its direction of travel, z, and at any time, t. We can now describe any electromagnetic wave by three generic parameters (an amplitude A, a velocity v and a wavenumber k) and two specific parameters (the distance, z, from some predefined co-ordinate origin, and a time elapsed from some predetermined moment, t) that describe what that wave is doing at any particular place and time. Equation (3.1) is therefore just a snap shot at time $t = 0$,

$$\psi(z, t)|_{t=0} = \psi(z) = A \sin kz. \tag{3.3}$$

Note that I haven't drawn your attention to any illustration yet. It is important to understand that simply on the basis of the oscillatory nature of the sine function, we have developed a mathematical function that has all the same properties of a travelling wave. The manner in which we picture that wave comes *after* the mathematical description. We can then use that function to consider some special cases and gain some insight into the properties of such waves even before we start to visualise what is happening.

For instance, if we keep either z or t fixed, we still have a sinusoidal disturbance and so the wave is periodic in both space and time — fixing one still leaves a disturbance that can be described by a sine function. The spatial period, the distance over which our function repeats itself, is known as the *wavelength* and is denoted by the Greek letter *lamda*, λ. By defi- *Wavelength* nition, an increase or decrease in z by the spatial period should give the same value of ψ, i.e.,

$$\psi(z, t) = \psi(z \pm \lambda, t).$$

For a harmonic wave, this is equivalent to altering the argument of the sine function by $\pm 2\pi$, since $\sin(x \pm 2\pi) = \sin(x)$, so that,

$$
\begin{aligned}
\sin k(z - vt) &= \sin k[(z \pm \lambda) - vt] \\
&= \sin[kz \pm k\lambda - kvt] \\
&= \sin[k(z - vt) \pm 2\pi],
\end{aligned}
$$

giving

$$(kz - kvt) \pm k\lambda = (kz - kvt) \pm 2\pi.$$

k and λ are both positive numbers so that the wavenumber k must be related to the wavelength by

$$k = \frac{2\pi}{\lambda} \qquad [\text{m}^{-1}]. \qquad (3.4)$$

We have now defined the constant, k, in (3.2) in terms of the spatial period of the wave.

In exactly the same way, we can consider the temporal period, T. This is the amount of time it takes a complete wave to pass a stationary observer. It is left to the reader to show that using the same method as above,

$$kvT = 2\pi, \qquad (3.5)$$

or, since we can substitute Equation (3.4) for k,

$$T = \frac{\lambda}{v} \qquad [\text{s}]. \qquad (3.6)$$

This gives us a relationship between the spatial period of the wave, the temporal period, and the wave velocity. For the rest of the book we will only be dealing with electromagnetic waves, so we may replace v with c, the conventional symbol for the velocity of EM radiation.

The period T is the number of units of time per wave cycle, but we often use its inverse,

$$\nu = \frac{1}{T} \qquad [\text{s}^{-1}], \qquad (3.7)$$

Frequency   which is the *frequency* denote by the Greek letter *nu*, ν, the number of waves (or cycles, or oscillations) per unit of time. The unit of frequency is the number of cycles per second and is measured in *Hertz*, named after Heinrich, who was introduced in Chapter 2.

We now have the simple, but extremely important, relationship

$$c = \lambda\nu \qquad [\text{ms}^{-1}]. \qquad (3.8)$$

There is one further quantity that is used to describe wave motion: the *Angular Frequency*   angular frequency, denoted by the *omega*:

$$\omega = \frac{2\pi}{T} = kv \qquad [\text{radians s}^{-1}]. \qquad (3.9)$$

This is a rate of change of phase angle and so is measured in units of radians per second, rather than the more common waves-per-second for "frequency". The additional 2π term is simply the radian equivalent of one full cycle.

This leads to the most common way of expressing (3.2):

$$\psi(z, t) = A\sin(kz - \omega t). \qquad (3.10)$$

The reason for using an angular frequency is more apparent if we consider a particular way of geometrically representing a sinusoidal wave — with a circle. This may not seem an obvious choice at first, but Figure 3.1

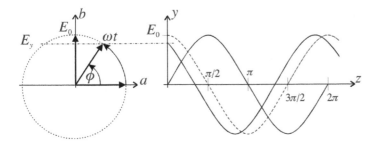

FIGURE 3.1 This figure illustrates how a sinusoidal waveform can be represented by a pointer (of length E_0) rotating anticlockwise. The amplitude of the wave is the projection of the pointer onto the y-axis. The phase, ϕ, denoted by the Greek letter *phi*, is then the angle that the pointer makes with the horizontal axis. The initial phase, ϕ_0, is the angle at time zero.

illustrates how the oscillatory nature of a wave can be related to the cyclic motion of a rotating vector. If we consider a pointer, or vector, of length A turning anti-clockwise about the origin of a Cartesian coordinate system with a constant angular frequency (rotations per second) of ω, it will trace out a circle over time[16]. If we project the end of the pointer onto the y-axis (i.e., $A\sin(\phi)$) and then plot how this changes with time (so that $\phi = \omega t$), we find that the projected function is identical to the function given in Equation (3.10), for a static location ($z = 0$). We now have a very convenient way of describing a wave as a function of time — as a rotating vector with length A and angular frequency ω.

This is not quite yet a complete description of the wave since at time $t = 0$, and position $z = 0$ (the origin of the wave), the pointer may start at any angle, and not necessarily so that $\psi(z,t) = 0$. We therefore need an extra parameter that we call the *initial phase*, ϕ_0. *Phase*

The *phase* of the wave describes at what stage of the cycle the wave is in and by definition we make it equal to the angle ϕ in Figure 3.1. Mathematically this is the entire argument of the sine function of Equation (3.10), so that,

$$\phi = (kz - \omega t).$$

The initial phase defines the stage of the cycle that the wave starts from, so that the most complete description of the wave is

$$\psi(z,t) = A\sin(kz - \omega t + \phi_0). \qquad (3.11)$$

A complete description of an electromagnetic wave travelling at velocity c therefore requires a measure of amplitude A and absolute phase ϕ, in addition to a measure of the frequency or wavelength.

---

[16] You should recall that one full cycle is 2π radians. If you don't, then refer to Appendix A.

In fact, even this is still not entirely the whole story since we must also describe the direction of the wave. For instance, if we have a collection of waves each travelling in a different direction then we no longer have the convenience of being able to define the z-axis as being "along the line of propagation". We therefore need a way to describe the direction of travel

Wave vector

for the wave. A convenient way to do this is to define a wave *vector*, **k**, rather than just a wave *number* (which is a scalar quantity and has no direction). The wave vector can be further broken down to give a unit vector $\widehat{\mathbf{k}}$ which has unit magnitude and so only defines the direction of the wave, and the original wavenumber, k, which characterises the magnitude of the wavenumber so that

$$\mathbf{k} = k\widehat{\mathbf{k}}, \qquad \left|\widehat{\mathbf{k}}\right| = 1 \qquad [\text{m}^{-1}].$$

The vertical bars indicate that it is the length of the vector (the *magnitude*). The wavevector now characterises the direction of the wave as well as the original wavenumber.

3.1.2   Complex Wave Description

As was intimated earlier in this chapter, Equation (3.11) is only one way in which to mathematically represent a wave. There is another more condensed way to represent a wave using the properties of complex numbers. If mathematics is not your forte, do not panic, it is not imperative that you completely understand this terminology, but it is such a convenient (and common) way to describe a wave that I would urge you to make the extra effort to get to grips with the complex format — it is by far the most useful for microwave remote sensing. (A primer on complex numbers is given in Appendix A which should help you get started.)

The one relationship that is required to make the link from sinusoidal waves to complex numbers is

$$e^{i\theta} = \cos\theta + i\sin\theta$$

where $i = \sqrt{-1}$, e is Euler's constant[17] and θ is some angular value (such as a phase). Although a validation of this relationship is given in Appendix A, you needn't worry about where it comes from — just accept it as given.

In this case, we call the cosine term the *real* part and the sine term the *imaginary* part. Note that the right hand side of this equation simply relates to the vector of length A we used to describe a wave in Figure 3.1. The cosine term is the projection of the vector onto the x-axis and the sine term the y-axis projection. The use of the imaginary term i can be thought of as merely a convenience for keeping the two components separate. We now call the y-axis the imaginary axis with units of $i = \sqrt{-1}$, so that the axes in Figure 3.1 can be labelled "real" and "imaginary", rather than x and y.

---

[17] e is an irrational number that is equal to 2.718281828... You will find it on any good calculator.

θ then corresponds to the phase, ϕ, and A the amplitude of the wave, so that Equation (3.11) can be re-written as

$$\psi(z,t) = Ae^{i(\omega t - kz + \phi_0)} = Ae^{i\phi} = A(\cos\theta + i\sin\theta). \qquad (3.12)$$

This includes everything we need to describe a travelling wave: the amplitude, the initial phase and a wavenumber. By convention, when we want to consider the actual wave profile, we simply consider the real component only[18].

The complex representation looks unnecessarily ... well, unnecessarily complex, but later on we will find that this makes much of the future mathematics far easier. In conjunction with Figure 3.1 it also provides a visually straightforward way to picture waves of different amplitudes and phases by considering rotating vectors of different lengths and starting angle. Each wave has an amplitude (which is related to the power of the wave) and a phase angle (which describes at what stage the cycle of the wave is at) and so each wave can be represented by a unique vector. The phase angle can be any number within the complete cycle of 0 to 2π (or 0–360°) and in some circumstances, rather than necessarily representing its true *initial* phase, it could simply be arbitrarily assigned as the phase angle relative to some other pre-defined wave (and is therefore a phase *difference*), which is often a more convenient way to describe phases in a practical measurement context.

From Figure 3.1 it should be apparent that an amplitude (or magnitude) and an angle between 0 and 2π (polar co-ordinates) is equivalent to giving a real and imaginary component (Cartesian co-ordinates). Each vector (and therefore wave) only requires two values to describe it within this context because we have assumed the velocity and the frequency are constant. Differences in angular frequency, ω, would correspond to different rates of rotation of the vectors, which is difficult to represent in graphical form. These diagrams therefore always assume a constant angular frequency and wave velocity.

3.2   Energy and Power of Waves

All of the discussion so far is relevant to all types of wave phenomena. In this book, however, we are explicitly interested in electromagnetic waves so we will now introduce a rather more specific terminology for the amplitude. Instead of the general term A, we want a term that is more physically meaningful for EM waves, which, you will recall, are actually a combination of electric and magnetic field waves. It is convention to always consider *only* the electric field vector in our description of the wave and use its amplitude, E_0, to characterise the wave, as introduced above. There is

---

[18]   Clearly the description of a sinusoidal wave can be done equally well using a sine function or a cosine function, the only difference being a quarter cycle ($\frac{\pi}{2}$) difference on how the phase is described.

good reason for this as for most natural substances it is only the electric field that is altered directly when EM waves interact with them.

The amplitude of a wave is a useful parameter, especially when we are interested in the exact form of a wave and when considering the rotation vector description in Figure 3.1. However, in practical situations a detecting instrument will often measure the incident power, P, which is defined as the energy per unit time (giving units of Watts, W). Power-like properties are also what we will use when considering wave interaction with materials and objects, so it is important to make the link between amplitude and power. The power of a waveform is given by the square of the amplitude divided by a constant, η (the Greek letter *eta*) that is determined by the material through which the wave is travelling, so that

$$P = |E(z,t)|^2 = \frac{E_0^2}{2\eta} \text{ [W]}. \qquad (3.13)$$

where η is termed the *impedence*. This is a straightforward relationship, but a significant one because it is important to know that power scales differently from amplitude. The distinction being important when we consider the combination of waves in the next section, and in later chapters on how to record data and how to represent data in an image.

Areas

So far the discussion has centred around waves along a specific direction, but in remote sensing we are often dealing with distributed areas — antennas, ground targets, cross-sections, and so on — so we need to define a power per unit area, i.e. a *density*, given in units of Wm^{-2}. There are two special cases for the power density: the first is the *irradiance*, E which is defined as the total amount of radiant power incident on a unit area. The second is the radiant *exitance*, M, which is the total amount of outgoing radiated power (emitted or reflected) from the unit area. Since the distinction between these two terms is the direction of flow of the power, and because in remote sensing the unit area is usually on the Earth's surface, it is often convenient to label these terms *downwelling* and *upwelling* radiation, respectively.

Directional Power

The exitance and irradiance refer to the total power leaving or arriving over an area, but it is often necessary to distinguish radiant power coming from different directions — for instance, to characterise the radiation incident upon a microwave antenna. To this end we require some kind of "unit direction", which by convention we use the steradian, the unit of solid angle (see the Appendix for a definition and explanation of solid angles). From a defined location, we can define the *radiance* as the measure of how much radiative energy comes from a given direction, and has units of $Wm^{-2}sr^{-1}$.

It might also be necessary to distinguish between different frequency ranges also, in which case we need to define a *spectral* radiance, i.e. radiance per unit frequency, or $Wm^{-2}sr^{-1}Hz^{-1}$. Note that this can also be defined as "per unit wavelength" in which case the units are $Wm^{-2}sr^{-1}m^{-1}$.

It is this spectral density per unit solid angle incident on a point that we

more conveniently call "brightness", or "intensity". In remote sensing, the
incident point of special significance is the detector, so that brightness becomes a measurement quantity as opposed to the spectral radiance, L, of
the object being observed. If we do not consider the spectral characteristics of the detector, then B and L are equivalent. This may seem strange
at first since the energy leaving the target must be greater than the energy
arriving at the instrument, but the inclusion of the "per unit solid angle"
balances out the "per unit area" so that the two are equivalent. A white A4
page of paper, for instance, does not look less bright simply because you
move it further away. The total amount of energy leaving the page remains
constant, but as you move the page away two things happen: the total energy at your eye (the detector) drops because of the increasing distance,
but the energy that does arrive is now condensed into a smaller solid angle
(the apparent angular dimension of the page). Both the brightness at the
detector and the radiance at the object remain constant. The use of brightness as a means of quantifying observations is therefore particularly useful
as it says something about the target, rather than about the way in which it
is being measured.

*Brightness and Inten-
sity*

3.2.1   Polarisation

There is one final property that we need to consider for EM waves: polarisation. This is because EM waves are *transverse* (the oscillations are
perpendicular to the direction of travel of the wave — "up and down" like
a rope) unlike sound waves, which are longitudinal (the oscillations are
in the same direction as the wave vector — "to and fro" like a spring).
A transverse wave therefore has an extra property that longitudinal waves
lack: a parameter that describes the direction in which the oscillations are
taking place. If you were to send a wave down a piece of rope, for instance, in order to describe the wave completely you should also define the
way in which the rope is waving (up-down or left-right, for instance). For
EM waves, this property is known as the *polarisation*.

Many wave effects, such as interference and diffraction, which are discussed later, can occur with any kind of wave, including sound waves and
surface waves on a pond of water, as well as EM waves. These effects are
generally independent of whether the waves are longitudinal or transverse.
Polarisation, however, is a property of wave phenomena that is meaningful
only for transverse waves.

Any transverse wave can be represented as a superposition of two component waves, one having displacements only along the x-direction, the
other only along the y-direction (where we are again taking the direction
of propagation as being along the z-axis).

The wave having only y-displacement is said to be linearly polarised
in the y-direction, and the one with only x-displacements is linearly polarised in the x-direction. Linear refers to the fact that the oscillation is
in a straight line. The perpendicular polarisation planes are usually arbi-

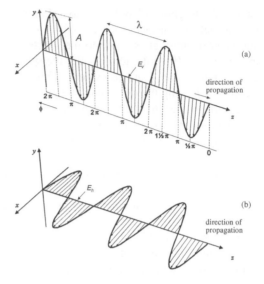

FIGURE 3.2 When the component of E in the x-direction is zero, $E_x = 0$, there is only oscillation of the electric field vector in the y-direction, and the wave is said to be vertically polarised (a). Similarly, when the component of E in the y-direction is zero, $E_y = 0$, there is only oscillation of the electric field vector in the x-direction, and the wave is said to be horizontally polarised (b). (After van der Sanden, 1997).

trarily called *horizontal* and *vertical*, usually with reference to the Earth's surface. By convention, the x-direction is usually taken as the horizontal and the y-direction the vertical. This situation is illustrated in Figure 3.2. The radiation can then be said to be horizontally polarised, or vertically polarised. We will discover in the following chapter that polarisation may also be circular, or elliptical, but that every polarisation can still be described using a combination of only two (complex) terms, usually a horizontal and a vertical component.

For EM radiation, by convention, the direction of polarisation is taken to be that of the electric-field vector, not the magnetic field. This is because the most common manifestations of EM radiation are due chiefly to the electric field, not the magnetic field. The wave equation used for EM radiation therefore takes on the general form

$$E(z,t) = E_0 \sin(kz - \omega t + \phi_0) = E_0 e^{i\phi} \tag{3.14}$$

where E_0 is the amplitude of the electric field vector (rather than the more general term, A). However, because we are now dealing with polarised waves we need to describe the amplitude in the x-y plane. We can do this by describing the x and y motions independently, so that we now have

$$E_x(z,t) = E_{0x} \sin(kz - \omega t + \phi_0) = E_{0x} e^{i\phi}, \tag{3.15}$$
$$E_y(z,t) = E_{0y} \sin(kz - \omega t + \phi_0 + \varepsilon) = E_{0y} e^{i(\phi + \varepsilon)}. \tag{3.16}$$

The term ε describes the difference in phase between the x and y waves.

Only the basic concept of polarisation is introduced here, with a more complete examination of the topic given in Chapter 4.

3.3 Combination of Waves

One question that is very important when dealing with waves is: what happens when we combine two, or more, waves together? Like many cases in physics, there is a simple answer using some basic simplifying assumptions, which then becomes increasingly more complicated as the concept is developed. One thing is for sure: the concepts relating to the superposition of waves (*coherence* and *interference*) are critical for understanding active microwave imaging and some advanced types of passive imaging. It is therefore extremely important to have an understanding of these principles.

The superposition of waves in called *interference*. At any given time   *Interference* and place, the wave that results from the addition of two waves of identical frequency and amplitude A, is the sum of the instantaneous amplitudes of the two waves. These waves can be at different relative stages of their oscillation cycle, i.e. they can have different initial phase ϕ_0. It is the phase difference between the two waves (which we will represent as $\delta\phi$) that will determine the final wave amplitude, which may be anything between 0 and $+2A$. This is illustrated in Figure 3.3. When the phases are identical, the phase difference is 0, and the waves combine maximum-to-maximum and minimum-to-minimum. In this special case, the waves are said to interfere *constructively*, resulting in a final amplitude of $2A$. A phase difference of one half cycle (one half wavelength or π radians) results in maxima combining with minima and vice versa, giving *destructive* interference and an amplitude everywhere equal to 0 — the waves cancel each other out, resulting in no wave at all!

Visualising the combination of waves is relatively straightforward for these two special cases, but when the phase difference is one quarter cycle ($\pi/2$ radians) it is less obvious that we get a wave of amplitude $\sqrt{2}A$.

More generally, what about the situation where more than two waves overlap, all with different amplitudes and phase values. This can be calculated analytically, using either the trigonometric or complex expressions, but it is not easy to picture what would result from such a process using the traditional sinusoidal wave profiles used in Figure 3.3. A far easier representation is to use the vector approach introduced in Figure 3.1, whereby we add together (nose-to-tail) each vector that represents an individual wave. The vector that results from this addition represents the new wave. Using this method, it is possible to get a real feeling for what interference entails, even for many waves. For instance, let us reconsider the two simple cases introduced above. For constructive interference of two identical waves, the vectors are the same length, but they also have the same direction, so when they are added together the result is a wave that

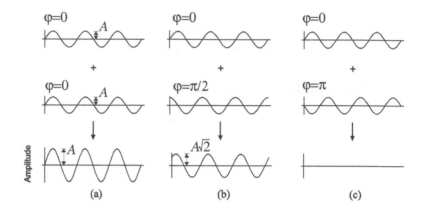

FIGURE 3.3   The resulting wave form from a superposition of waves depends upon the relative phase of the original waves. When they are in-phase, $\delta\phi = 0$, and the result is constructive interference with a wave form twice the amplitude of the original waves. When the waves are completely out of phase, $\phi = \delta\pi$, and the waves destructively interfere, resulting in a wave with zero amplitude; i.e. no wave at all.

has the same phase as the originals, but twice the amplitude ($2A$). On the other hand, if the waves are exactly out of phase, the two vectors point in opposite directions and a vector addition gives you 0 — no wave at all, as determined before. This is illustrated in Figure 3.4.

The case of the quarter cycle phase difference can then be found from vector addition and Pythagoras, so that the amplitude would be $\sqrt{2}A$ and the phase would be half way between the phases of the original waves. In a similar manner we can visualise the addition of many waves, as shown in Figure 3.5. Such a visual aid is an important tool when trying to understand interference effects. Those readers who are not mathematically inclined are particularly encouraged to become familiar with this graphical representation, as it can provide a means of representing quite complicated situations without recourse to the mathematics.

3.3.1   Coherence

A prerequisite for waves to interfere is that they are *coherent*. In fact, coherency may be defined as that condition which allows interference, but that is clearly not such a helpful definition when you are trying to understand what it means, so let me be more specific.

Two waves with a phase difference that remains constant over time, are said to be coherent. This implies that their frequencies are identical even if their amplitudes and initial phases are different. It is this condition that has allowed us up to now to consider waves as *stationary* vectors (as in Figure 3.4) and ignore the fact that they are really rotating with time (as in Figure 3.1). If they rotated at different angular frequencies the result of the vector

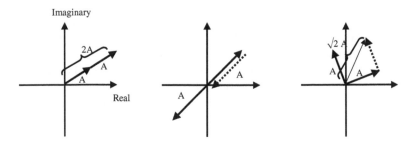

FIGURE 3.4 This figure is equivalent to Figure 3.3 but using the vector representation. In the first diagram, the two waves are in-phase and they have the same amplitude, A. They are therefore identical. When we add these two vectors together we get a new vector, with the same initial phase, but twice as long. When these two waves are out of phase by a full half cycle, as in the centre diagram, the two vectors add together to get a zero (null) vector. In the final case, the phase difference is a quarter cycle ($\frac{\pi}{2}$). The vector addition then gives a wave that has an amplitude of $\sqrt{2}A$ (from Pythagoras) and a phase angle half way between the two initial phases. Note that these diagrams will be the same (albeit rotated) even if we change the absolute phases of the vectors — it is the relative phases between the two vectors that is the important consideration.

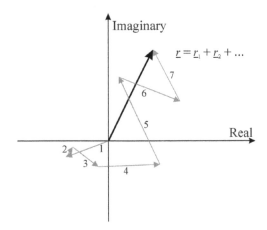

FIGURE 3.5 The vector representation for a wave is a convenient way to visualise the process of interference. Each original wave is represented by a vector and the resulting wave is determined by adding (in a vector fashion) the individual contributing waves. The resulting vector gives the amplitude and phase of the wave that is the result of the interference.

addition would vary with time, meaning the interference effects are less predictable. In this context coherence can be thought of as a measure of predictability — the higher the coherence of two waves, the easier it is to predict the properties of one of those waves given knowledge of the other.

Mathematically, we quantify this predictability by using a measure of the similarity of two waves measured over some finite interval of space or time. It makes no sense to consider coherence at some infinitesimal *point* in space and time — coherence is only meaningful when taken over some finite interval of space or time (in the same way as predictability is also only meaningful when predicting over space or time).

The complex coherence between two waves, E_1 and E_2, is denoted by a term using the Greek uppercase gamma, Γ_{12}, and is defined by a complex cross-correlation[19] of the two waves such that as

$$\Gamma_{12} = \langle E_1 E_2^* \rangle. \tag{3.17}$$

Here the $\langle ... \rangle$ brackets denote the *ensemble average*, which strictly speaking means that the values are averaged over different *realisations* of the same measurement – i.e. many times under the same conditions. The star indicates the complex conjugate, so that the sign of the imaginary part is changed.

In (3.17) Γ_{12} can vary in value if the amplitudes of the wave change. A more general parameter that is not dependent upon the absolute amplitudes of the waves, is a normalised version of Γ_{12}, usually expressed using lower case gamma, such as γ_{12}, and defined by a complex cross-correlation of the two waves such that

$$\gamma_{12} = \frac{\Gamma_{12}}{\sqrt{\Gamma_{11}\Gamma_{22}}} = \frac{\langle E_1 E_2^* \rangle}{\sqrt{\langle |E_1|^2 \rangle \langle |E_2|^2 \rangle}}. \tag{3.18}$$

Ensembles   In practice, remote sensing measurements cannot always measure exactly the same thing many times, as an ensemble average strictly requires, so the ensemble averaging is often approximated by an average over measurements that are very nearly identical. This may be by making measurements very close together in time, or measurements that are very close spatially. The average is then taken over a series of measurements or over a number of locations, the assumption being that each measurement effectively represents a new observation of the same target area.

A more practical alternative way of expressing coherence can therefore be given by

$$\gamma = \frac{\sum_N E_1 E_2^*}{\sqrt{\sum_N |E_1|^2 \sum_N |E_2|^2}}, \tag{3.19}$$

where for convenience we now drop the subscripts on γ, and the upper case sigma, Σ, signifies a sum over N elements. In this case, N is the

---

[19] It is a *cross*-correlation because we are correlating one wave with another, and it is complex because the waves are described using complex numbers.

number of samples over which the coherence is being estimated, either in time or space (for instance, over a collection of pixels). Note that the denominator in each equation simply normalises the measurement using the magnitudes of the two waves so that the magnitude of γ will range from 0 (not coherent, or *incoherent*) to 1 (completely coherent). The coherence is only equal to 1 when $E_1 = E_2$. Note that γ is a complex number, so it also has a phase value.

In the real world, the coherence of EM radiation is rather complicated. The molecular or electron transitions that cause radiation occur over finite (albeit very small) time scales, for instance, and this may result in minute variations in radiation frequency. Minute, but enough to disrupt the coherence. Radiation originating from different locations within a source will also lose coherency because of their spatial separation. Similarly radiation that was generated at the same location but at different times will also be less coherent. In general, the more you separate the source of the waves in either space or time, the less coherent the waves.

We will see later that a key component of imaging radar systems is that they are able to maintain coherency between consecutive pulses, and because of this it is possible to assume the waves are always coherent. In passive systems, it is not so obvious that the source may have any degree of coherency, but under particular conditions (i.e. short time intervals and over small spatial scales) passive sources will also maintain a degree of coherence.

3.4   The Most Important Section in This Book

The discussion on interference above rather abstractly discussed the combination of two waves travelling in the same direction — a simple one-dimensional case. In this section we now consider the effects of interference in a two dimensional plane. In order to do this, we must first look again at what information is contained within the phase of a wave.

I call this the most important section of this book because the concepts introduced here are so rarely dealt with in remote sensing text books, yet they form the foundation of many of the more difficult topics found in later chapters — antenna sensitivity patterns, aperture synthesis, image speckle, interferometry, to name a few.

3.4.1   Phase as a (Relative) Distance Measure

In a fairly straightfoward manner, the phase of a wave changes with distance. If we have two identical waves but allow one to travel a fraction of a wavelength longer, then both from Figure 3.1 and Equation (3.11) we can see that the phase will also change. Similarly, a wave travelling from a source, say at point A, to a detector at point B_1, will travel a distance that can be divided up into some number of whole wavelengths, plus some additional fraction of a wavelength.

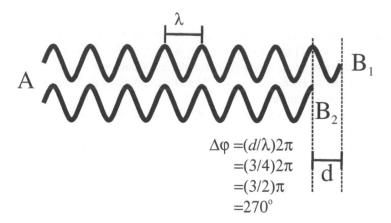

FIGURE 3.6 Measuring the phase difference measured at the two detectors B_1 and B_2 gives you information on the separation distance, d. In this case, a phase difference of $3\pi/2$ equates to a path length difference of $3/4$ of the wavelength. Knowing the wavelength then gives a value for d, as long as $d < \lambda$.

If the phase of the received wave is measured at the detector it does not tell us about the absolute distance (it does not count all the whole wavelengths) but it does tell us something about the remaining fraction of a wavelength.

Unfortunately, unless you know the initial phase of the emitted wave at A, and the absolute number of whole waves, there is nothing useful you can actually do with this phase information on its own. However, imagine that we have another detector at a second location B_2, a small (<1 wavelength) distance, d, closer to the emitter at A (see Figure 3.6). In this case, we will measure a different phase at B_1 compared to B_2, and this difference in phase is a direct measure of the difference in the length of the path from A to B_1, versus A to B_2, i.e. the distance d. A phase difference of $3\pi/2$ (270°) for instance, indicates a path length difference of three quarters of a wavelength (whatever that may be for the particular frequency of waves being used). In this case it does not matter what the initial phase is at A, since it is the same wave that both detectors measure – it is the phase *difference* that is the key piece of information that tells you about the different path lengths and hence the distance, d.

Notice that I specified that d was less than one wavelength. If it was greater than this, we start to compare the phase of two different parts of the wave and our phase difference can be out by whole numbers of wavelengths.

Let me explain that a different way.

Imagine d is slowly increased: 0, $\pi/2$, π, $3\pi/2$,... Eventually we reach

2π. But in terms of measuring the wave, there is no way to tell the difference between 2π and 0. It is a cycle (cf. Figure 3.1). We therefore have an ambiguity in our measure of phase — we know the relative difference in that extra fraction of a wavelength, but we still do not know the number of whole wavelengths. If we shift d in steps of λ we would always measure the same phase difference. This phase ambiguity can always present itself as a limiting factor whenever we are considering phase information.

This relationship between path length and phase is an important one, and is fundamental to the following discussion on interference patterns.

3.4.2   Combining Two Waves in 2-D

Figure 3.7 illustrates the case where there are two sources, S_1 and S_2, both lying in the same horizontal plane and emitting waves of the same frequency and amplitude A in all directions. To begin with we will consider the case where the emitted waves are both "in-phase" (i.e. the two transmitters are synchronised so that at any given instant the emitted waves have the same phase). The distance, d, between them is larger than a few wavelengths. Since we have defined the waves as being at exactly the same frequency and amplitude, we can assume that these waves are coherent, so at any location where the two wave influences meet, they will exhibit interference.

It is also worth clarifying that for this discussion the two sources are vertically polarised. In fact the following description of how the waves interfere is similar no matter what polarisation the sources are — as long as they are the same. This discussion would equally apply to longitudinal waves, such as sound waves. However, by keeping the discussion to a horizontal plane, with vertically oscillating sources, allows us to use a convenient analogy — that of surface water waves — as some of the graphics will imply.

Qualitatively we can say that at each point in the plane in front of the sources there is one wave arriving from S_1 and one from S_2. In each *Qualitative Description* case there is a phase difference between the two arriving waves as a consequence of the path length difference between each point and the two sources. In some locations the waves will arrive in-phase, with a phase difference of 0 or some multiple of 2π, so that they interfere constructively to give an amplitude of $2A$. Over time, a detector at this location would see a wave with a frequency of the original source, but an amplitude of $2A$, or a power proportional to $4A^2$.

In those locations where the two path length distances differ by half a wavelength the waves are out of phase by π (one half wavelength or $180°$) and the waves interfere destructively resulting in zero amplitude. There are also locations where the phase differences are somewhere between these two extremes and the amplitude lies somewhere between 0 and $+2A$ (refer back to Figure 3.3). The space in front of the two sources, therefore, has resulting wave patterns with amplitudes lying somewhere between 0 and

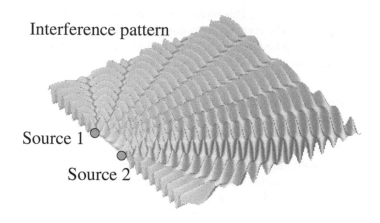

FIGURE 3.7 This figure illustrates the interference pattern lying in front of two identical (coherent) wave sources that are in phase with each other. The surface is drawn to help you see the nature of the pattern. It should also help you make the analogy to water waves, for instance. The same pattern would exist using sound, in which case the two sources would be loudspeakers and the amplitude of the pattern would indicate loudness. Note that this image is a snap-shot — the waves continuously move outwards from the sources.

2A. Let us now consider the nature of the pattern of distribution of these highs and lows.

Figure 3.7 was a simulated result for an example that has $d = 10$ wavelengths. The figure shows the resulting field at some *instant* (remember the waves are moving) in a representation that relates back to the water wave analogy mentioned above. At a distance of a few wavelengths away from the transmitters we can see the emergence of a regular pattern of peaks (highs) and troughs (lows) associated with the constructive interference. They make a distinct pattern in the plane around the two sources, which is clearer if we look at the amplitude pattern in plan view shown in Figure 3.8, where the grey scale goes from $-2A$ to $+2A$. The pattern seems to "radiate" out from the location of the two sources in a number of discrete directions. This is known as an "interference pattern" and in the case of

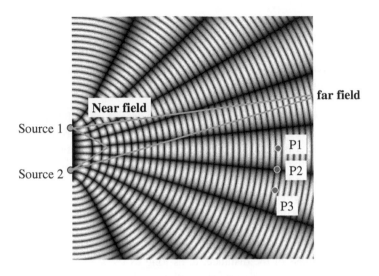

FIGURE 3.8 If two nearby coherent sources generate the same wave signal, an interference pattern is generated. With electromagnetic radiation, this pattern exists in the space infront of the sources rather than as something you actually see. If a sensor is placed at P_1 or P_3, a signal maximum is detected, whereas at P_2 a signal minimum is detected.

two sources the pattern is well defined.

Before looking in more detail as to why this happens and what governs the shape of the pattern it is important to note one distinct feature of the pattern — near to the sources (i.e. within a handful of wavelengths) the pattern is very much more complicated than it is when you get much further away and the pattern is simply a collection of radiating beams. This is a general trend in coherent processes and so it is often important to make the distinction between what is referred to as the "near field" and the "far field". The key distinction is whether or not you are dealing with spherical wave fronts or plane, straight parallel wave fronts, over some small region (which may be another aperture, or a measuring instrument). When the difference between the actual wavefront and a plane wave front is less than one eighth of a wavelength $\frac{\lambda}{8}$, over the region in question, we say that we are dealing with the far field case. This is known as the "Fraunhofer criterion". The distinction between near and far fields is therefore a combined effect of distance from source and the extent of the region of interest. In Figure 3.8 the wavefronts can be considered straight and parallel when considering *either* a very small portion of the image, or at a great distance from the sources. The key factor is therefore that the region of interest must be very much smaller than the distance from the sources.

Near field vs far field

In the simple case that we are dealing with here, determining the solution of the near field is not an intractable problem, but as we introduce more factors into this scenario it should become apparent that dealing with the near-field is a very different ball game and one that is worth trying to avoid whenever possible. Besides, in a remote sensing context we are almost exclusively dealing with the far field condition (since by definition sources and detectors are very much farther apart than the size of the regions of interest, i.e. pixels). On a number of occasions throughout this book we will utilise some of the simplifications that can be made by assuming we are dealing with the far field.

The Interference Pattern

Let us now be a bit more explicit about the properties of the interference pattern by first considering the resulting wave field at some arbitrary points. What happens at point P_1 in Figure 3.8, for instance? To answer this we must consider the phase difference between the signal arriving from S_1 and that arriving from S_2. From symmetry arguments alone we can see that the path length (the distance travelled) from either source to P_1 is the same, since it lies on the axis of symmetry. The phase between each signal will therefore also be equal, resulting in constructive interference. In fact any point along this central axis of symmetry will have constructive interference regardless of the distance, d, as long as the sources are in-phase. It is therefore convenient to refer all other angles to this direction — i.e., this direction is defined as having a "look angle" of 0°. This is sometimes referred to as the *pointing angle*, or *boresight angle*, depending on the physical context with which we are dealing[20].

Now consider another location, this time below the main axis, such as P_2. From the diagram, it is possible to say that at P_2 the path length from S_1 is longer than that at P_1, but that the path length to S_2 is now slightly shorter. From Section 4.1 we know that a difference in path length equates to a difference in phase, since a path length, say, one half wavelength longer will mean the wave goes through a half cycle extra. If at P_2 the path length from S_1 is a whole half cycle, or π radians, longer than that from S_2 then this will correspond to a region of destructive interference.

If we look further than P_2, eventually the path length difference gets long enough that they are one full cycle, 2π, out of phase, and the waves interfere constructively again. If we go to even larger look angles we find that the phase difference becomes greater than 2π ... then it reaches 3π (destructive interference), then 4π (constructive interference, again) and so on. This is the reason we get many "beams" of constructive interference radiating in different directions.

How many of these beams there are, and the direction into which they radiate, must be related to the geometry between S_1 and S_2 and the size of the wavelength. If we halve the wavelength of the transmitted waves, for instance, do we get the same pattern? No. We get a similar pattern, and its shown in Figure 3.9, but this time we have a maximum (constructive interference) at P_2 where before we had a minimum. This is because although the physical path lengths are the same, there are now twice as many wave-

---

[20]   It is also important to note at this stage that all the discussion here applies equally well to the left hand side of the sources, with the solutions identical although mirrored along the vertical axis between the sources. For now we simply ignore this as it is independent of what happens on the right hand side.

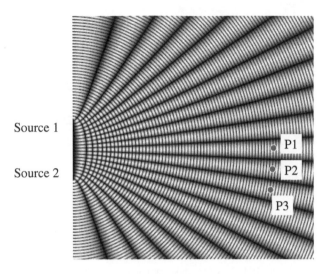

Source 1

Source 2

P1

P2

P3

FIGURE 3.9 Same figure as 3.8 except using waves with half the wavelength. Notice that there are now twice as many "beams" radiating from the sources.

lengths that can fit into that distance. If the pathlength difference was half a wavelength in the first case, then this now equates to a full wavelength (one full 2π cycle). By halving the wavelength, we double the number of beams (but each beam is now half as wide).

3.4.3   Quantifying the Interference Pattern

To provide a quantitative description of what is going on, we will invoke the fact that we are mainly interested in the far field solution. This allows for some very useful simplifications. Firstly, since we are now only considering the interference pattern very far away from the sources, we can assume that the pathlengths from both sources are parallel. In such a case we no longer want to think about P_1 as a specific location, but as a *direction*. This far-field case is illustrated in Figure 3.10 where the diagram has been redrawn to emphasise the directional peaks of the interference patter. For the rest of this chapter, and in fact most of this book, this far-field simplification will be the assumed unless otherwise stated.

We can determine the path length difference by considering the geometry *at the sources*, as illustrated in Figure 3.11. The far-field condition allows us to assume that the rays are in a specific direction and therefore parallel, which greatly simplifies the geometry. We can then derive an expression for the path length difference δl based on the separation d, and the look angle θ, such that

$$\delta l = d\sin\theta,$$

which we can use to determine the phase difference, $\Delta\phi$, by relating δl

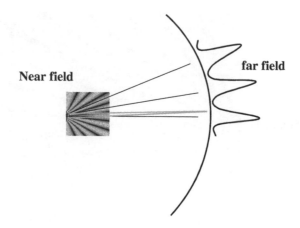

FIGURE 3.10 In this diagram we consider the same interference patterns from the previous figures but in the far field. It is then more appropriate to consider directions rather than locations within the field.

as a proportion of a wavelength,

$$\Delta\phi = \frac{\delta l}{\lambda}2\pi = \frac{2\pi d \sin\theta}{\lambda}, \quad \text{with } \delta l < \lambda. \tag{3.20}$$

Note that δl may be equivalent to many wavelengths, but in considering our interference pattern we are only interested in the *relative* phase difference, not the absolute phase difference — you will get constructive interference whether the phase difference is 0, 2π, 4π, or any integer number, n, of 2π's.

If we consider only those angles that give a peak in the pattern, i.e., constructive interference, then the phase difference must be equal to an integer multiple, m, of 2π. A maximum in the peak will therefore only occur if θ, λ and d satisfy the following:

$$\frac{2\pi d \sin\theta}{\lambda} = m2\pi, \tag{3.21}$$
$$d\sin\theta = m\lambda,$$

or

$$\theta = \sin^{-1}\left(\frac{m\lambda}{d}\right). \tag{3.22}$$

Physical Context   Within the context of this book, one relevance of the scenario described is in the context of active remote sensing — i.e. the transmission of microwaves for use in radar systems. From a practical point of view, the interesting thing about this scenario is that we have gone from a simple antenna that radiates equally in all directions, to two simple antennas working together to direct the wave energy into a small number of discrete directions. Its not perfect, because ideally we would like to radiate all the energy into one narrow direction, but it is a significant improvement on the pattern

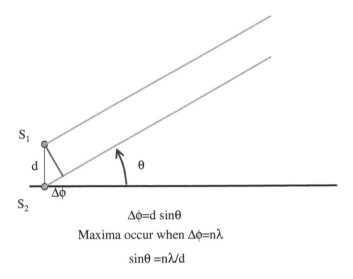

$\Delta\phi$=d sinθ

Maxima occur when $\Delta\phi$=nλ

sinθ =nλ/d

FIGURE 3.11 When considering the far field case it is appropriate to consider only the direction. This diagrams illustrates the geometry at the sources that influences the directionality of the interference pattern.

from a single source. We will show below how further improvement can be made in this regard by adding some more simple antennas.

Another impressive thing that can be done with just these two sources is to change the direction, or "steer", the interference pattern. This is done by introducing a phase difference to the signals transmitted from S_1 and *Changing Phase* S_2. If, for instance, we make S_1 lag S_2 by half a cycle, then the waves at P_1 are no longer in phase, but are exactly out of phase, giving destructive interference. A peak in the pattern is no longer along the line towards P_1, but is towards P_2, which now has constructive interference. If we have control over the phase difference between the transmitted signals we can then point the pattern in any direction we choose. We will return to this concept in Chapter 6.

3.4.4   Passive Case

Despite the comments above, it is important at this stage to appreciate that the analysis we are doing is not restricted to the active case where S_1 and S_2 are transmitters. We can also ask the question: what happens if S_1 and S_2 are now detectors and we place a source in the direction of point P_1? The discussion so far has only been one of geometry — the path length differences determine the phase difference and the path length difference is the same whether it is coming or going, so all the same principles and geometry must apply. A source in the direction of P_1 will result in the two measurements made at S_1 and S_2 being in phase, while a source at P_2 will

Reciprocity

result in the two measurements being out of phase by half a cycle (π). A process that works the same either "forwards" or "backwards" is referred to as reciprocal. We will return to the idea of reciprocity when we look at antennas, and when we consider polarimetry.

In general, the phase difference between the detectors will be given by the equation:

$$\delta\phi = \frac{2\pi\delta l}{\lambda} = \frac{2\pi d \sin\theta}{\lambda}. \tag{3.23}$$

Not surprisingly (since the geometry is the same) this is directly equivalent to Equation (3.20).

Direction

This is another important result because it means we can now use the phase of *measured* signals to tell us something about the *direction* of the source. If we have detectors that are sensitive to phase, and a method of combining the signals measured at each detector in a coherent manner (i.e. the measurement system is said to be "coherent") then we have a way of making measurements that have some sensitivity to direction even though our individual sensors are not! For instance, if we only consider received signals that are in-phase, we will only measure sources that lie within the maxima of the pattern in Figure 3.8. If we only consider measurements that have a phase difference of π then we are now sensitive only to a set of directions that lie within the minima of that pattern. The phase difference therefore tells us something about the direction of the source, albeit very crudely at this stage. In a remote sensing context, we have now achieved a degree of directional sensitivity for detection, using only two non-directional detectors, so long as we can measure the phase *difference* of the detected signal. Note that it is the difference that counts, not the absolute phase of the wave. In fact for both transmission and detection, we have now shown how even just a simple modification — two simple sources/detectors instead of just one — has given us some capability of directivity, so long as we have control over the phase.

IPD

This technique is exactly what mammals (including humans) do with their two ears to determine the horizontal direction of sound sources. The use of two ears provides a vital indicator of the direction of low frequency sound sources by measuring an Interaural Phase Difference (IPD). The IPD is simply the phase difference associated with path length difference between a waveform arriving at the ear nearest the sound source and arriving at the ear furthest away. This path length difference results in a phase difference between the ears of a fraction of a cycle. Such a phase shift between sound arriving at each ear can provide the information needed to find the corresponding azimuth angle of the sound source, just as described above for the detectors at S_1 and S_2.

Note, though, that there is a limitation to this technique: the ambiguity that exists between points such as P_1 and P_3, which both exhibit constructive interference. A constructive signal at the detectors gives an ambiguous result because it cannot distinguish between a source at P_1 and P_3. If we detect a zero phase difference it does not tell us from which of the many beams the signal originated. Mammals get around this because most sound sources cover many different frequencies (wavelengths) and Equation (3.23) is dependent upon wavelength— for a given wavelength, a measured phase difference will correspond to a different set of maxima in the interference pattern. If you have a sufficient number of frequencies, then you are left with only one direction that satisfies the phase differences for all the measured frequencies. This combined process can provide an azimuthal (horizontal) resolution of about 5–10° for humans, and as little as 1.5° for echolocating bats (Masters et al, 1985).

This is perhaps best explained through an example. If we reduce the wavelength of the signal used in Figure 3.8, but keep the same separation distance d, then we get the pattern shown in Figure 3.9. Note that the pattern is similar, but the beams are narrower and there are more of them. This makes sense, as a shorter wavelength means you do not have to increase the angle by as much to still result in a path length change of 2π. Now, if we make measurements at two frequencies or wavelengths from points S_1 and S_2, then as before, P_1 gives a high combined response, and it does so for both frequencies. P_2 on the other hand, has a high response at one frequency and a low response at the other, so we can now distinguish between signals from P_1 and those from P_2. By measuring across a wide range of frequencies, you can, in principle, distinguish the direction to any source using only two coherent detectors — such as a pair of ears[21].

Unfortunately, in most microwave remote sensing, both passive and active, we have to deal with a narrow bandwidth of frequencies, and so we need a different method of restricting the directional sensitivity. Mechanical methods such as the use of parabolic reflectors to focus the waves from a narrow direction are a suitable response to such a problem, and we look at those in Chapter 6 but for now, we will continue to concentrate on the more generic role of interference patterns.

3.4.5 Multiple Source Interference Pattern

Our aim in this section is to see if we can make our interference pattern such that we can focus the sensitivity to a single direction. For both the transmission case and the receiving case, we have the problem of having too many maxima and minima within the interference pattern. How can we improve upon this to get a single narrow beam?

The first improvement we can make is to add a third source, half way between S_1 and S_2. (This same argument applies for adding another detector, but for simplicity we will concentrate our discussion to sources rather than detectors.) Adding another source helps because at direction P_1, in the far field, all three sources combine constructively to give a total ampli-

---

[21]   It is for this reason that emergency vehicles are best served by wide-band sirens, rather than the traditional dual frequency "nee-naw". A wide band of frequencies allows other drivers to more accurately locate the emergency vehicle and thus take action to prevent hindering its path.

tude of $3A$, corresponding to a power of $9A^2$. At P_3, on the other hand, while we originally had two waves adding constructively to give a maximum in this direction, the contribution from the third source S_3 at point P_3 is actually half a cycle out of phase with the contributions from S_1 and S_2. S_3 lies exactly half way between S_1 and S_2, and so is half way between 0 and 2π. The result is that whereas S_1 and S_2 are pulling one way at point P_3, S_3 is pushing the other, giving the total signal here to be

Sidelobes

$(2A - A)^2 = A^2$. This is tending towards what we want — more amplitude in the central beam and less in the side beams, or "sidelobes". We can minimise these sidelobes even more by adding a few more sources between S_1 and S_2 — with each additional source we reduce the amplitudes in the sidelobes.

It still does not completely solve the problem however, because if we go to a large enough pointing angle eventually we find a direction where all the sources combine constructively again, giving another direction with a maximum amplitude. The only way to completely reduce the number of maxima is to have enough sources to make the spacing between each less than a wavelength — there is then only one direction (the boresight direction) in which every source can be exactly in phase. By way of example, the pattern resulting from ten such sources across a total distance of 6λ is shown in Figure 3.12. By using a combination of many identical, in-phase, point sources, we have now managed to direct most of the energy into a single beam — we have now got a narrow sensitivity range, even though each individual source has no directivity. There is still evidence of some sidelobes, albeit with very small amplitudes compared to the central beam, so not all the energy is radiated within the main beam, but it is still a vast improvement on two sources.

Whether dealing with transmission or reception by such an array of point sources/detectors it is clearly going to be a key question to ask about the width of the central beam and the amplitude of the sidelobes. We deal with that in the following section.

3.4.6   Beamwidth and Angular Resolution

Let us take the scenario of multiple source interference to an extreme limit by considering the number of sources across the distance d as increasing towards infinity (and all in-phase). In the near field condition, the pattern becomes much more ordered, with almost linear wavefronts parallel to the distance d, as illustrated in Figure 3.12. These linear parallel wavefronts correspond to a narrow beam in the far field, since they are now only travelling in a very narrow range of angles. It is not an infinitely narrow beam — it has a finite angular width and there are still some sidelobes.

Aperture

This scenario is of extreme importance because it can be used to describe the pattern that is formed by an *aperture* or an *antenna* — a transmitting pattern when there is a source at the aperture, but as also as a sensitivity pattern when the aperture feeds into a detector. It is convenient at

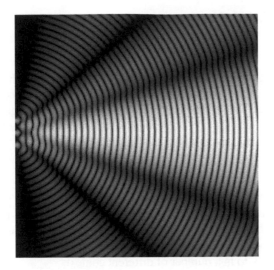

FIGURE 3.12 This interference pattern is created using ten sources, each closer than a wavelength to its neighbour. Note that most of the energy now ends up in the central beam, although there is still some energy in the sidelobes.

this point to generalise by talking about an "aperture" (of which an antenna is a specific example) because the same physics describes the properties of all apertures, including optical collecting devices.

For the passive optical case, the beamwidth corresponds to the angular resolution of a collecting aperture. It is therefore a very important question to ask, "how narrow is the beam in Figure 3.12?" as this will ultimately be important for defining the angular resolution of the system (a topic that will be dealt with in more detail in Chapter 6).

Firstly it is necessary to decide how to describe the width of the beam — there is no definitive way to do that. We could, for instance, choose the angular width when the intensity has dropped to half the maximum — this is the so-called *full width at half maximum* (FWHM) and is a common measure of beamwidth. *FWHM*

However, a simpler definition is to consider the look angle at which *resolution* the first minimum occurs. A convenient way to consider this question is to consider the vector addition of all the sources and picture what is going on using the vector diagrams introduced in Figure 3.5. We can represent the situation in the farfield as being a combination of n vectors each of amplitude A, each with a relative phase difference of $\delta\phi$. Minima occur when the vectors add in such a way as to bring the point of the last vector back to the origin: i.e., a zero vector result. By definition, this must be when the cumulative angle is 2π, so that $n\phi = 2\pi$. This is equivalent to saying that the first minimum is when the two ends of d (first to last n) are exactly in phase again, so the angle from the centre line to the first minimum, θ', is given by

$$\theta' = \sin^{-1}\left(\frac{\lambda}{d}\right). \tag{3.24}$$

Two things to note: the first is that the total beamwidth is twice this angle, which is the total angle between the two minima on either side of the main lobe. The second is that as long as d is much larger than λ and we use radians for the angles, then for small angles, $\sin\theta \simeq \theta$, so we can then write the anglular width of the beam to be,

$$\theta \simeq \frac{2\lambda}{d}. \tag{3.25}$$

The important thing to remember about this relationship is that if we increase the wavelength or decrease the distance d, we get a broader main beam. Decreasing the wavelength or increasing the total width of the line of sources gives a narrower beam. This is a very important result that we will come back to in a number of different contexts.

This may seem slightly different to forms of this equation you may have come across before. This is because it is more common to consider the *angular resolution* of the aperture, rather than the beamwidth. The angular resolution in this context is found by considering when two overlapping beam patterns are clearly two beam patterns, and not a single broad one. Again, there is some flexibility in how this distinction might be made. One criterion is to say that the two beamwidths are separable if the peak of one is located over the first minimum of the other, which would be given by

Angular resolution

$$\theta_r \simeq \frac{\lambda}{d}. \tag{3.26}$$

You will find that the angular resolution is more often quoted for a circular disk of diameter d, rather than a line of dipoles. For a circular disk of diameter d, the width is given by

$$\theta_{cd} \simeq \frac{1.22\lambda}{d}, \tag{3.27}$$

defined by the so-called *Airy disk*.

Other shapes of disk, or aperture, can give different beam angles, but they are never better than this result.

3.4.7   Huygens' Wavelets

Another important principle worth mentioning here is that introduced by Christiaan Huygens (1629-1695) a Dutch physicist and astronomer who made key developments in understanding the planets during the 17th century. Within the framework of his wave theory of light, which he first introduced, he took the idea of an infinite number of sources one step further by demonstrating that we can generalise the concept to any surface that is scattering or reflecting a wave. Any surface can be considered to be composed of an infinite number of closely spaced "secondary isotropic

Secondary sources

sources" — i.e., although they are really reflections of finite wavefronts they can be represented by a new set of sources that re-radiate a new wave of intensity and phase governed by the incoming wave.

The principle can be further extended to describe individual wavefronts so that any wavefront, of whatever shape, can be represented at any instant by a collection of imaginary sources across the wavefront each emitting isotropically, but with a phase and amplitude governed by the properties of the local wavefront. This is not meant to be a physical explanation of what it happening, merely an explanatory and analytical convenience that allows one to explore and explain wave phenomena — waves always act *as if* this is what is happening.

The collection of sources in a straight line described above, for instance, can now be thought of as representing a wavefront at an *aperture* of diameter d, instead of a row of sources. In this case we assume there is a sequence of plane (parallel) waves incident on some surface parallel to the wavefronts, which has a hole in it of dimension d. The infinite collection of small sources lying along d now represent the straight wavefront as it arrives at the aperture.

What then, is the difference between a straight wavefront at an aperture of size d and a collection of many small point sources lying along a line of length d? The answer is that there is no difference. The pattern that emerges at the far side of the aperture is identical to the pattern we got when considering the many source pattern of Figure 3.12. Equations (3.25) and (3.27) now equally well apply to a continuous aperture — in fact, you will often see these terms described as the "aperture limit". We will revisit this concept when discussion diffraction in Chapter 5.

3.4.8   More on Coherence

The discussion on coherence in Section 3.1 can now be illustrated using the approach above to describe multiple source interference.

One case we might consider is that each of the sources, $S_1...S_n$ are not antennas but individual atoms of, say, a gas. In this case the locations of each atom are not in a regular straight line, but will be randomly located. In order to simplify things, let us take the case of the ten sources given in Figure 3.12 but randomly move those sources off the line by anything up to a wavelength. The interference pattern shown in Figure 3.13 is what results. There is now no regular differences in the path lengths from each individual source and so the result is a rather chaotic pattern that possesses no obvious regularity. Without knowledge of the precise locations of each source, it would be impossible to predict the nature of this pattern — knowing the properties of the pattern in one location is no guarantee of predicting it in another.

Coherence is a comparative quantity — it is the measure of the similarity between two waves separated in space or time. The pattern is Figure 3.13 could be said to be incoherent because there is no regularity across the

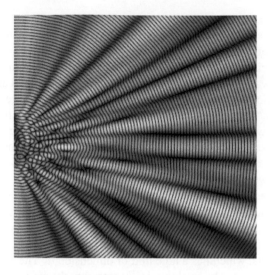

FIGURE 3.13 This is the interference pattern that results when the ten scatterers from Figure 3.12 are moved in random directions (but by less than a wavelength). The interference pattern now looks rather chaotic and irregular, although it still maintains some qualities similar to Figure 3.8. It can therefore be considered to be only partially coherent.

space it occupies — if we compare the waves in one direction with those in a marginally different direction, we find that they do not behave in similar ways. By contrast, the regularity of Figure 3.8 means that the behaviour is very similar between adjacent directions and is therefore predictable.

Natural sources like gas molecules have an added level of incoherence because firstly they generate waves that are slightly different in frequency, and secondly they are moving. The resulting pattern appears to be extremely random, changing in time as well as space. However, because such changes happen over such a short time scale, the average power in the far field evens out to some average, unchanging, value. Only measurements made in time intervals shorter than these fluctuations would measure anything other than a constant intensity. This is what we expect from an incoherent source, such as a light bulb, viewed by the naked eye (which integrates over too long a time to observe the fluctuations).

It should be noted here that if we were to be pedantic, it could be argued that there is *some* regularity in the pattern shown in Figure 3.13. The reason for this is that the sources are not truly random in 2-D space, since they still lie approximately along the original line (but with some perturbation). This pattern might be better described as "partially coherent" as some of the features (the beam-like nature of the pattern, for instance) are similar to Figure 3.8.

3.5   Propagation of Microwaves

So far the discussion has considered propagating waves, but very little consideration has been given to what they are propagating through, and what properties of that medium influence the EM waves. Consider the effect that different materials can have on the electromagnetic radiation we are most familiar with – visible light. Some materials are transparent (*e.g.*, glass, water), others opaque (*e.g.*, brick, wood), and some materials are highly reflective, such as polished metal. The transparent materials may also bend, or refract, the light (take a look at a ruler dipped into a bowl of water, for instance) and some glass may be transparent only to a small range of visible wavelengths, so it appears as coloured glass.

All of these effects are a consequence of the electromagnetic properties of the material. Since we are considering the interaction of electromagnetic waves with materials, we must be able to describe and quantify their electromagnetic properties. These are the properties that are important for understanding and interpreting the measurements made in microwave remote sensing, and can change significantly across the electromagnetic spectrum.

The three terms used to characterise the electromagnetic properties of a material are: the *electric permittivity*, represented by the Greek letter *epsilon*, ε; the *magnetic permeability*, with the letter *mu*, μ, and *electric conductivity*, given by g.

It turns out that for most cases in remote sensing it is only the first of these that we need consider in detail, but for completeness here is a description of all of them.

The electrical conductivity of a material is important because we are dealing with waves that consist of electric (and magnetic) fields. Electrical conductivity can be understood on the basis of the mobility or lack of mobility of electrons in a material. In metals, for instance, the electrons are not bound to a single atom, but are free to move through the metal. For this reason metals are good conductors of electricity. The electric conductivity, g, for metals is therefore very high. *Electrical conductivity*

EM waves cannot propagate in a conducting material because the electric field component induces currents in the material that result in the dissipation of the energy of the wave. For an ideal conductor the electric field will be zero everywhere inside the material. When an EM wave strikes such a material, it is totally reflected. A polished metal surface is therefore a good reflector of EM waves, and as a result metals are opaque to most EM radiation, including microwaves.

There are no *perfect* conductors under the conditions we encounter in remote sensing, so there is always some penetration and only partial reflection, but in the microwave region some materials come very close.

For a vacuum $\mu = \mu_0$ (the *permeability of free space*) and has a value of $4\pi \times 10^{-7} \mathrm{Ns^2 C^{-2}}$. In a material medium, the permeability is usually higher, so for convenience we relate it to μ_0, by defining a dimensionless *Magnetic permeability* quantity, the *relative permeability* μ_r, such that:

$$\mu = \mu_r \mu_0 \quad \text{[-]}.$$

For nonmagnetic materials, such as the Earth's atmosphere and most objects on Earth[22], the value of μ_r is very close to unity, so that $\mu \approx \mu_0$ for most practical applications. We therefore do not deal with it again in this book.

Electric permittivity

In a similar manner to the magnetic permeability, we have $\varepsilon = \varepsilon_0$, the *permittivity of free space*. From experiment, this is found to be $8.8542 \times 10^{-12} \text{C}^2\text{N}^{-1}\text{m}^{-2}$. In a material medium the permittivity is also usually higher and is related to ε_0 by defining a dimensionless quantity, the *relative permittivity* ε_r, such that:

$$\varepsilon = \varepsilon_r \varepsilon_0 \quad \text{[-]}.$$

This is an important parameter when using microwaves to look at the surface of the Earth, and has a real part that varies between 1 to 80. Most naturally occurring materials on the Earth's surface are at the lower end of this limit.

Complex Notation

It should be noted that in general all of these electromagnetic terms, μ, μ_0, μ_r, ε, ε_r, ε_0, g, are complex numbers, so for example,

$$\varepsilon_r = \varepsilon_r' - i\varepsilon_r''$$

where ε_r' is the real part, and ε_r'' the imaginary part (and $i = \sqrt{-1}$)[23]. This may seem rather complicated and abstract at the moment, but the physical implication of this will become apparent later in this chapter.

Dielectric Media

Most solid materials encountered in remote sensing are non-conducting. Such a material is known as a *dielectric*, and it is usual to hear, from radar scientists in particular, expressions such as "dielectric properties" when talking about a target or an object. In general they are referring only to the relative permittivity of the material.

Dielectric Constant

The electric permittivity is sometimes used interchangeably with the term "dielectric constant". In fact, the dielectric constant should really be reserved as a description of the real part of the complex electric permittivity, ε_r'.

More on Maxwell

It is also interesting at this stage to put these parameters into the context of Maxwell's theory of electromagnetic radiation. As mentioned in the previous chapter, Maxwell's Equations, quite unexpectedly, formed the ground work of Einstein's Theory of Special Relativity. This was because while his theory described the EM radiation as waves, this was only satisfied if,

$$v_0 = \frac{1}{\sqrt{\varepsilon\mu}} = \frac{1}{\sqrt{\varepsilon_0\mu_0}} \quad \text{[ms}^{-1}] \quad (3.28)$$

---

[22]   In fact, most substances, with the exception of ferromagnetic materials, are only weakly magnetic; none is actually nonmagnetic.

[23]   It seems a common convention that physicist tend to use i as the symbol for $\sqrt{-1}$ whereas electrical engineers prefer j. You should be aware of both these terms but realise that the meaning and use are identical.

where v_0 was the group velocity of the waves in free space. Using the results of other laboratory experiments that had determined ε_0 and μ_0, Maxwell evaluated v_0 and found that it was approximately $3 \times 10^8 \mathrm{ms}^{-1}$ — a theoretical determination of the speed of light! What is remarkable is that it predicts that the speed of light in a vacuum is constant — a result that was later shown by Einstein by a completely different approach.

It is customary to designate the speed of light in a vacuum by the symbol c, a recent value of which is

$$v_0 = c = 2.997924562 \times 10^8 \quad \pm 1.1 \quad [\mathrm{ms}^{-1}].$$

I was careful to specify that this is the speed of light in a *vacuum*. A further interesting consequence of Equation (3.28) is when we consider what happens when the wave is not in a vacuum but in a dielectric material, so that $\varepsilon = \varepsilon_r \varepsilon_0$ and $\mu \approx \mu_0$: *Waves in Media*

$$v = \frac{1}{\sqrt{\varepsilon_r}\sqrt{\varepsilon_0 \mu_0}} = \frac{c}{\sqrt{\varepsilon_r}}.$$

The wave velocity in the medium is therefore less than in a vacuum by a factor proportional to the square root of the dielectric constant. The wavelength in a dielectric medium is also reduced, corresponding to the reduction of the wave velocity:

$$\lambda = \frac{v}{\nu} = \frac{1}{\sqrt{\varepsilon_r}} \frac{c}{\nu}.$$

The velocity v and wavelength λ of the wave are reduced by the factor $\sqrt{\varepsilon_r}$. This factor is more commonly termed in optics the *index of refraction* (or *refractive index*) n. The frequency of the wave remains constant.

In Chapter 5 we will see how these parameters influence the progress of EM radiation at boundaries.

3.5.1 Through Lossy Media

As an electromagnetic wave travels through a dielectric medium, it will inevitably lose energy. This is called *attenuation*.

For the present purposes it is sufficient to treat the attenuating medium *Attenuation* as a homogeneous lossy dielectric, meaning that we deal with the bulk properties of the medium rather than with absorption or scattering by individual molecules or particulates, which will be discussed in the next chapter. As above, such a medium can be characterised by a complex relative permittivity $\varepsilon_r = \varepsilon_r' - i\varepsilon_r''$. In this notation, the real part corresponds to the lossless dielectric constant, while the imaginary part ε_r'' describes the energy losses.

The attenuation acts exponentially, so that there is an exponential decay of the wave amplitude as it propagates through the medium. This is characterised by the constant of attenuation κ which can be used to determine the *penetration depth*, δ_p. The penetration depth is defined as the *Penetration Depth* distance at which the power is reduced by a factor e, and is related to the relative permittivity, ε_r, through

$$\delta_p \approx \frac{\lambda\sqrt{\varepsilon_r'}}{2\pi\varepsilon_r''}. \tag{3.29}$$

This is a good approximation as long as $\varepsilon_r''/\varepsilon_r' < 0.1$ and the scattering within the medium is negligible. For most microwave applications some penetration does occur, except for liquid water or wet snow. Long wavelength microwaves (L and P-band), for instance, can have a penetration depth of a few metres in very dry soil.

3.5.2   Moving Sources

Doppler Shift

One important characteristic of waves that must also be mentioned here, is the so called Doppler effect, named after the Austrian physicist and mathematician Christian Doppler (1803-53) who first described the effect in the middle of the 19th century. When a wave source and a detector move relative to each other the detector will measure a wave frequency slightly different to the original wave. This shift in frequency is termed a "Doppler shift" and results in an increase in the detected frequency when the source and detector are moving towards each other and a decrease in frequency when they are moving apart. This is the reason for the audible change in pitch that you can hear when a car (or siren on an emergency vehicle) drives past you at high speed — "eee-owww" — the pitch being higher as it approaches than when it recedes.

We can quantify this effect by considering the change of geometry associated with a moving source and describing it using the rotating vector approach of Figure 3.1. If a wave source S, is moving towards a detector at D, the distance it travels in a time interval t is given by $d = Vt$, where V is the relative velocity in the direction between the source and the detector. This change in distance will have an impact on the phase that is measured at the detector, such that there will be additional phase change given by:

$$\phi_D = \frac{d}{\lambda}2\pi = \frac{Vt}{\lambda}2\pi \qquad \text{[rads].} \tag{3.30}$$

This is the proportion of one wavelength that constitutes the travelled distance $d = Vt$, and then converted to phase angle by making it a proportion of 2π. When we consider the rotating vector description, we see that this phase change constitutes an additional rotation of the vector. This is much like the situation in Figure 3.6, but the distance d is now the distance travelled.

Now, it is not the phase change itself we are interested in, but its *rate of change* with time. We saw in describing Equation (3.9) that the rate of change in time of a phase is the angular frequency, ω. For a unit time interval t, the rate of change of ϕ_D is given by

$$\frac{d\phi_D}{dt} = \omega_D = \frac{V}{\lambda}2\pi \qquad \text{[rads s}^{-1}\text{],} \tag{3.31}$$

where ω_D is the Doppler angular frequency shift on a signal of wavelength λ as a result of a constant relative velocity V. We can convert to a Doppler frequency shift by dividing through by the 2π radians that correspond to one cycle so that:

$$f_D = \frac{V}{\lambda} \qquad [\text{s}^{-1}], \qquad (3.32)$$

where f_D is now the Doppler frequency shift, and V the relative velocity.

It should be noted that this is nothing to do with Special Relativity. This is the classical derivation of Doppler shift, and is sufficient for the purposes of this book.

One important thing to note here is that for an individual measurement of a wave, a change of frequency is equivalent to a change of phase. You might picture a wave travelling between two points, with the source at point A and the detector at point B (as in Figure 3.6). A second wave with a very small increase in frequency would result in marginally more wave cycles within the distance between A and B, resulting in a change of phase being measured at B for this second wave.

3.6   Where Do Microwaves Come From?

If we are to observe microwaves in the natural world, it is important to know what is causing them. Furthermore, if we wish to generate our own microwaves for active remote sensing, such as radar, we must be able to produce microwaves artificially.

3.6.1   How Are They Produced in Nature?

It is traditional at this stage of the proceedings to introduce the concept of *blackbody radiation*. However, it might be argued that for microwaves it is perhaps more appropriate to talk about *thermal* radiation, since we are dealing with a part of the EM spectrum where concepts such as "black" have very little meaning.

Heat energy is kinetic energy of (usually) random motion of particles in matter, and can result in different forms of excitation: electronic, from electrons getting knocked into higher energy levels, or in the form of kinetic (rotational or vibrational) energy levels of the molecules themselves. If random, this type of transformation leads to emission over a wide spec-   *Blackbody Emission* tral band, much like a blackbody, the hypothetical, ideal radiator that totally absorbs and re-emits energy incident upon it.

The important conclusion from this is that anything that has a physical temperature will emit some EM radiation.

It was Max Planck (1857–1947) who in 1900 first derived the law for   *Planck Function* the intensity of the radiation emitted at different frequencies by any sufficiently opaque body of temperature T. In fact, this derivation was to

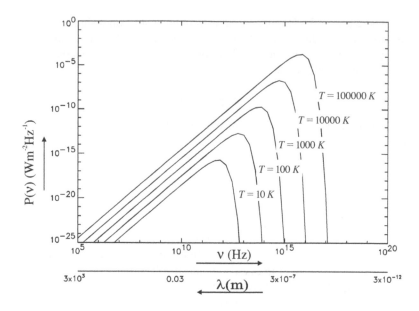

FIGURE 3.14 The Planck Function for five different thermodynamic temperatures. As the temperature increases the total amount of radiated energy increases and the peak also moves to higher frequencies (higher energy).

initiate the study of quantum mechanics, and was the first appearance of Planck's constant, h, that has been a fundamental constant in quantum physics ever since. Fortunately, quantum theory is well beyond the scope of this book, as is Planck's original derivation of his thermal emission law, so it is just quoted here.

Planck's law is usually written as the *Planck function*, $B_\nu(T)$, which represents the intensity of radiation per unit frequency range (at frequency ν), per steradian (i.e., unit direction) from a blackbody at physical temperature T and is given by:

$$B_\nu(T) = \frac{2h\nu^3}{c^2} \frac{1}{e^{\frac{h\nu}{kT}} - 1} \qquad [\mathrm{Wm^{-2}sr^{-1}Hz^{-1}}] \qquad (3.33)$$

where h is Planck's constant, c is the speed of light and k is Boltzmann's constant.

The Planck function for different values of T is shown in Figure 3.14. Note that temperature is given in units of Kelvin, K, named after Lord Kelvin (1824–1907) who designed the scale, in which 0 K is defined as absolute zero, equivalent to $-273.16°$C. The size of one degree increment in Kelvin is the same as the size of one degree Celsius.

Brightness and Intensity

B_ν is sometimes considered as a surface brightness, which is the flow of energy across a unit area, per unit frequency, from a source viewed through free space in an element of solid angle $d\Omega$. The specific intensity I_ν, on the other hand, is the instantaneous radiant power that flows at each point in a medium, per unit area, per unit frequency interval (at a specified frequency), and in a given direction per unit solid angle. Note that brightness and intensity have the same units, and in the case of a blackbody are equivalent.

A blackbody is a useful theoretical concept for describing radiation principles, but real materials do not behave like blackbodies, rather they tend to be "greybodies". We therefore need to define how "grey" a body is, and for this we define a term *emissivity*, ϵ, that describes how efficiently *Emissivity* an object radiates energy, at a certain frequency, compared to a blackbody, such that[24]:

$$\epsilon \stackrel{\text{def}}{=} \frac{\text{brightness of an object at temperature } T}{\text{brightness of a blackbody at temperature } T}. \tag{3.34}$$

Strictly speaking a greybody has a constant emissivity over all frequencies. A more likely case for natural objects is an emissivity that varies over frequency — it may resemble a blackbody more closely at some frequencies than others. These are termed *selective radiators*.

What about radiation that is incident on a blackbody or greybody? Or more exactly, what will happen to the energy of the incident radiation? It can either be reflected at the surface, absorbed by the medium or transmitted through the medium. We therefore define a number of other parameters that quantify these processes:

The *reflectivity*[25], symbolised by the Greek letter *rho*, ρ, is the ratio of the power reflected from a surface to the incident power in a given direction. Complete reflection gives $\rho = 1$.

The *transmissivity*, symbolised by the Greek letter *upsilon*, Υ, is the ratio of the power transmitted through a medium to the power incident on the surface of the medium. A transparent medium therefore has $\Upsilon = 1$, whereas an opaque medium has $\Upsilon = 1$.

And the *absorptivity*, symbolised by *kappa*, κ, is the ratio of power absorbed by a medium to the incident power. When $\kappa = 0$ the material is said to be "lossless".

These properties may all be directional quantities, and all of them will be frequency dependent.

The conservation of energy implies that all the incident energy must end up somewhere — it is either reflected, absorbed or transmitted through the material, so that,

$$\rho + \Upsilon + \kappa = 1, \tag{3.35}$$

since all of these parameters are fractions of the initial incident power.

Here we introduce another radiation law named after a famous scientist from history: *Kirchoff's Radiation Law*. Simply stated, it implies that *Kirchoff's Law* "good absorbers are good emitters". Specifically it says that the spectral

---

[24] Note that we use a different version of *epsilon* here to distinguish it from the permittivity, ε.
[25] Reflectivity is sometimes confused with albedo. The former includes a directional component, whereas albedo encompasses a whole hemisphere and so accounts for all incoming and outgoing radiation.

emissivity of an object equals its spectral absorptance, so that,

$$\epsilon = \kappa. \qquad (3.36)$$

While this is strictly true only for conditions of thermal equilibrium (when energy in = energy out). This relationship holds for most conditions encountered in Earth observation. This situation is also often referred to as "lossless" since no energy is lost in the interaction.

Similarly, in most remote sensing situations measuring the surface, the objects of interest are assumed to be opaque to thermal radiation, i.e. $\Upsilon = 0$, so that Equation (3.35) becomes,

$$\rho + \epsilon = 1, \qquad (3.37)$$

making it possible to derive surface emissivity characteristics from measurements of reflectivity and vice versa.

In fact the absorption and emission are usually not continuous over all frequencies — a blackbody is only a hypothetical model. In general, all these parameters are *very* frequency dependent and we refer to them as *Selective Absorption* selective absorber or emitter.

3.6.2   Radiation Laws

For completeness, three further radiation laws are given here.

Stephan-Boltzman Law   The first determines the frequency-integrated radiance of a blackbody surface, such that the total emitted power density, S, can be written as

$$S = \sigma T^4 \qquad [\text{Wm}^{-2}]$$

where σ is a constant and is equal to 5.6697×10^{-8} $[\text{Wm}^{-2}\text{K}^{-4}]$. This is known as the *radiation law of Stephan-Boltzman,* and σ is the Stephan-Boltzman constant [26]. It is found by integrating Equation (3.33) over all frequencies. It is important because it tells us that as the temperature increases, so does the intensity of the emitted radiation (to the fourth power of the temperature!).

Wein's Law   The second Radiation Law is known as *Wein's Law* and describes the relationship between the temperature of a blackbody and the frequency (or wavelength) at which the peak of the Planck function occurs. In terms of frequency, it is stated as

$$\nu_{\text{max}} \approx T \times 5.876 \times 10^{10} \qquad [\text{K}^{-1}\text{s}^{-1}].$$

This tells us that as the temperature of a body increases, the thermal radiation it emits will peak at higher frequencies (smaller wavelengths means higher energy). Qualitatively this relationship is characterised in common expressions describing degrees of heat — for instance "red hot" or "white hot". Imagine taking a rod of metal from a furnace that is so hot the bar

---

[26]   Sigma (σ) is the traditional symbol used to represent the Stephan-Boltzman constant, but do not get it confused with the scattering cross section that is introduced later, which by convention is also represented by σ.

is glowing white —i.e., emitting EM radiation mostly in the middle of the visible region of the EM spectrum (where our eyes are most sensitive). Being so hot the radiation is also very intense (from the Stephan-Boltzman Law) but as it cools, the intensity drops and the bar glows less brightly. As it cools the $\nu_{\max}$ also drops, so the colour appears to change, from white, through yellow and orange, to eventually a much dimmer red glow. This glow may fade away completely to our eyes as the temperature drops, but only because our eyes are not sensitive to the infrared radiation it is now emitting (although you could probably feel the effect of the radiated energy with your heat sensitive skin). $\nu_{\max}$ will continue to drop (going to longer and longer wavelengths) as the rod cools until the rod reaches the ambient temperature of the air[27].

It is left as an exercise for the reader to determine what temperature would have a radiation peak in the middle of the visible region.

The third radiation law relates specifically to microwaves. At the low frequency end of the microwave spectrum we can consider the *Rayleigh-Jeans limit* where $h\nu \ll kT$. This reduces the Planck function to

Rayleigh-Jeans Law

$$B_\nu(T) \approx \left(\frac{2\nu^2 k}{c^2}\right) T, \qquad (3.38)$$

which gives $B_\nu(T)$ as a linear relationship with physical temperature (since for an observation at a given frequency the expression within the brackets remains constant). It is then convenient to define a *microwave brightness temperature T_B* as

$$T_B(\nu) \stackrel{\text{def}}{=} \left(\frac{c^2}{2\nu^2 k}\right) B_\nu = \left(\frac{\lambda^2}{2k}\right) B_\nu \qquad [\text{K}]. \qquad (3.39)$$

This is the *Rayleigh-Jeans equivalent brightness temperature*, rather than the *thermodynamic brightness temperature*. The latter refers to the temperature that a blackbody would have to produce the observed brightness, and so is an approximation to the real temperature of the object. The former is a defined term that provides a convenient unit for characterising the intensity of the radiation — measurements of radiation intensity can now be given in units of temperature (as a brightness temperature), instead of, say, the voltage measured by a radiometer. However, it can't be stressed enough that although the measurements are given in temperature, this need not explicitly imply any assumption about the physical properties of the object.

Equivalent Brightness Temperature

However, as a good approximation for most microwave remote sensing cases, we can relate the microwave brightness temperature to the physical temperature by subsituting (3.38) into (3.39) and using (3.34) to equate the

---

[27] Merely out of interest it is worth noting that experienced craftsmen who use furnaces (glass blowers or metal workers, for example) are often able to judge the temperature of a furnace simply be observing its subtle changes in hue. The makers of samurai swords, for instance, must look for the "colour of the moon in august" at a key stage in the metal hardening process.

observed brightness to the blackbody brightness, to get

$$T_B = \epsilon T, \tag{3.40}$$

where T is the physical temperature and ϵ is the emissivity. Note that the convention followed is to use upper case subscripts on T to indicate brightness temperatures, and lower case subscripts to identify physical temperatures.

Note also that Equation (3.39) is for unpolarised measurements. When a single polarisation, p, is being measured it loses the factor of 2 so that,

$$T_B(\nu, p) \stackrel{\text{def}}{=} \left(\frac{\lambda^2}{k}\right) B_\nu \quad [\text{K}], \tag{3.41}$$

where p refers to a particular polarisation state (discussed in Chapter 4).

3.6.3   How Are Microwaves Produced Artificially?

Without going into the technical detail of the electrical engineering, it is worth summarising how microwaves can be generated artificially. The key point is that, as with naturally occurring EM waves, artificially produced ones are also generated by transformation of energy from other forms, such as kinetic, chemical, thermal, electrical, magnetic or nuclear. In all cases these forms of energy cause the movement of electrical charge, and it is this motion (or more precisely, acceleration, which is a change in motion) of the charge that is the fundamental initiator of all EM fields. In the case of artificially producing them, we must have some way of controlling this transformation of energy.

Dipole antenna

The simplest case is referred to as a *dipole antenna*. This is simply a conducting rod that has an alternating current sent through it (it is alternating since we do not want the current to go anywhere, merely oscillate back and forth).

The transformation mechanisms are generally different in character over the EM spectrum. Some mechanisms are very ordered, like a laser, and produce very coherent light with a very fine range of frequencies (small bandwidth), while others are chaotic or random and produce incoherent (wide-band) light.

Lasers

Lasers (light amplification by stimulated emission of radiation) use the excitation of electrons in molecules and atoms to enforce a very selective emission of radiation with very narrow bandwidth EM radiation that is consequently very coherent. Lasers can produce light over a range of visi-

Masers

ble and infrared frequencies. The *maser* is the equivalent for microwaves, the term standing for "microwave amplification by stimulated emission of radiation". The maser was invented in 1953 by Charles Townes. They differ from lasers in that instead of exciting electrons in atomic energy levels, they excite different levels of rotational energy in a molecule. The molecule is excited from one level to another and then when it drops back to the original level it radiates the excited energy as an EM wave — in this case, a microwave.

Microwaves can also be generated using electron tubes which use the motion of high speed electrons in specially designed structures to generate a variable electric/magnetic field, which is then guided by hollow metal tubes (waveguides) to a radiating structure (i.e., an antenna). The magnetron is the best-known such device, utilising a magnetic field to force electrons to rotate — such rotation means the electrons are being accelerated, resulting in the generation of EM radiation in the microwave region of the spectrum. The magnetron was notable for its high efficiency at converting electrical power to radiated microwave power, hence its importance in the development of practical radar devices. *Microwaves*

For radar purposes one of the most important requirements when generating microwaves is that they are coherent — or more specifically, you can generate a stream of coherent pulses that all begin with the same phase.

Generally, the more organised the transformation mechanism, the more coherent the resultant radiation.

Radio frequency waves, which are longer than microwaves, extending from a few Hertz up to 10^9Hz (λ from many kilometres to 0.3m, or so) are generally emitted using an assortment of electric circuits with periodic currents of electric charges in wires. The alternating current in domestic power lines, for instance, generate EM radiation. There is no theoretical upper limit to the EM radiation that could be generated — you could even swing a charged ball at the end of a string and generate an extremely long wavelength wave, albeit a very weak one. TV and radio use the higher frequency end of the radio frequency band. *Radio Waves*

Waves emitted by a radio transmitter are usually linearly polarised; a vertical rod antenna[28] of the type widely used on cordless telephones, emit waves which, relative to a horizontal plane around the antenna, are polarised in the vertical direction (parallel to the aerial). A receiving dipole antenna that is oriented perpendicular to the emitter will, in the idealised case, receive none of the incoming polarised waves. *Radio Transmission*

Light from ordinary sources — the sun, light bulbs, etc — is generally not polarised. The radiation originates from the vast number of molecules that make up the light source, acting like small "dipoles", but positioned and oriented randomly. The result is that while individual molecules may radiate a polarised wave, the total emitted radiation is made up of a random mixture of waves linearly polarised in all possible directions. It should be noted that the process of reflection, or transmission through a polarising medium, can select out only the waves of a specific polarisation. The sun may emit unpolarised EM radiation, but the reflected light from the sky, sea, snow, etc, may be distinctly polarised. This is the reason for using polaroid sunglasses when driving or skiing — the "glare" from the horizontal *Sources of Polarisation*

---

[28]　Such an antenna is called a *dipole antenna*, since the rod has two "poles". It is also worth noting here that in radar nomenclature, the plural of antenna is usually written antenn*as*, not antenn*ae*, a term which is normally reserved for insects.

surface is primarily horizontally polarised. By allowing only vertically polarised light to pass through, the sunglasses will selectively extinguish the
glare (you can compare the difference yourself by looking through your
glasses rotated through 90°).

3.7   Further Reading

This chapter was inspired by three fundamental texts that I would urge
you to follow up after (or during) reading this book: Richard Feynman's
"Lecture Notes in Physics" (1963), Schanda's "Physical Fundamentals of
Remote Sensing" (1986) and Hecht's "Optics" (2001). These three books
cover in much greater detail the broad topics introduced here, and for those
readers with a confident mathematics background are definitely recommended reading.

4
POLARIMETRY

"If you are going through hell, keep going."
— Sir Winston Churchill (1874-1965)

Polarisation describes the path of the tip of the electric field vector of an electromagnetic wave. As was the case for the general description of a wave given in Section 1.1, we can describe the polarisation in terms of the pattern carved out in space by the tip of the vector at a fixed time, or we can describe how the tip of the vector appears to move in time at a fixed location. The latter method is the more usual in remote sensing and was the framework used to introduce the basic concept of polarisation in Chapter 3. For waves of a fixed frequency, the observed polarisation at a fixed location takes the form of simple geometric shapes such as lines, ellipses and circles.

This chapter focuses on the properties of polarised waves, how to describe them mathematically, and what they can tell you about a distant object. The use of polarimetric information is important in both active and passive microwave remote sensing. In remote sensing, if we are able to describe the polarimetric properties of the object or target under observation, then it may be possible to determine some of its physical properties. The first basic principle of polarimetry is defining some system of quantitatively describing the polarisation state of a wave.

Polarimetry is one of the most challenging aspects of microwave remote sensing. It not only requires an ability to visualise in three dimensions and to then conceptualise those three dimensional attributes changing in time, but it also requires mathematical notation and techniques that are often beyond even the more numerate applied scientist. This chapter attempts to capture some of the flavour and to summarise the main features — the hidden wonders of polarimetry are extensive and ominous and this chapter only touches the surface. Readers may choose to skip this chapter if they feel it is beyond them — the rest of the book has been written with the expectation that not every reader will delve into polarimetry. The rewards are great, however, for those who keep going and work at understanding even a small portion of the theory presented here.

4.1   Describing Polarised Waves

In Chapter 3, a description was given for an individual wave and how it might be polarised, and some arbitrary set of axes were used to describe horizontally and vertically polarised waves. How and why those particular axes were chosen was not discussed. The coordinate system used is merely a frame of reference and is entirely arbitrary, although the plane in which polarisation is described is always perpendicular to the direction of the wave since polarisation is a result of the transverse nature of the waves. This plane of reference is fixed in space, so that in microwave polarimetry the polarisation properties are described by considering the apparent motion of the electric field vector as the wave moves through the x-y plane.

For convenience, in remote sensing the horizontal axis, x, is defined to be parallel to the surface of the Earth when we are taking oblique (off-nadir) measurements. The vertical axis, y, is then defined as being perpendicular to this so that the x-y plane is defined at the instrument, not the ground (i.e. it need not be "vertical" with respect to the Earth's surface). When viewing towards the nadir, the choice is completely arbitrary since there would be no distinction between horizontal and vertical at the ground surface. In such a case the instrument designers would choose the axes based on the instrument configuration. If the field of view is large enough (or if the instrument is scanned) so that part of the data is effectively low-oblique, then the surface in this direction can be taken as the horizontal axis. Otherwise, the horizontal is taken as parallel to the motion of the platform.

Basis

Since it is a 2-D property (it is in a plane) the only requirements for a polarimetric coordinate system, or "basis", is that it is composed of two polarisation states that are combined together to form the actual polarisation, and these states must be orthogonal. Qualitatively, orthogonality means that each state does not contain any element of the other. The basis described above uses two linear polarisation states, which means that they must be perpendicular — i.e. horizontal does not contain any "verticalness", and vice versa, so they are orthogonal. Since it uses linear polarisations as the two fundamental polarisations, it is referred to as the "linear basis".

You could equally choose $+$ and $-$ 45° as the pair of linear axes, or you could even choose clockwise and anticlockwise circular polarisations as the two orthogonal states. It is this arbitrary definition of the axes, or polarimetric basis, that has led to many (but not all) of the key approaches to polarimetric analysis being preferentially chosen to be "basis invariant", meaning that they concentrate on polarimetric properties or descriptions that are not dependent upon the choice of coordinate system. It is always most advantageous to choose a measured property of a target that is based on the properties of the target alone and not the manner in which it is observed or in which the data is recorded.

A further thing to consider is the definition of the z-axis, which runs parallel to the direction of travel of the wave. Which way should it point — towards the direction of travel or away from it? It is important to consider this as the description of the polarisation will be different depending on which one is used. If a right-handed system[29] is used in both cases (which is generally the convention in physics) then the horizontal component of the wave would be reversed — the diagonal component and the handedness would then be reversed. *Handedness*

Most optical-based polarimetry uses the forward direction of the travelling wave as the positive z-direction, taking the perspective of the wave itself. In this case we can imagine sitting on the wave and looking in the direction of travel – whether travelling from or to an instrument – when the polarisation is defined. This is referred to as the *Forward Scattering Alignment* (FSA), as it is most useful when considering waves scattered, or originating, from some object. However, when making a microwave measurement, an antenna is used and so it is often more convenient to use the antenna coordinate system, regardless of whether the wave is coming or going (i.e. being received or being transmitted). This is achieved by a seemingly clumsy arrangement whereby a regular right-handed system (like the FSA) is used when transmitting a wave from an antenna, but a left-handed system is used when receiving waves. This is known as the *Back-Scatter Alignment* (BSA) — the x and y-axes remain as before, but the direction of the z-axis is chosen to be parallel to the direction of propagation of the wave. *FSA* *BSA*

The reason for this alignment is that it allows for some simplifications in the way in which data are described.

The coordinate system does not influence the nature of the polarisation, only the manner in which we describe it. However, this can lead to some confusion, especially when considering the changes of polarisation due to some interaction at a boundary as differences in the chose of basis may seem to alter the direction of oscillation of the horizontal component. This change in direction would be apparent as a half-cycle (π) phase shift in that component (since $\sin(\theta + \pi) = -\sin(\theta)$). *Mirror images*

4.1.1 Summary of Linear Basis

Within a defined coordinate system using two orthogonal states, all the possible polarisation states can be described by using some combination

---

[29] You can see the difference between right- and left-handed co-ordinate systems by defining your thumb and first two fingers as follows: x-axis=thumb, y-axis=first finger, z-axis=second finger. By pointing each digit in a direction 90 to its two neighbours, you will form a set of right-handed axes with your right hand and left-handed axes with your left hand. The right-handed system gives the conventional set of axes that you will see in most text books. If you compare the two systems you will see that you can only make two pairs of axes point in the same direction at any one time — the third pair always point in opposite directions.

of these two base states. This text will focus on the linear basis and consider some of the possible combinations. The simpler ones are when the polarisation is linear (the electric field vector moves in a straight line within the x-y plane). If the oscillation is only in the x-direction, we say the wave is horizontally polarised, while if it is only in the y-direction we refer to it as vertically polarised. Before expanding this concept further in order to consider how we describe the full range of polarisation states, it is first necessary to discuss in more detail what happens if we add together two polarised waves.

4.2   Superposition of Polarised Waves

The superposition of waves was considered in quite some detail in Section 3 when discussing interference, but the assumption throughout was that the individual waves were all of the same polarisation. In polarimetry this must be generalised to consider the superposition of waves of different polarisation. It is important to consider the superposition of polarised waves for a very practical reason — polarimeters only measure (or generate) microwaves with two orthogonal polarisations. These two polarisations are then combined to infer the actual polarisation of the measured wave.

As was the case for interference, when two polarised waves with the same amplitude are combined coherently the key consideration is the phase difference between them, but in this case rather than resulting in a new wave of different phase and amplitude the result of adding two orthogonally polarised waves is a wave of a different polarisation.

What happens, for instance, if a horizontal, E_x, and vertical, E_y, are combined? Firstly, consider the case when they are of the same amplitude ($|E_x| = |E_y| = A$) and are in-phase so that when E_x reaches a maximum so does E_y — they will therefore also cross the origin and reach their minima at the same time. Remember that E_x and E_y are describing the motion of the electric field vector as the wave travels past a point in space — at any point in time the resulting electric field vector is therefore the vector sum of E_x and E_y. Do not lose sight of the fact that E_x and E_y are continuously oscillating so the resulting electric field vector will also oscillate[30]. In this case, the new vector traces out a straight line at $+\frac{\pi}{4}$ ($+45°$) to the two axes, as is apparent from Figure 4.1.

The amplitude of this new linearly polarised wave is given by $\sqrt{E_x^2 + E_y^2}$ = $\sqrt{2}A$. Note that the amplitude only tells us about the maximum wave amplitude available — the amplitude of the wave does not tell us anything about the polarisation state. This is an important practical consideration when dealing with remotely sensed data — the polarimetric information is independent of the variations in total measured intensity (or backscatter).

---

[30]   Do not confuse the diagrams in this chapter with the complex wave vector diagrams of Section 3 which represent properties of different waves but with the same polarisation.

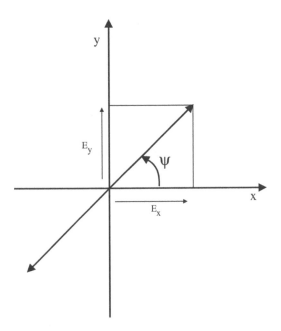

FIGURE 4.1 A linearly polarised wave with an orientation angle (ψ) 45 can be generated by the combination of one horizontally and one vertically polarised wave, with the two waves having the same initial phase.

We can generate linearly polarised waves at any angle to the x-axis by changing the relative amplitude of the two original waves. As the y-component tends to zero, the angle of oscillation gets smaller (when measured from the x-axis) and the wave tends towards being horizontally polarised. This gives us our first true polarimetric property: the *orientation angle*, ψ, defined as the angle between the maximum amplitude of the polarised wave and the horizontal axis[31], symbolised by the Greek letter *psi*, and defined as

Orientation angle

$$\psi = \tan^{-1} \left(\frac{|E_y|}{|E_x|} \right), \tag{4.1}$$

given in units of degrees or radians, whichever is more convenient for the particular problem at hand.

The two basis states are independent of each other (by definition) so it is also possible for them to have a relative phase difference, δ, between the E_x and E_y components, say, by making E_x lag behind E_y by a small fraction of a wavelength. Now they do not reach their maxima and minima together, rather E_y gets there fractionally ahead of E_x. The resulting pattern traced out by the new E-vector is now a very narrow ellipse, with its central axis oriented at $+\frac{\pi}{4}(+45)$ and with the point of the vector moving

---

[31]   This is not always the case — on occasion you will still find that some data defines the orientation angle relative to the vertical, rather than the horizontal.

Elliptical polarisation

Circular polarisation

Ellipticity angle

Handedness

around the ellipse in a clockwise direction.

This is *elliptical polarisation*. As the magnitude of the phase difference increases, the resulting ellipse gets progressively fatter, until eventually, at $\delta = 90$ the two axes of the ellipse are the same and the wave vector traces out a circle — the wave is now *circularly polarised*.

When E_x reaches a maximum, E_y is crossing the origin and vice versa. Note that the value of ψ remains the same throughout these changes, since the amplitudes have not changed, only the relative phase.

We also want to describe the degree of ellipticity, which can be done through the use of the angle

$$\chi = \tan^{-1}\left(\frac{b}{a}\right), \tag{4.2}$$

where a and b are the major and minor axes, respectively, as illustrated in Figure 4.2. We refer to χ (the Greek letter *chi*) as the *ellipticity angle* and it ranges from $-\frac{\pi}{4}$ to $+\frac{\pi}{4}$, where the sign denotes the handedness of the polarisation — positive for left-handed and negative for right-handed. The two extreme values denote circular polarisation, whereas an ellipticity angle of zero indicates linear polarisation.

The handedness of the polarisation is a difficult thing to conceptualise, for it depends on the system being used to describe the polarisation and the direction of observation. If we stop the wave at a given moment, we could imagine that the point of the electric field vector for a circularly polarised wave carves out a spiral with a right-handed sense[32] — in this case the polarisation is said to be "right-handed circular". But if you use the system whereby polarisation is described by fixing the location and tracking the vector in time as it passes through a reference plane, then the tip of the vector traces out a circle. For a right-handed circular wave the circle is tracked-out in the anti-clockwise direction if you are looking along the direction of travel. The same wave, travelling through the same reference plane, would trace out a *clockwise* circle were you to observe it from the other direction (opposite to the direction of travel). Well, I did say it is difficult to conceptualise!

For now we consider the system of looking opposite to the direction of travel.

If the phase difference is increased further, the polarisation will again become a clockwise ellipse, but this time with $\psi = +\frac{3}{4}\pi$ (135°). Once the phase difference reaches one half wavelength (π), E_x will be reaching its maximum when E_y is at a minimum, and vice versa — the result is a linearly polarised wave but with $\psi = +\frac{3}{4}\pi$.

To complete the full cycle of polarisation states, the phase difference must increase further, from π to 2π. This will take the polarisation through

---

[32]   The handedness of a spiral, or screw, can be determined by taking either hand, pointing the thumb along the line of travel of the wave, then curling your fingers into a fist. As your hand moves along the direction of the wave, your fingers should curl in the direction of the spiral for your right hand, when its a right-handed screw, and for your left hand for a left-handed screw.

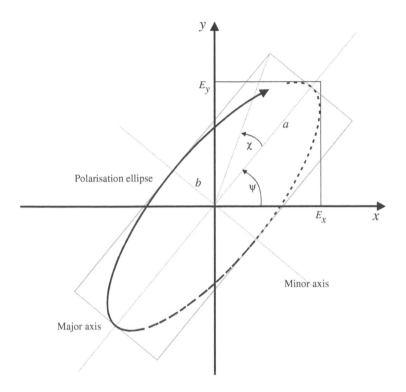

FIGURE 4.2 The general pattern of polarisation (looking towards an incoming wave out of the page) is elliptical. The shape of the polarisation ellipse is described by two parameters: the ellipticity angle, χ, and the orientation (rotation) angle, ψ. The size of the ellipse is governed by the maxima of the x and y (horizontal and vertical) electrical field components.

all the left-handed elliptical conditions, including left-handed circular, and back to linear polarisation at $+\frac{\pi}{4}$. Remember, a relative phase shift of 2π is equivalent to no phase shift at all.

This sequence of polarisations generated from two orthogonal linear waves is summarised in Figure 4.3. This is an important figure as it illustrates what was discussed above: any polarisation state can be described by the addition of two orthogonal linearly polarised waves. It is important because it means that any microwave system need not generate or detect all possible polarisation states, but instead can get away with only dealing with two orthogonal polarisations, so long as the phase of the signal is also included. The most common, but not exclusive, method is to use linear horizontal and linear vertical polarisations. For transmitters, these two linear states, could, in principle be used to generate any polarisation state simply by controlling the relative amplitude and phase of the two waves, as above. The converse is true for receivers: as long as we measure the phase difference as well as the amplitude in both linear polarisations, we can determine the complete polarisation state of a wave.

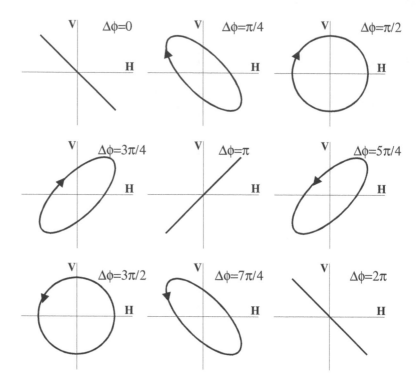

FIGURE 4.3 Changing the phase difference between two linearly polarised waves gives a variety of resultant polarisations. When the phase difference is 0 or π, the result is another linearly polarised wave. Phase differences of $\pi/2$ and $3\pi/2$ both result in circular polarisations, but with opposite handedness. Phase differences inbetween these special cases results in the generalised elliptical polarisation. (After van der Sanden, 1997).

In some radar systems only one polarisation is transmitted and one polarisation is received. The ERS SAR transmitted vertical and received vertical (VV), for instance, whereas the JERS-SAR and Radarsat both transmit and receive only horizontally polarised waves (HH)[33]. Polarimetric radars will measure more than one channel. For instance, they may transmit a single polarisation but record two polarisations in receive mode. The ASAR instrument on Envisat can receive both H and V, having transmitted either an H or V pulse. Fully polarimetric, or "quadpol" radar systems transmit both H and V and receive H and V. The PALSAR on the ALOS platform, and Radarsat 2 are both designed to have quadpol modes.

The horizontal and vertically waves are actually transmitted independently — they are transmitted on alternating pulses, but with such a short time delay that the waves can be considered to have been backscattered

---

[33]   Note that here the upper case H and V are used as we are referring to instrument defined properties, rather than arbitrary directions.

from exactly the same part of the surface being observed[34]. But as long as the phase differences are accurately recorded this is not a problem since all the possible polarisation states, for reception as well as transmission, can be synthesised after the data collection, even though they were not actually recorded at the time. This is the principle of a technique used in radar imaging called *polarisation synthesis*, whereby the backscatter can be determined for all possible transmit and receive configurations, even though only horizontal and vertical waves are used for transmission and reception. The mathematics of this process will be dealt with later in this chapter.

Polarisation Synthesis

4.3   Representing Polarisation

The previous section gave an explanation of those parameters necessary to describe a polarised wave. In this section the question is how can these parameters be efficiently represented, either graphically or mathematically, to aid in interpretation or data analysis.

4.3.1   Poincaré sphere

It is useful to introduce here the idea of the Poincaré sphere, named after Jules Henri Poincaré (1854–1912). The idea of using a sphere to represent polarisation arises from the realisation that ψ and χ act in much the same way as latitude and longitude, in that the ellipticity angle spans half the range of the orientation angle and as the magnitude of the ellipticity increases, the orientation becomes less significant. In this way different polarisations can be represented on the surface of a sphere using 2ψ as longitude and 2χ for the latitude[35]. Around the equator of the Poincaré sphere lie all the linear polarisations each with varying orientation angle (longitude) and with increasing latitude we have polarised waves with increasing ellipticity. At any one latitude there are the family of ellipses around the globe, but with different orientations. At the poles we have the two circular polarisations — conventionally with the left-handed circular at the top (North) pole and right-handed at the bottom. Note that orientation is no longer relevant when the polarisation is circular. Additionally, the radius of the sphere can be defined to be a measure of the amplitude (or intensity) of the polarised wave, so that the Poincaré sphere is a convenient conceptual picture for all polarisations and intensities.

Another important feature of the Poincaré sphere is that orthogonal (opposite) polarisations lie on antipodal points on the surface of the sphere — e.g. horizontal linear polarisation lies directly opposite vertical linear polarisation, and right handed circular lies on the opposite point of the sphere to left handed circular.

Orthogonal Polarisations

---

[34]   Notice that if the waves were not transmitted *independently*, then the system would actually be transmitting a linearly polarised wave with $\psi = \frac{\pi}{4}$.
[35]   Ellipticity and orientation only range over ± 45 and ± 90 instead of 90 and 180 for lat/lon so that 2ψ and 2χ are used instead of ψ and χ.

The Poincaré sphere is merely one way to represent polarisation state. Its usefulness is mostly as a conceptual tool, rather than as a method of graphical display, since visualising a complete sphere is not possible to do effectively in two dimensions. It also does not provide any convenient method of analysis of the polarimetric information. For that, a mathematical description of polarised waves is required.

4.3.2   Mathematical Description

The most straightforward mathematical description of a polarised wave is to use the two components of the electric field vector, E_x and E_y, and represent together in a *wave vector* defined as,

$$\mathbf{E} = \left(\begin{array}{c} E_x \\ E_y \end{array} \right). \tag{4.3}$$

However, it is not always practical to measure the components of the electric field vector directly. Historically, the initial quantitative study of polarisation in the early 1800's was within the realms of the visible and unlike microwave detectors, visible light detectors do not measure the polarisation directly by determining the electric field components — the frequencies in the visible region are far too high. Determining the polarisation of visible light was first achieved by a more deductive approach whereby you measure the intensity of the wave after it has passed through a linear polarising filter (a filter that only allows linearly polarised waves to be transmitted) that is oriented at different angles.

In 1852 George Gabriel Stokes (1819-1903) introduced four parameters based on such optical measurements that allow a complete description of a polarised wave. His four parameters are normally collected together as a vector in order to make them easier to deal with. The Stokes vector is still regularly used for representing polarimetric data from radiometers and some radars and so it is important to know the origin and relevance of the parameters.

4.3.3   Stokes Vector

The classical Stokes vector is composed of four elements, which for the sake of consistency with the majority of the literature will be referred to by their more common symbols, I_0, Q, U and V.

The first Stokes parameter, I_0, is chosen so that it is proportional to the total intensity of the wave, so that,

$$I_0 = \left\langle E_y^2 + E_x^2 \right\rangle = \left\langle E_y^2 \right\rangle + \left\langle E_x^2 \right\rangle. \tag{4.4}$$

The $\langle ... \rangle$ brackets denote averaging over time which is necessary when the wave is not completely polarised. I_0 is a measure of the total amount of energy in the wave and so does not actually tell you anything about the polarisation at all — it is a measure of how much energy you would measure with a detector that is not sensitive to polarisation. If you think back to the polarisation ellipses in Figure 4.2 you will see (through application of Pythagoras' Theorem) that I is the square of the amplitude of the electric field vector, which we know from Section 2 is equal to the power.

The other three Stokes terms describe the state of the polarisation. The second term

$$Q = \langle E_y^2 \rangle - \langle E_x^2 \rangle, \tag{4.5}$$

reflects the tendency of the polarisation to be more vertical ($Q > 0$) or horizontal ($Q < 0$).

The third and fourth terms jointly represent the phase difference, δ, between the vertical and horizontal components of the wave. The third term,

$$U = 2E_y E_x \cos \delta = 2 \operatorname{Re} E_y E_x^*, \tag{4.6}$$

expresses the tendency to be polarised at $+45°$ ($U > 0$) or $-45°$ ($U < 0$). The superscript $*$ refers to the complex conjugate of the number.

The handedness (left or right) is described by the fourth term

$$V = 2E_y E_x \sin \delta = 2 \operatorname{Im} E_y E_x^*, \tag{4.7}$$

so that $V < 0$ for right-handed polarisations (lower hemisphere of the Poincare sphere) and $V > 0$ for left-handed polarisations (upper hemisphere). U and V are sometimes referred to as the *in-phase* (real) and *quadrature* (imaginary) covariances, respectively, between the vertical and horizontal field components.

These might seem rather arbitrary definitions, but they originate from *Stokes Vector* the fact that each of the Stokes parameters directly relate to the transmittance through different polarising filters and are therefore based on quantities that could be measured in a laboratory.

The Stokes parameters can then be written as a Stokes Vector, g, such that

$$\mathbf{g} = \begin{bmatrix} I_0 \\ Q \\ U \\ V \end{bmatrix} = \begin{bmatrix} \langle E_y^2 \rangle + \langle E_x^2 \rangle \\ \langle E_y^2 \rangle - \langle E_x^2 \rangle \\ 2\operatorname{Re}\langle E_y E_x^* \rangle \\ 2\operatorname{Im}\langle E_y E_x^* \rangle \end{bmatrix} = I_0 \begin{bmatrix} 1 \\ \cos 2\psi \cos 2\chi \\ \sin 2\psi \cos 2\chi \\ \sin 2\chi \end{bmatrix}, \tag{4.8}$$

where the relationship between the four Stokes parameters and the angles of the polarisation ellipse have been included[36]. We can now refer to one parameter, namely the Stokes vector g, rather than the individual components. It is convenient in polarimetry to represent the polarimetric properties in vector form (recall that a vector is simply a column of numbers that can be dealt with in a number of well-defined ways) since it allows for the application of matrix operations to modify this vector. Such modifying operations can then be used to characterise the impact of transmission or

---

[36] The Stokes vector is not always used as a record of the polarisation. When it is, for example with the NASA JPL AirSAR airborne radar imaging system, readers should also be aware that the placement of the horizontal and vertical component within the Stokes vector can be switched around. The underlying advice is always to check in detail the format of the data you are using!

Modified Stokes Vector

scattering events on the wave. This modification will be described in more detail in Section 3.6.

In the context of using matrix operations for solving radiative transfer problems it turns out that the classical Stokes vector is not in the most convenient form. An alternative to the four elements described above can be used by noting the following relationships:

$$I_y = |E_y|^2 = (I_0 + Q)/2, \qquad (4.9)$$
$$I_x = |E_x|^2 = (I_0 - Q)/2.$$

The modified Stokes vector is then represented by

$$\mathbf{g}_m = \begin{bmatrix} I_v \\ I_h \\ U \\ V \end{bmatrix} = \begin{bmatrix} \langle E_y^2 \rangle \\ \langle E_x^2 \rangle \\ 2\operatorname{Re}\langle E_y E_x^* \rangle \\ 2\operatorname{Im}\langle E_y E_x^* \rangle \end{bmatrix} = I_0 \begin{bmatrix} \frac{1}{2}(1 + \sin 2\psi \cos 2\chi) \\ \frac{1}{2}(1 - \sin 2\psi \cos 2\chi) \\ \sin 2\psi \cos 2\chi \\ \sin 2\chi \end{bmatrix}.$$
$$(4.10)$$

Readers should be aware of this subtle difference as many authors are not always explicit as to which form of the Stokes vector they are using. The two are related by a matrix operation such that,

$$\mathbf{g} = \mathbf{U}\mathbf{g}_m$$

where

$$\mathbf{U} = \begin{bmatrix} 1 & 1 & 0 & 0 \\ 1 & -1 & 0 & 0 \\ 0 & 0 & 1 & 0 \\ 0 & 0 & 0 & 1 \end{bmatrix}.$$

4.3.4   Brightness Stokes Vector

In passive polarimetric measurements it is often convenient to rephrase the Stokes vector representation in terms of brightness temperature using the Rayleigh Jeans equivalent brightness temperature, defined in Equation (3.41) for a polarised measurement at vertical and horizontal polarisations. The Stokes vector is then in units of temperature and is given by,

$$\overline{T}_B = \begin{bmatrix} I_0 \\ Q \\ U \\ V \end{bmatrix} = \begin{bmatrix} T_V + T_H \\ T_V - T_H \\ T_{45} - T_{-45} \\ T_L - T_R \end{bmatrix} = \frac{\lambda^2}{k} \begin{bmatrix} \langle E_v^2 \rangle + \langle E_h^2 \rangle \\ \langle E_v^2 \rangle - \langle E_h^2 \rangle \\ 2\operatorname{Re}\langle E_v E_h^* \rangle \\ 2\operatorname{Im}\langle E_v E_h^* \rangle \end{bmatrix} \quad [\mathrm{K}]$$
$$(4.11)$$

where the subscripts L and R refer to left and right handed circular polarisation respectively and k is Boltzman's constant. A modified brightness Stokes vector can then also be defined in the same way as above.

4.3.5   Partially Polarised Waves

An important benefit of the Stokes vector is that it can describe partially-polarised waves as well as completely polarised waves. When discussing the polarisation ellipse of Figure 4.2 it was assumed that the wave was completely polarised, meaning that the polarisation ellipse is composed of waves that are of exactly the same frequency (monochromatic) and so would remain unchanged over a period of time (i.e., over the measurement interval). In practice, a microwave system will not be truly monochromatic since they cover a finite range of frequencies. The polarisation parameters of the target may also vary relatively quickly with time (over an ocean for instance), or spatial location (since it is always being measured from a moving platform). Natural objects tend to emit radiation with very random polarisation (although they will have some polarisation due to the orientation of its surface).

In a radar system, the measured signal is made up of a collection of echoes over both the integration time of individual pulses, as well as integration over a series of pulses. Each echo can have a slightly different polarisation.

In all practical cases, therefore, a measured wave will only be partially polarised, so a method for describing partially polarised waves is required.

For completeness the previous section already used the $\langle ... \rangle$ brackets to denote averaging over time, but this is only strictly necessary when the waves are not completely polarised.

In the case of a completely polarised wave the Stokes parameters are not independent, but satisfy

$$Q^2 + U^2 + V^2 = I_0^2. \qquad (4.12)$$

Waves that satisfy this equation are therefore called *completely polarised waves*. Completely polarised waves correspond to a single point on the surface of a Poincaré sphere, since the length of the Stokes vector equals the total power of the wave (i.e., the radius of the Poincaré sphere). *Completely polarised waves*

For the other extreme case of *completely unpolarised waves*, the magnitudes of E_x and E_y will be equal, and since the phase angle between the two components will be random, the time average over the cosine and sine terms in (4.8) will be zero, so that

$$I_0 = 2 \langle E_y^2 \rangle = 2 \langle E_x^2 \rangle, \text{ and } \quad Q = U = V = 0.$$

This is unsurprising, since the latter three terms define the state of the polarisation and there is no polarisation state for an unpolarised wave. Completely unpolarised waves therefore have a Stokes vector with zero length and so are located at the centre of the Poincaré sphere. Being at the centre, it has no direction so does not correspond to any single location on the surface of the sphere.

Fortunately, most instruments use a range of frequencies that is small compared to the centre frequency so we can assume we are always dealing with *quasi*-monochromatic waves. Measurements are made with short enough integration times that we can assume we are dealing with partially polarised waves and so the $\langle ... \rangle$ brackets remain. For the general case, *Quasimonochromatic waves* therefore,

$$I_0^2 \geq Q^2 + U^2 + V^2. \tag{4.13}$$

Degree of polarisation   Another important property that we should be able to quantify is the *degree of polarisation*: how much of the power in the wave is polarised and how much of it is unpolarised? This is given by the degree of polarisation, m, defined by

$$m = \frac{\text{polarised power}}{\text{total power (polarised+unpolarised)}} \tag{4.14}$$

$$= \frac{\sqrt{Q^2 + U^2 + V^2}}{I_0} \quad \text{[-]}.$$

For completely polarised waves, $m = 1$ and for completely unpolarised waves, $m = 0$.

Additivity   We can now make use of another important property of the Stokes vector : for independent waves travelling in the same direction their Stokes vectors are *additive*. This means that any Stokes vector can be represented by an addition of two or more other Stokes vectors. This property can be used to describe partially polarised waves, by rewriting the partially polarised wave **g** as a combination of a completely polarised wave $\mathbf{g}_p$ and an unpolarised wave $\mathbf{g}_u$ so that

$$\mathbf{g} = m\mathbf{g}_p + (1 - m)\mathbf{g}_u, \tag{4.15}$$

where m is the degree of polarisation. This is important when trying to focus on the polarised wave that is believed to originate from some target of interest, as opposed to other intervening factors (such as instrument noise) that contribute only an unpolarised signal.

4.3.6   The Stokes Scattering Matrix

The Stokes vector is a convenient way to characterise the polarisation of any EM wave (at any frequency over the entire EM spectrum) and when used in the antenna coordinate reference frame (BSA) it is a good way of representing the measurements of a polarimeter. In microwave radiometry, this is quite sufficient as it encapsulates everything there is to know about the received radiation. For active microwave instruments, however, it is possible to choose the polarisation of the transmitted (illuminating) radiation. The measured Stokes vector describing the received polarisation state is then dependent upon the polarisation of the transmitted wave *as well as* the scattering properties of the target. The same conditions are relevant when simulating a passive microwave response, since as well as emitted radiation, the target may also be scattering ambient microwave signals from other parts of the scene. For this reasons, a more elaborate scattering description is also useful in a passive microwave context.

Target scattering   Suppose that an arbitrary polarised wave represented by a Stokes vector $\mathbf{g}_i$ interacts with a target and results in a wave given by the vector $\mathbf{g}_s$. The target can then be thought of as "operating" on the incoming wave to produce a new wave. For now the scattering is considered to be in some arbitrary direction so that $\mathbf{g}_s$ is described in the forward scattering alignment (i.e. in the direction of scattering). This scattered wave is a result of some of the incident waves scattering from some object or target area (the exact nature of which does not matter at this stage). The scattering process transforms one vector, $\mathbf{g}_i$, into another vector, $\mathbf{g}_s$, a process that can be described very efficiently using a 4×4 real matrix that we will term $\mathbf{M}$. Recall that a matrix is just an array of numbers which has prescribed addition and multiplication operations (a summary of matrix algebra is given in Appendix A).

If $\mathbf{M}$ represents the scattering transformation matrix, then

$$\mathbf{g}_s = \mathbf{M}\mathbf{g}_i. \qquad (4.16)$$

$\mathbf{M}$ relates the polarisation properties of incident and scattered waves, not of transmitted and received waves. If we want to rewrite this in terms of the wave arriving at an antenna, an additional factor must be included to account for the drop in power associated with the distance from the target to the receiver, R. We may then consider the incident wave, $\mathbf{g}_i$, which is the wave that arrives at the target, and the received wave, $\mathbf{g}_r$, such that

$$\mathbf{g}_r = \frac{1}{4\pi R^2}\mathbf{M}\mathbf{g}_i. \qquad (4.17)$$

The form and name given to $\mathbf{M}$ is dependent upon the nature of the coordinate system used. In various forms it may be referred to as the *Stokes scattering matrix* or the *Stokes scattering operator* or the *Mueller matrix*, the last one being the version most commonly used in optics. A more useful version for remote sensing purposes is the *Kennaugh matrix*, $\mathbf{K}$, in which the scattered wave is described in the receiving antenna reference frame (BSA) so that both the incident wave and the scattered wave are described from the point of view of the antenna.

Mueller matrix

Kennaugh matrix

It is this matrix operator that therefore contains the information we are looking for with regard to the target's effect on a polarised wave. It is the 4×4 real matrix $\mathbf{M}$ (or $\mathbf{K}$) whose elements we will denote by m_{ij} $(i, j = 0, \ldots, 3)$, that usually forms the central concern of polarimetry since it describes the complete polarimetric response of the target. In a radar context especially, this is a much more interesting parameter than either $\mathbf{g}_s$ or $\mathbf{g}_r$, which describe the waves, not the target.

From now on we need only consider the properties of $\mathbf{M}$ (or $\mathbf{K}$) and investigate how we can make most use out of the 16 numbers it contains. Note that while the Kennaugh matrix is most often employed, the colloquial use of the name Mueller matrix is often loosely used to describe this operator whether it is in FSA or BSA.

Note that another matrix transformation allows for the conversion between Kennaugh and Mueller matrices.

4.3.7 The Scattering Matrix

For radar polarimetry, the Stokes vector is not the most effective way to characterise the data since there are effectively two measurements of polarisation to quantify — one for each of the orthogonal transmitted pulses. In the linear basis, the radar systems transmits a horizontal signal, measures the echo polarisation, then transmits a vertically polarised wave and measures that polarisation of that echo. At least two Stokes vectors would then be required. Since the polarimetric measurement of the echoes are made as orthogonal measurements it is convenient to define an alternative scattering matrix, $\mathbf{S}$, such that

$$\mathbf{S} = \begin{pmatrix} S_{VV} & S_{VH} \\ S_{HV} & S_{HH} \end{pmatrix}, \tag{4.18}$$

where the subscripts refer to the pairs of transmit and receive measurements (vertical and horizontal for the linear basis). Each of the elements S_{pq} are complex numbers describing the phase and amplitude of the q-transmit and p-receive wave[37]. When the subscripts are the same the measurement is described as co-polarised (often abbreviated to "co-pol") and when different, cross-polarised ("cross-pol"). In the general case, p and q can be any pair of orthogonal polarisations, such as R and L circular.

The advantage of this matrix is that it describes the relationship between incident and scattered *wave fields*, rather than Stokes vectors, such that

$$\begin{pmatrix} E_v^s \\ E_h^s \end{pmatrix} = \frac{e^{-ik_0 r}}{R} \begin{pmatrix} S_{VV} & S_{VH} \\ S_{HV} & S_{HH} \end{pmatrix} \begin{pmatrix} E_v^i \\ E_h^i \end{pmatrix}. \tag{4.19}$$

A radar system that measures the amplitude and phase of each of these four terms is described as "fully polarimetric", whereas if it only measures a sub-set of the scattering matrix it is "partially polarimetric".

Radar reciprocity   In most conditions encountered in Earth observation the principle of *reciprocity* can be invoked, which in polarimetry implies that $S_{HV} = S_{VH}$ (compare this with Section 4.4). This is a practical convenience since the cross-polarisation term will be of much lower intensity than the co-pol terms and so will be influenced more by background or instrument noise. To provide a more accurate estimate, it is often assumed that $S_{HV} = \frac{1}{2}(S_{VH} + S_{HV})$. Since this is an average the effect of noise is reduced.

4.3.8 Target Vector

Once reciprocity has been assumed, there are effectively only three complex measurements in a fully polarimetric system. It is therefore often convenient to define a target vector, instead of a matrix, as it is easier to

---

[37] In matrix notation the order of the operations is reversed so that for pq, q operates before p. It is rarely a point of confusion, but be aware that it is now commonplace for either order to be used when describing transmit and receive pairs.

deal with. In linear coordinates the target vector is defined as

$$\mathbf{k} = \begin{bmatrix} S_{VV} & S_{HV} & S_{HH} \end{bmatrix}^T \qquad (4.20)$$

where the superscript T refers to the transpose of the vector[38]. One means of visualising polarimetric data is simply to display the three components of this vector into a red-green-blue colour composite.

The linear basis is not always the most efficient for dealing with the analysis of polarimetric data, and so sometimes the data will be transformed into what is know as the "Pauli" basis, which is composed of the *Pauli basis* sum and difference of the co-pol terms, and twice the cross-pol term. The target vector in the Pauli basis is therefore given by

$$\mathbf{k}_P = \frac{1}{\sqrt{2}} \begin{bmatrix} S_{HH} + S_{VV} & S_{HH} - S_{VV} & 2S_{HV} \end{bmatrix}^T \qquad (4.21)$$

with the factor $\frac{1}{\sqrt{2}}$ added to normalise the result. For many applications this is a much more useful method as it helps to emphasise the phase difference between the HH and VV terms (the addition and subtraction being of complex numbers). In a colour composite this also serves to increase the contrast between different types of scattering, as will be discussed further in Section 6.4 below.

Direct scattering is dominated by the first term; double-interaction terms are dominated by the second term; and multiple scattering dominates the third term.

4.3.9   Covariance Matrix

From the target vectors a number of other matrices can be defined that characterise the similarity of the polarimetric channels. This is similar to the measure of coherence, introduced in Section 3.1, as they record the statistical inter-relationship between the different channels, rather than the channels themselves.

The most common is the *covariance matrix*, **C**, which is generated from the matrix multiplication of the target vector with its complex conjugate, such that

$$\mathbf{C} = \mathbf{k}.\mathbf{k}^{*T} = \begin{bmatrix} S_{VV} \\ S_{HH} \\ S_{HV} \end{bmatrix} \begin{bmatrix} S_{VV}^* & S_{HH}^* & S_{HV}^* \end{bmatrix} \qquad (4.22)$$

$$= \left\langle \begin{bmatrix} |S_{VV}|^2 & S_{VV}S_{HH}^* & S_{VV}S_{HV}^* \\ S_{HH}S_{VV}^* & |S_{HH}|^2 & S_{HH}S_{HV}^* \\ S_{HV}S_{VV}^* & S_{HV}S_{HH}^* & |S_{HV}|^2 \end{bmatrix} \right\rangle . \qquad (4.23)$$

The $\langle ... \rangle$ brackets denote ensemble averaging, which as in Section 3.1 is approximated by averaging over a set of measurements (usually pixels). *Coherency matrix*

---

[38] All vectors are column vectors by default. In the interests of space saving it is convenient to simply give vectors as row vectors, but for completeness we use the transpose superscript to indicate that the vector is really a column.

An alternative to this is the coherency matrix, $\mathbf{T}$, that is formed in a similar way, but using the Pauli target vector, such that

$$T = \mathbf{k}_P.\mathbf{k}_P^{*T}. \tag{4.24}$$

It is left as an exercise for the reader to determine the individual matrix elements.

4.4   Passive Polarimetry

A polarimeter is any instrument that can measure the polarisation of an EM wave. In general atmospheric sounding does not require a polarimeter since there is very little influence on polarisation from the atmosphere. An important exception is measuring from spaceborne systems, where the ionosphere may introduce Faraday rotation (described in Section 1.5). However, at the higher microwave frequencies used in atmospheric sounding the Faraday rotation is minimal.

Far more important in the context of passive microwave sensing is the signal from the Earth's surface. Even when measuring the atmosphere, the underlying Earth signal may make a significant contribution to the measured signal. The relationship between polarisation and emissivity (or reflectivity) will be considered in Chapter 5 where it will be shown how it depends on incidence angle. The only angle where there is no difference in the polarisation is at zero degrees (nadir) which would not allow measurement scanning to increase the swath.

Increased surface roughness also minimises the difference in polarisation and the extreme case of volume emission and scattering, such as a vegetation canopy or rain, will completely depolarise the signal. Making polarised measurements can therefore be important for discriminating between targets, or at least the source of the microwaves.

4.5   Polarimetry in Radar

4.5.1   Radar Polarimeters

A radar polarimeter is an instrument that not only measures the polarisation of the returned echo, but can also transmit the equivalent of a full range of polarisations. In practice, a radar polarimeter will only transmit two orthogonal polarisations, usually horizontal and vertical, and then receive these two polarisations. Since the phase information is known (by control of the transmitted wave and measurement of the received wave) as well as the amplitude, the information contained within these two polarisations is sufficient to synthesise the response from all possible combinations of transmit and receive polarisations.

If a system does not determine the full polarimetric response for some reason, it is said to be only partially polarimetric. Some systems may have the ability to both transmit and receive two polarisations, but may do so with some restrictions that provide only partially polarimetric data — for example, it may simply be that data telemetry rates are not fast enough to handle fully polarimetric data, or that there are power limitations. Similarly, if the phase information is not sufficiently well calibrated, or if echoes from different polarisations do not correspond to exactly the same target area, then the data is not fully polarimetric. *Partial Polarimetry*

Polarimetric data from radar systems are potentially very powerful as it contains information on the orientation of the scattering targets, and as such can be used to distinguish between different types of targets. Different agricultural crops, for instance, have different polarimetric responses, whereas full forest canopies tend to be completely depolarising. Chapter 5 considers polarimetric responses from real objects in more detail, and will include a description of how these effects depend upon wavelength.

4.5.2   Polarimetric Synthesis and Response Curves

One might expect that if a target has a preferred shape or orientation, then differently polarised incident waves will result in differently polarised echoes. Imaging radar systems cannot completely control the illumination polarisation on a pixel-by-pixel basis, but by measuring the full polarimetric data, it is actually possible to process the data *as if* you could change the transmit and receive polarisation for each pixel. This technique is referred to as "polarisation synthesis" and can be used to simulate the response for any arbitrary combination of transmit and receive polarisations. These responses are often referred to as polarimetric "signatures" however this implies a unique one-to-one correspondence between the pattern and the target, which is not the case. The response for a surface, for instance, is identical to that from a trihedral corner reflector. *Polarisation synthesis*

Analytically, the synthesis is achieved by multiplication of **K** by the Stokes vectors **g** for the transmit and receive polarisations (subscripted t and r respectively), so that the backscattered power $P(...)$ for a given combination of transmitted and received ellipticity and orientation angles, is given by

$$P(\chi_r, \psi_r, \chi_t, \psi_t) = \mathbf{g}_r \mathbf{K} \mathbf{g}_t. \qquad (4.25)$$

If **K** is appropriately calibrated, then P may be replaced with some real quantities that correspond to the target characteristics (rather than the signal measured at the antenna). The range of 2χ and 2ψ are polar coordinates within a Poincaré sphere so that the synthesised polarisation response P, for any given $\mathbf{g}_r$ and $\mathbf{g}_t$, can be considered as an intensity pattern across the sphere. Alternatively, if the magnitude of P represents the range from the centre of the sphere, then the result can be thought of as an irregular surface that characterises the polarimetric response.

To visualise such a response is not straightforward. Initial visualisations, developed by Huynen, plotted contours of equal power on the Poincaré sphere, which were termed *gamma spheres*. This requires the use

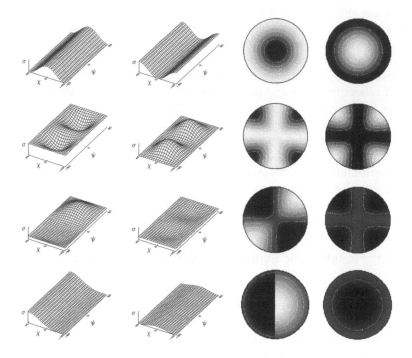

FIGURE 4.4 A comparison of the two techniques for a number of idealised responses. Example (a) demonstrates the responses of a flat surface, showing the change in rotation that occurs as a result of such interactions. Note that this becomes less relevant as the ellipticity of the impinging wave decreases. Example (b) shows the characteristic cross shape produced by a dihedral corner reflector. The orientation of such scatterers is more clearly visible using the new technique. Finally, examples (c) and (d) illustrate the benefits of the new technique for examining the orientation and helicity of scattering objects.

of two projections to map the left- and right-handed polarisations (van Zyl et al., 1987) . This technique was subsequently redeveloped to form what might be referred to as the "classical" polarisation response graph. This consists of a 3-D surface (or 2-D contour plot) of P using what amounts to a latitude-longitude representation of the gamma sphere (i.e., using simple linear axes of χ and ψ ranging between $\pm45°$ and $\pm90°$)[39] (van Zyl et al., 1987, Zebker et al, 1987). Normally two graphs are produced: one each for the co-pol and cross-pol response (where the two states are orthogonal corresponding to the antipodal point on the Poincaré sphere).

Some examples are demonstrated in Figure 4.4. Despite now being a familiar method in the literature, this presentation of the polarimetric re-

---

[39] Be aware that confusion can sometimes arise because of inconsistent labelling of the orientation axis such that in some publications it ranges from 0 to 180° and in others from -90 to 90°.

sponse is not optimum. It fails to optimally present the data in an intuitive and straightforward manner, making it difficult to interpret and analyse and so hinders its widespread use. One alternative visualisation method that tries to maintain the integrity of the Poincaré space, while at the same time optimising the ease of interpretation of the polarimetric response is describe by Woodhouse and Turner (2003). The examples in Figure 4.4 show both methods for comparison. In contrast to the simple lat-long plot of the traditional method, the alternative uses an equal area, polar azimuthal (Lambert's) projection of the Poincaré sphere. As such, we are effectively looking at one of the poles, with ellipticity (χ) measured from the pole to the equator (i.e., along lines of longitude), and orientation (ψ) measured around the circumference (i.e., around lines of latitude). The method also utilises the fact that the orientation angle is measured modulo 180° so it is possible to represent both the left handed and right handed polarisation states (the "Northern" and "Southern" Poincaré hemispheres) on the one circle. Such an approach also has the added advantage of mapping the horizontal and vertical polarisations to the horizontal and vertical directions, respectively, making them straightforward to interpret.

The complete structure of the projection is given in Figure 4.5 with the lat-long projection shown for comparison.

The interest in the use of polarimetric response graphs (or more usually, pairs of co- and cross-pol graphs) is that they offer the potential of an almost instant overview of the key features of the polarimetric data. They do not display the full range of information contained within polarimetric data (they do not display information on phase, for instance) but they do characterise the full range of polarimetric response of a symmetric target. In this way idealised scatterers such as spheres, diplanes and dipoles exhibit distinctive patterns in the graphs.

Note that these graphs do not show the physical ground — this is merely a symbolic representation of the response, not a visual representation of the target.

4.6   Important Polarimetric Properties

The important thing about polarimetry is its relationship to the target properties. If different targets did not respond differently to different polarisations, then polarimetry would not be very beneficial. However, a complete discussion on all the possible parameters with which polarimetric data can be characterised and how these relate to target properties is beyond the scope of this text and readers are referred to Cloude and Pottier (1996) and Ulaby et all (1990) for further reading. Here the most important, or most frequently used, parameters are described, with a brief qualitative summary of some others, whereas Chapter 5 will consider in more detail how different features of the Earth's surface effect polarimetric measurements.

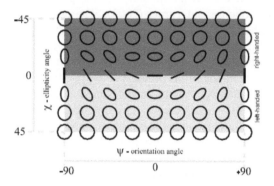

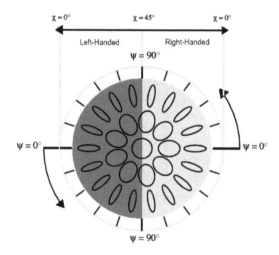

FIGURE 4.5 The polarimetric response corresponds to the surface of the Poincaré sphere, so can be respresented by using different projections much like maps of the world. Above is a latitude-longitude projection which is the most common form of the polarimetric response plot (even though it is often represented as an isometric surface). Beneath that is a polar plot that preserves the orientation information. Some examples polarisations are plotted on the projection for reference.

4.6.1   Unpolarised Power

Ironically, the first polarimetric parameter to consider is actually a "non-polarimetric" parameter. For any single polarisation state, the total power is the first term of the Stokes vector, I_0, such that

$$I_0 = \langle E_v^2 \rangle + \langle E_h^2 \rangle.$$

This is the unpolarised signal — the signal that would be measured if the detector was not sensitive to the polarisation. The term "total power" *Total power* is usually confined to radar where it is used to describe the unpolarised response for a radar — i.e. if the radar was unpolarised, what would be the power of the returned echo? The total power, T_u, is given by

$$T_u = \frac{1}{4}(|S_{HH}|^2 + 2\,|S_{HV}|^2 + |S_{VV}|^2), \qquad (4.26)$$

where S_{pq} is the complex scattering amplitude for a q-polarised transmit wave and p-polarised receive wave. It is also often called the *span* because it equals the span of the covariance matrix (Equation 4.23). Note that this is *not* a measure of how much of the signal is unpolarised, but a measure of the total power scattered from the pixel or target area back to the antenna, without taking the polarisation into account at all. It is therefore the direct radar equivalent of the first Stokes parameter, I_0, which would be used in passive polarimetry to characterise the intensity of the signal.

The total power radar image has been used in a number of applications and analysis methods. Despite the importance of the polarimetric information, the unpolarised power is still one of the most important sources of scene information primarily *because* it is not polarimetric, and so is more closely related to scene parameters such as local incidence angles, shadowing, or the dielectric properties of the target — i.e. it is related to the unpolarised brightness of the target, rather than shape or orientation.

As one might expect for an non-polarimetric parameter, the total power is independent of the polarimetric coordinate system — it is basis invariant.

4.6.2   Degree of Polarisation and Coefficient of Variation

For a polarised wave, it is useful to be able to describe how much of the wave is polarised and how much is unpolarised. In the context of the Stokes vector, this was given by, m, defined in Equation (4.14), where

$$m = \frac{\text{polarised power}}{\text{total power}}. \qquad (4.27)$$

It is useful to define a radar equivalent to m, and the polarimetric synthesis described earlier, gives a visual way of explaining terms such as the "coefficient of variation" defined by van Zyl (1997), which tries to quantify the proportion of the signal that is *un*polarised in a radar system. It is reasonable to assume that the maximum received power in polarimetric synthesis will correspond to a polarised signal, while the minimum, or background power, will be unpolarised. It can then be defined as

$$v_u = \frac{\text{minimum received power}}{\text{maximum received power}} = \frac{P_{\min}}{P_{\max}}, \qquad (4.28)$$

where $P_{\min}$ is often called the "pedestal height" (the underlying unpolarised background signal) in the polarimetric response curve. Similarly, the proportion of polarised signal is given by

$$v_p = \frac{P_{\max} - P_{\min}}{P_{\max}}. \qquad (4.29)$$

4.6.3   Polarimetric Ratios

In both active and passive polarimetry, it is often worthwhile to consider polarimetric ratios. By ratioing, the emphasis is on the relative properties of different polarisations, rather than the absolute intensity. Such parameters should therefore be more indicative of geometric or shape properties of the target, rather than dielectric or local illumination conditions. Ratios are calculated using temperature units for radiometry and power (or equivalent) units for radar.

The ratio of vertical to horizontal signal, r, is important because it relates directly to the expressions for reflectivity (for passive case) or backscatter (active case), that will be considered in Chapter 5. It is given by

$$r = \frac{T_V}{T_H} \qquad (4.30)$$

Co-pol ratio

for passive sensors, and for active sensors it is termed a *co-pol ratio* and is given by

$$r_{\text{co-pol}} = \frac{|S_{VV}|^2}{|S_{HH}|^2} = \frac{P_{VV}}{P_{HH}} = \frac{\sigma_{VV}^0}{\sigma_{HH}^0}, \qquad (4.31)$$

where the P terms correspond to backscattered power and the σ^0 terms are a measure of backscatter (the *normalised radar cross-section*) that is not defined until the next chapter but is included here for completeness and future cross-referencing.

Vegetation tends to be both a depolarising emitter and scatterer, so that horizontal and vertical contributions become progressively similar as the proportion of signal from the vegetation increases. Flat (horizontal) surfaces tend to have $T_H > T_V$ but $P_V > P_H$. The ratios can therefore be used as a crude indicator of, among other properties, the degree of vegetation cover.

Another property in radar systems that indicates volume scattering (in-
Cross-polarised ratio   cluding vegetation) is the *cross-polarised ratio*

$$r_{\text{cross}} = \frac{P_{HV}}{P_{HH}}, \qquad (4.32)$$

since the cross-pol contribution will increase with vegetation cover.

The horizontal term has been used as the normalising factor, but there is no universal agreement as to whether the H or V term should be used as the denominator[40].

4.6.4   Coherent Parameters

The concept of coherence was introduced in Section 3.1 within the context of the interference of waves, although it was explained that coherence can be interpreted as a measure of similarity between waves, or as a measure of predictability. The same approach can be applied to polarimetry, and in particular to quantify the similarity of waves at different polarisation. Note that the elements of the covariance matrix correspond directly to the cross-correlations used to define complex coherence in Equation (3.17). The normalised complex coherence for polarimetry can then be defined as

Polarimetric coherence

$$\gamma_{pq, \, p'q'} = \frac{\sum_N S_{pq} S^*_{p'q'}}{\sqrt{\sum_N |S_{pq}|^2 \sum_N |S_{p'q'}|^2}}, \qquad (4.33)$$

where the Coherence is now measured between the signal for a q-polarised transmit wave and p-polarised receive wave, and the signal for a q'-polarised transmit wave and p'-polarised receive wave. This is a complex number so it has both amplitude and phase.

The amplitude of (4.33) gives a measure of the degree of polarimetric coherence and lies between zero (incoherent) and one (total coherence). As before, a high coherence implies that knowledge of one polarisation will allow you to predict the other, so that the more random the spatial elements of a target, the less knowledge will be common to both polarisations and hence the lower the magnitude of the polarimetric coherence.

Degree of polarimetric coherence

One special case of $\gamma_{pq, \, p'q'}$ that has been used often in the literature is the coherence of the two co-polarised states, HH and VV. The magnitude of $\gamma_{HH, \, VV}$ can be an indicator of the degree of polarisation since low values indicate there is no correlation between the two co-pol states, which usually implies the target is not strongly polarising. Rougher surfaces tend to have a low values of $|\gamma_{HH,VV}|$ with volume scatterers (like vegetation canopies) having very low coherence.

Note that the coherence can be between any two polarisation states, so that we could also determine the coherence between $(HH + VV)$ and $(HH - VV)$ — this would be a different result to that between HH and VV. Polarimetric coherence therefore depends on which basis is being used and what polarisations are chosen.

The phase of (4.33) represents the phase difference between the two polarisation states. The phase of $\gamma_{HH, \, VV}$ is an important parameter as it has physical meaning related to the number of interactions and may also be

Polarimetric phase difference

---

[40]   An important point to remember is not to ratio decibel data. If your values are in decibels, then a ratio of a/b is actually found by the difference $a - b$.

an indicator of displacement or transmission effects. It is usually termed the polarimetric phase difference (PPD).

The PPD has been used in a number of applications to classify the scattering mechanisms of the target scene, whereby a PPD near $0°$ is interpreted to mean single bounce scattering and PPD near $180°$ is interpreted as "double bounce" scattering, which refers to the interaction between two adjacent, but perpendicular, surfaces. PPD values that lie between these two extremes are interpreted as multiple scattering if the distribution across the range of phase angles is approximately uniform. This interpretation of PPD, however, is not always applicable, since $HH - VV$ phase angle differences may arise from the HH and VV backscatter sources being at different range distances, or when the H and V waves propagate through an object at different velocities (e.g. Ulaby, et al, 1987).

The PPD can be estimated from the scattering matrix or covariance matrix from the following relationships:

$$S_{HH}S_{VV}^* = A_{HH}A_{VV}\exp(\phi_{HH} - \phi_{VV}) \qquad (4.34)$$
$$= A_{HH}A_{VV}(\cos\Delta\phi + i\sin\Delta\phi)$$
$$\tan\Delta\phi = \frac{\mathrm{Im}(S_{HH}S_{VV}^*)}{\mathrm{Re}(S_{HH}S_{VV}^*)}. \qquad (4.35)$$

Polarimetric entropy   A variant on the polarimetric coherence is the polarimetric *entropy*, which can be interpreted as a basis-invariant "total" polarimetric coherence. This provides a generic coherency measure, rather than one based on an arbitrary choice of polarisations of basis. As with the normalised coherence, the polarimetric entropy H is a real number between 0 and 1, although in this case low values indicate a high degree of polarisation, whereas depolarising targets have values of H approaching unity. The entropy is found by an eigenvector analysis of the coherency matrix, and interested readers are referred to Cloude and Pottier (1997) for a detailed derivation.

4.6.5   Polarimetric Decomposition

A very different approach to parameterising the information contained within a polarimetric radar measurement is *polarimetric decomposition*. The principle behind this technique is that a polarimetric response can be characterised by a combination of idealised scatterers. What separates the methodologies is the nature of these idealised base scatterers. From a purely theoretical approach, for instance, the polarimetric radar response from a target can be described as a combination of a diplane (double bounce), a sphere and a helix (which possesses a "handedness"). By changing the proportional contribution from each of these three targets, any measured response can be replicated (to within measurement error and noise).

Unfortunately, while theoretically meaningful, in the context of remote sensing of the Earth, this decomposition is not physically meaningful. There is no evidence that strong helical scatterers exist in nature, for instance, and dipole scattering (which is the predominant contribution in vegetation canopies) is not characterised at all.

A more physical-based approach was described by Freeman and Durden (1998) , whereby an imaged land area was characterised using varying contributions of dihedral (double bounce), direct surface (which equates to spherical scattering) and depolarised signal that corresponds to volume scattering from vegetation. These three terms then provide a useful means of generating an RGB colour composite for visualising the data.

4.7   Further Reading

General information on polarisation will be found in Hecht (2003). For more polarimetry in a radar context, readers should seek out Ulaby and Elachi (1990), while a thorough review of decomposition methods can be found in Cloude and Pottier (1996).

5
MICROWAVES IN THE REAL WORLD

"So far as the laws of mathematics refer to reality, they are not certain. And so far as they are certain, they do not refer to reality."
— Albert Einstein, *Geometry and Experience*

In this chapter we consider how microwaves interact with materials and objects (both natural and artificial). If we are to make meaningful measurements using microwaves it is important to know what governs their interaction with the natural world. One material of particular interest is the atmosphere — because we want to measure its properties, but also because if we are to measure the Earth's surface, we need to understand the influence of the intervening atmosphere.

It is this chapter that forms the underlying theoretical justification for using microwave remote sensing instruments to observe the Earth since it makes the link between measurable microwave properties and the physical attributes of natural objects.

5.1 Continuous Media and the Atmosphere

Figure 1.1 shows the transmissivity of a nominal clear atmosphere over a frequency range from radio and microwaves, to the visible. The transparent (windows) regions in the visible, infrared, low frequency microwave and radio bands can be used to observe the surface, the lower atmosphere and aerosols from satellites, since the attenuation by the atmosphere is minimal. In the case of microwaves with frequencies less than about 10GHz, we may even go so far as to ignore the impact of the atmosphere altogether. As the frequency increases, however, radiation is increasingly attenuated and by 100GHz the atmosphere is almost opaque — i.e. it becomes an atmospheric wall. As the atmosphere becomes increasingly opaque we begin to see less of the surface, but can make increasingly effective measurements of the atmosphere. In the true "wall" region, we only see the top of the atmosphere, since even the lower parts of the atmosphere can then no longer be seen through the upper opaque layers.

Whether looking at the surface or the atmosphere, it is important to understand the theory of how radiation is altered as it travels through a homogeneous medium. This is known as radiative transfer theory.

5.1.1   Radiative Transfer Theory

The original form of the equation of radiative transfer was developed by Chandrasekhar for describing the radiative properties of stellar atmospheres. It describes the intensity of radiation propagating in a media that simultaneously absorbs, emits and scatters the radiation. The key difference with the lossy medium described in Section 5.1 is that we can now include the case where the medium itself emits its own radiation rather than simply being a passive medium. In remote sensing of the terrestrial atmosphere, we are always dealing with media with a high enough temperature to emit significant microwave radiation. Thermodynamic temperatures within the atmosphere are well above 200K for most of the lower and middle atmosphere corresponding to a significant microwave emission (see Figure 3.33). At low frequencies (long wavelengths) we can usually assume the medium to be locally non-scattering, non-refractive and in thermal equilibrium[41]. This greatly simplifies the problem as there is no need to model precipitable water vapour or aerosols because they are so small by comparison to the wavelength. At the higher end of the microwave frequencies, however, it is necessary to start considering the scattering arising from water droplets (referred to as "hydrometeors").

For imaging applications (both active and passive) we can also use a radiative transfer approach, but this time to describe the interaction with layers on the ground — e.g. layers of snow or vegetation. This requires local scattering to be considered as well as emission. For active sensors the local emission is so small in comparison to the incident microwaves it is usually ignored. For simplicity we keep the discussion in this section confined to the atmosphere, with other radiative transfer applications discussed later, as appropriate.

The Problem   The main question in radiative transfer theory is this: given an incident radiation intensity onto a slab atmosphere, what is the emergent radiation intensity from the other side? The key to solving the radiative transfer problem is that you want to determine the intensity of radiation at an arbitrary point s along the path of the radiation (where the incident radiation is defined to be at $s = 0$). However, the radiation that is incident is not the only observed radiation since at each point along the path from 0 to s the

---

[41]   The important aspect of thermodynamic equilibrium is that the gas is neither warming up nor cooling down — there is a balance between absorbed and emitted radiation — so we can use Planck's equation to describe the relationship between radiation and temperature.

atmosphere itself radiates energy according to the Planck function (3.33). Additionally, both the atmospheric radiation that is emitted and the initial radiation are subject to an exponential decay due to absorption. Figure 5.1 illustrates this situation.

We therefore have the following circumstance:

Instantaneous intensity at s = incident intensity scaled by absorption

+accumulated emission over path

(again, scaled by absorption).   (5.1)

We know the form of the localised emission at an arbitrary point s' — this is the blackbody radiation $B_\nu\left(T(s')\right)$ modified by a term that accounts for the selective absorption, which for this case we take as the same as the selective emission, utilising Kirchoff's radiation law described in Section 6.2. In radiative transfer, since this is a property of a volume, and can also change within the medium, rather than being a property of an object, we use an absorption *coefficient*, instead of an absorptivity. Effectively they *Absorption Coefficient* describe the same physical property. We use the term $\kappa_\nu(s')$ to represent the volume absorption coefficient[42] at a frequency ν and position s'. For a continuous medium like the atmosphere, we only consider absorption and not other forms of energy loss, such as scattering.

In radiative transfer we also need a measure of the absorption over a path length — in particular, the absorption between our arbitrary point s' and the instrument location at s. We call this the *optical depth*, or *opac-* *Optical Depth* *ity*[43], which is defined over the distance s' to s, for a given frequency ν. It is denoted by the function $\tau_\nu(s', s)$, symbolised with the Greek letter *tau*, and is given by:

$$\tau_\nu(s', s) = \int_{s'}^{s} \kappa_\nu(s'')\,ds''. \qquad (5.2)$$

The integral merely tells us that we are summing up all the κ_ν's from s' to s, at each point s'' along the path. The opacity acts exponentially, so that the initial intensity in Equation (5.1) has to be modified by the term $e^{-\tau_\nu(0,s)}$, which accounts for the total absorption from 0 to s. The emitted radiation at each point s' must also be modified by a similar term for absorption between s' and s. The form of the radiative transfer equation, *Radiative Transfer Equa-* can then be given as: *tion*

$$I_\nu(s) = I_\nu(0)e^{-\tau_\nu(0,s)} + \int_{0}^{s} \kappa_\nu(s')B_\nu\left(T(s')\right)e^{-\tau_\nu(s',s)}\,ds', \qquad (5.3)$$

where $I_\nu(s)$ is the intensity of radiation at frequency ν and position s along some ray path, $I_\nu(0)$ is the initial intensity at the start of the path (behind the slab of atmosphere) and the integral sign this time is adding up all the emission along the path between 0 and s, as well as modifying

---

[42] The volume absorption coefficient differs from the mass absorption coefficient used by Chandrasekhar by a factor equal to the reciprocal of the density of the absorbing substance, *i.e.*, ρ^{-1}.
[43] The opacity is the degree of "opaqueness".

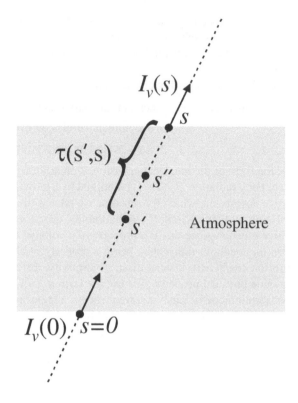

FIGURE 5.1 Illustration of the formulation of the radiative transfer equation through a slab of atmosphere.

it for absorption. The $\kappa_\nu(s')$ term has been included in order to modify the blackbody emission in correspondence to its selective emission (since we cannot assume it is a pure blackbody) . We can use the absorption coefficient because we utilised Kirchoff's radiation law — the spectral absorption should equal the spectral emission.

5.1.2   Microwave Brightness Temperature

In practice, the radiative transfer equation is usually presented in terms of brightness temperature, rather than intensity. The microwave brightness temperature, T_B was defined in Equation (3.39) by taking advantage of the Rayleigh-Jeans limit of the Planck function. This affords a linear relationship of the Planck function with physical temperature, so that it makes sense to use the brightness temperature instead of intensity as the quantity that is recorded and analysed. The brightness temperature has units of Kelvin. Formulations of the radiative transfer equation can then be done using brightness temperatures instead of intensity by using a linear conversion given by

$$I_\nu = \frac{2k}{\lambda^2} T_B(\nu). \tag{5.4}$$

For a fixed wavelength of interaction, the fraction factors out of the radiative transfer equation given in (5.3) so that it can be rewritten as

$$T_B(s) = T_B(0)e^{-\tau_\nu(0,s)} + \int_0^s \kappa(s')B\left(T(s')\right)e^{-\tau(s',s)}\,ds', \tag{5.5}$$

where the subscript ν has been dropped for clarity, even though the variables remain frequency dependent. The one point to remember here is that the brightness temperatures are not the the physical temperatures of the objects — brightness temperature is an observational, rather than a physical, property. We can simplify even further by writing (5.5) as:

$$T_B = T_{B0}\Upsilon + \int_0^s \kappa(s')T(s')e^{-\tau(s)}\,ds \tag{5.6}$$

where the brightness temperature of the background is given by T_{B0}. Note that we now have the actual temperature, $T(s)$, in the integral, but still modified by the absorption coefficient. The total absorption term, $e^{-\tau_\nu(0,s)}$, has been replaced by the transmissivity of the entire path length from 0 to s, such that

$$\Upsilon = e^{-\tau(0,s)}, \tag{5.7}$$

where $\tau_\nu(0,s)$ is the opacity along the total path length, giving a value of Υ that ranges from 0 (totally opaque) to 1 (totally transparent).

Equation (5.6) is a far more convenient form of the radiative transfer equation when using microwaves, and is surprisingly accurate as long as the R-J equivalent brightness temperature is used throughout the measurement chain, including the in-flight calibration. Not only do we now have the convenience of using units of temperature instead of intensity, but we have a much clearer sense of how the measurement relates to the physical properties of the atmosphere: the temperature and the opacity.

One special case is when the atmosphere is homogeneous with constant values of $T(s) = T$, and $\kappa(s) = \kappa$, throughout, so that $\tau = \kappa s$. In this case, the equation becomes

$$\begin{aligned}
T_B &= T_{B0}\Upsilon + \int_0^s \kappa T e^{-\kappa s}\,ds \\
&= T_{B0}\Upsilon + \left[-\frac{1}{\kappa}\kappa T e^{-\kappa s}\right]_0^s \tag{5.8} \\
&= T_{B0}\Upsilon + T(1 - e^{-\kappa s}) \tag{5.9} \\
&= T_{B0}\Upsilon + T(1 - \Upsilon). \tag{5.10}
\end{aligned}$$

Note that the second term in (5.10) is the atmospheric contribution, whereas the first term is the contribution from "behind" the atmosphere. It is worthwhile at this stage to make the effort to understand the steps above, since this equation will later form the basis of modelling the microwave response from surfaces as well as atmospheres.

5.1.3   Spectral Lines

The absorption coefficient κ_ν was quickly introduced above without saying much about its relationship to real atmospheres. κ_ν is a macroscopic parameter that describes the interaction of incident radiation with an absorbing medium but its properties need to be understood at the atomic level. Up to now the emphasis has been on describing the electromagnetic phenomenon as a wave — for dealing with microwaves this is much more appropriate than the alternative, which is to consider EM radiation as composed of many small particles, known as *photons*. However, the wave description is inadequate to describe the selective absorption (and emission) we observe in atmospheres, such as the Earth's atmosphere in Figure 1.1. For this we need to consider again the mechanism of how radiation is absorbed or emitted by individual atoms or molecules.

Atomic excitation   Atoms and molecules may exist at different levels of "excitation". This idea was touched upon in Section 6 when discussing the origin of microwaves. Changes in this excitation occur in discrete jumps or quanta — this is the fundamental idea behind quantum theory (the development of which was partly a consequence of Planck's work). For high energies these states of excitation are manifest as electrons in different "shells" – or in the very simple atomic model, you can consider these shells as being akin to higher or lower orbits for the electron. An absorption of an EM wave results in an electron jumping to a higher level, while an electron that drops in level will lose some of its energy by emitting an EM wave. The idea of an individual "wave" doesn't fit very nicely within this model, so a quanta of EM energy is required, which is where the photon comes in. Generally in remote sensing this distinction is not crucial, as we are always dealing with the equivalent of very many photons, and so the wave model is adequate.

Energy States   The excitation or relaxation process need not always be electrons jumping between shells. It can also be kinetic energy states of a molecule — rotational or vibrational states. Changes in the state at the molecular level also happen in discrete steps, and so an absorption of an EM wave will result in a change to a higher energy state, while a drop to a lower state will result in the release of EM radiation. Kinetic states are lower energy states than electron shells, so that whereas electron transitions tend to be in the NIR and visible, microwaves are the result of changes in rotational and vibrational states. The relationship between the energy of an EM wave and its frequency is given by $E = h\nu$, where h is Planck's constant.

Theoretically, in a gaseous medium where the molecules are not bound to each other, the interactions of photons with a molecule only occurs when the photon has a distinct frequency ν_{ul} given by the Bohr equation,

$$\nu_{ul} = \frac{(E_u - E_l)}{h} \qquad [\text{Hz}] \qquad (5.11)$$

where $(E_u - E_l)$ is the difference in energy between the upper and lower molecular states involved in the energy transition. Since the values

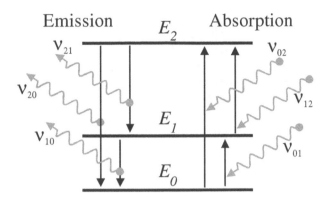

FIGURE 5.2   A simplified diagram showing three energy levels within a gas molecule. For microwaves these are likely to be rotational or vibrational energy states with $E_2 > E_1 > E_0$. Emission is when the molecules drops in energy level and emits a photon whereas absorption takes place when the molecule is excited to a higher energy level. Note that gases that emit well at certain frequencies will also absorb well at these same frequencies since $\nu_{21} = \nu_{12}$, $\nu_{20} = \nu_{02}$ and $\nu_{10} = \nu_{01}$.

of E_u and E_l are well defined values, the shifts in energy always corresponds to a discrete frequency. Figure 5.2 shows a diagrammatic representation of the energy levels and shows the process of absorption and emission. The result is that the absorption coefficient will be high for some discrete frequencies, but low for all the others, which is evident by distinct peaks, or *spectral lines,* in the atmospheric absorption shown in Figure 1.1. It should also be apparent from Figure 5.2 that frequencies where absorption is high, are also frequencies of (relatively) high emission. However, you can warm a gas with high energy radiation, but it may cool through a succession of much lower energy emissions from smaller jumps.

Molecular *collisions* can also cause shifts in the energy state. This *Molecular Collisions* is exactly what happens in the upper atmosphere — the atmosphere is warmed by the absorption of UV radiation by ozone, and the ozone molecules lose some of this energy through colliding with other molecules. These other molecules loose some of this thermal energy by emission of lower energy radiation in the microwave and far infrared (FIR) — high energy excitation is eventually lost through low energy emission.

This is also the process by which the "greenhouse" effect operates, *The Greenhouse Ef-* warming the Earth's atmosphere. High energy radiation from the Sun *fect* passes through the atmospheric window in the visible region (the Sun's peak radiation lies within the yellow part of the visible spectrum) and absorption at the Earth's surface causes the molecules to achieve a more energetic state. This energy is transferred, by collision, to the movement of the molecules and so the temperature of the surface rises. It then loses this energy, partly to the atmosphere by conduction, and partly by individual molecules emitting photons and returning to lower states of excitation. However, the lower energy microwave and thermal radiation it emits through this process of cooling down lies within an atmospheric wall. The result is that radiative energy reaches the surface very easily, but has a much harder time leaving through the atmosphere. The atmosphere itself then get progressively warmer until it is warm enough to re-radiate into space the equivalent amount of energy that is being absorbed at the surface.

Real Atmospheres   In a real atmosphere there are not just many transitions from a given molecular species, but also a number of different species. If the various transitions are taken as independent processes then the total absorption coefficient at any given frequency is simply the summation of the absorption coefficients for the individual transitions

$$\kappa_\nu = \sum_{\text{all transitions}} (\kappa_\nu)_{ul}. \tag{5.12}$$

In addition to calculating κ_ν for lines in the microwave region it is also necessary to add a *continuum* term, $\Delta\kappa_\nu$, for the water vapour transitions, such that

$$\kappa'_{H_2O} = \kappa_{H_2O} + \Delta\kappa_\nu. \tag{5.13}$$

The use of the continuum term is a result of empirical studies that show a discrepancy between the theoretical and measured values of water vapour absorption. Although the exact origin of the water vapour continuum is unknown, the common element among all the explanations is that it arises from the interaction between water molecules and either oxygen molecules, nitrogen molecules or other water molecules.

Ozone   Like water vapour, microwave interactions with the dominant isotopic form of ozone, $^{16}O_3$, are due to transitions between rotational states of the molecules, but O_3 has many more microwave transitions.

Microwave interactions with water vapour, molecular oxygen and ozone are the dominant absorption lines in the microwave region, and are due to electric dipole transitions between molecular rotational states. In their purest form, these lines would be extremely narrow, but in reality they are "broadened" by additional physical processes, such as pressure and temperature, which increase the width of the lines.

5.1.4   Line Broadening

Temperature Broadening   When the temperature increases, the individual molecules move faster. From any particular vantage point, there is therefore a greater spread in the range of velocities of molecules with an apparent trajectory towards or away from the observer. Emission (or absorption) from such moving molecules are therefore Doppler shifted (see Section 5.2) to slightly higher or lower frequencies. The result is a spreading of the observed spectral line, in proportion to the temperature, to cover a range of frequencies.

In the microwave region temperature broadening is significant only at low pressures, namely within the mesosphere ($\gtrsim$ 70km).

At the moderate to high pressures (in the context of Earth's atmosphere) *Pressure Broadening* in the stratosphere and troposphere, the line broadening is dominated by pressure (collisional) broadening, the details of which are not discussed here. It is sufficient to say that the broadening is a consequence of the perturbations of rotational energy levels by the increasing number of molecular collisions with increasing pressure.

5.1.5 Faraday Rotation

Faraday rotation is an important transmission effect that has an impact on polarised waves travelling through the ionosphere. The ionosphere is a region of ionised belts extending from about 100–500km above the surface of the Earth. The degree of the resulting electrification in this layer changes as a consequence of the solar illumination, and so varies diurnally, seasonally and with latitude. The importance of the ionosphere to microwave remote sensing is that satellite sensors have to make measurements of radiation that has passed through it (twice for radar) and is therefore a potentially significant practical limitation or, at the very least, a source of measurement error.

One of the discoveries made by Faraday was that an external magnetic field could influence the manner in which light propagated through a medium. In particular, it can rotate the polarisation of the light. Of course, visible light is just one form of electromagnetic radiation, so the Faraday effect applies equally well to microwaves, so that the local increase in the magnetic field within the ionosphere means that microwaves propagating through it are subject to Faraday rotation. The important thing to note here is that the degree of rotation drops off inversely with ν^2 so that the effect is greatest for lower frequencies (longer wavelengths). The reason Faraday rotation is a problem is if you are measuring only linear polarisations — a rotation around the line of site therefore impacts on the linear polarisation component of the measurement. This starts to become significant in L-band with maximum expected rotations up to about 100°, and becomes a severe limitation for the feasibility of spaceborne P-band systems which can suffer rotations of many hundreds of degrees.

For passive microwave sensing, the radiometric errors in linearly polarized brightness temperature measurements can be as large as 10 K. This error imposes a severe limit on the usefulness of data from daytime satellite passes for applications that require highly accurate measurements, such as soil moisture estimation or ocean surface salinity measurements. Fully polarimetric measurements are useful here since the effect only rotates the polarisation, rather than changing the nature of the polarisation. The basis-invariant information in the fully polarimetric measurement is therefore unaffected by the ionosphere.

5.2 Interaction With Discrete Objects

In radiative transfer we assumed that the dielectric properties, if they varied at all, only varied gradually and smoothly between the layers. We now consider what we might call "discrete objects", where we are really referring to the very general case of an interaction at a discrete *boundary* between two media of differing dielectric properties[44].

The key defining dimension for such interaction is the size of the elements that make up the boundary *in proportion to* the incident wavelength. When the dimensions of both media are large compared to the size of the wavelength we refer to the boundary as a *surface*, but when the dimensions of one of the media are smaller than, or similar in scale to, the wavelength we would talk about it being an *object*.

By way of example, consider the case of a smooth sphere: when the wavelength is much smaller than the diameter of the sphere (by about a factor of 100) then the boundary at the scale of the wavelength approaches being an infinitely flat plane (much like a flat Earth approximation when dealing with small scales on the Earth). In such a case we simply consider the boundary as a surface rather than consider it as an object, per se. As the wavelength gets longer eventually the size of the sphere begins to be important and we must take into account the boundary of the entire object (and more generally, its shape, since it need not be a sphere).

Key Properties   The nature of the interaction can therefore be considered to be governed by two key properties: the relative electromagnetic properties of the two media, and the shape and size of the boundary *in proportion to the wavelength*. This last point is repeatedly emphasised because various types of interactions that are often presented as being quite different can be the same fundamental process but at different scales. Partly because of this, the range of terminology can be confusing — diffraction, scattering and reflection, are all describing essentially the same phenomenon: the interaction of waves with objects or boundaries. In all cases it is the boundary that results in the incident waves being reflected (or refracted if the object is at all transparent) into different directions. These redirected waves then combine in the space surrounding the object and they will do so coherently if we are dealing with distances that are small compared to the wavelength.

When the boundary is very ordered (a smooth flat surface or a periodic surface) the resulting field can also be very ordered, whereas a randomly rough boundary, or a collection of randomly located small objects, will produce a chaotic and effectively unpredictable field (refer back to Section 4.8). Of course, concepts such as "small", "flat", "smooth" or "rough" are merely relative quantities and we will explore how to define these terms later in this chapter.

---

[44] Since dielectric properties govern the effects on electromagnetic radiation, there is no effect at a boundary between two media if they have the same dielectric properties. This is the basis of H.G. Wells' story, "The Invisible Man", where the main character is able to change the dielectric properties of his body to match that of the air around him — and therefore becomes transparent as there is no reflection or scattering from his body.

In the first instance, let us focus on discrete objects, or edges of objects, with surfaces considered in more detail in a later section.

5.2.1   Diffraction

Diffraction is the term generally used to describe the interaction of waves with any solid object (rather than a surface) but it is often used specifically to describe the interaction of a wave at an aperture. An aperture in this case can be considered as simply the space between two objects or a hole in a larger object.

The effects of what we call diffraction are an extension of the interference patterns we saw in Section 4.6 where we dealt with a vast number of superimposed waves. At the end of that section we introduced the idea that there was little difference in the interference pattern generated by a long continuous source and a collection of many closely spaced ($\ll \lambda$) isotropic point sources arranged in a line of the same length — this was the principle introduced by Huygens. The principle was also generalised not just to the source, but to any wavefront, so we can also apply it to the wavefront that exists across an aperture that is illuminated by a series of plane waves. An aperture should then act just as the many-source scenario shown in Figure 3.12 giving a generally plane wave beyond the aperture, but with sidelobes around the edges. The width of the main beam would therefore also be given by an expression similar to Equation (3.27), which reinforces the fact that the scale dimension in this case is the ratio of the wavelength to the diameter of the aperture. In Figure 5.3 we show a plane wave arriving at three different apertures — the first is much larger than the wavelength, the second of similar size and the third is much smaller than the wavelength.

If the wavefront at the aperture opening can be represented by a row of many point sources, then the waves behind the opening will combine in exactly the same way as the individual waves did in Figure 3.12 — in the diffraction case it looks like the wavefronts remains similar in the centre but are "bent" at the edges. This can be explained qualitatively by the fact that the waves at each edge of the aperture radiate much like a point source.

It is an important quality of waves that when they meet any kind of barrier, or aperture, they will exhibit this apparent distortion that we call *diffraction*. When the width of the aperture is comparable to the wavelength of the incident wave, the bent corners start to "join up" and the emerging waves become almost circular — the waves radiate as if they were almost emerging from a point source. As the aperture gets smaller still then the energy that gets through drops, eventually to zero.

The diagrams in Figure 5.3 are somewhat simplified, of course, since the strength of the actual pattern will vary with angle, as it did in Section 4 giving a strong central maximum and lower local maxima to either side — the sidelobes.

We are not familiar with the diffraction of EM radiation in everyday life because the wavelength of visible light is so small relative to everyday objects that the diffraction that *does* occur is hardly noticeable. However, there is one form of wave diffraction we are all familiar with: the diffraction of sound. Hearing sounds that originate from around a corner, for instance, is an example of wave diffraction.

5.2.2   Importance of Diffraction

Diffraction is important in microwave remote sensing for two reasons. The first is that an antenna, the element that is used to collect microwaves for detection, is an aperture (in the sense that it acts like a hole in the wave field it is measuring when in receiving mode as well as acting like a linear wavefront when transmitting) so that transmitted and incident waves are distorted by diffraction at the antenna. The distortion limits the angular resolving power of the antenna, just as we saw in Section 4.6. The theoretical limit of resolution due to diffraction is known as the *diffraction limit*. Optical systems are also limited by the diffraction of the aperture and this is often the factor that limits the angular resolution of the measurements. In principle, this is a physical limit to any optical[45] imaging system (although we shall see later when considering imaging radar that there are ways to make measurements that are not constrained by this limit).

Diffraction limit

The transmitted signal in a radar system is also diffracted. We can consider a similar set of cases to Figure 5.3, but as a sequence of transmitting antennas. In terms of the angular spread of the resulting waves, again the larger the aperture (antenna) the narrower the transmitted wave pattern.

The other reason diffraction is important is that many of the objects of interest on the Earth's surface are comparable in size to the wavelengths being used by the sensor (mm's to m's). Soil roughness elements, tree branches, wheat stalks, wind induced ripples over water and ocean waves, are all examples of features within this scale range. However, the theoretical analysis of diffraction is very difficult — an analytical determination of diffraction for anything more elaborate than a sphere or a long thin cylinder is so difficult as to be impractical to calculate.

Far Field

It is worth repeating that, as in the previous chapter, all the scattering and interaction considered here takes place in what was defined earlier as the *far-field* condition. In the case of scattering from objects this implies that the target object is small enough, and far enough from the source of the incident electromagnetic wave that the incident waves can be assumed to have straight, parallel wavefronts. This provides a simplified condition that allows for certain assumptions to be made, including that the intensity of the incident energy does not vary over the local target area (in any direction). This is a reasonable assumption in our case since, by definition,

---

[45]   Note that the term "optical" refers to the manner in which the waves are dealt with — usually by reflection or refraction through lenses — rather than referring to the visible part of the EM spectrum.

FIGURE 5.3 Parallel waves arriving at an obstruction will be diffracted. Diffraction is a coherent effect that is important both in system design and in modelling of scattering from objects.

remote sensing involves an instrument that is very much further away than the size of the target, or target area.

When the incident waves can be assumed to be plane waves, i.e., the wavefronts are straight and parallel, the diffraction is know as *Fraunhofer* diffraction, which is the case in remote sensing since we are almost always dealing with the far field. For the more general case when the waves are spherical, it is known as *Fresnel* diffraction. Near-field effects can become more significant, however, when dealing with multiple scattering between objects that are very close together — the needles on a conifer twig, for instance, or the closely spaced ice-crystals in snow.

5.2.3   Scattering

The general concept of *scattering* may be described as the *redirection of incident electromagnetic energy* by an object. "Scattering" and "diffraction" refer to the same physical process — a coherent distortion of an incident wave. In fact reflection, refraction, diffraction, can all be considered as essentially forms of scattering. Although somewhat of a simplification, it is often the case that when the term "scattering" is used explicitly, then we are usually referring to random distortion of waves by elements that are similar in size or less than the wavelength. Likewise, reflection is ordered scattering from surfaces that have features much less than the scale of the wavelength (i.e., "smooth") and diffraction is ordered scattering at discrete boundaries, such as well-defined edges and apertures.

When a discrete object scatters an electromagnetic wave the *diffracted field* is defined by the total field in the presence of the object, whereas the *scattered field* is defined as the difference between the total field in the presence of the object and the field that would exist if the object were absent, but with all the other conditions unchanged (Skolnik, 2002).

The effectiveness of a scatterer can be quantified by a term called the *scattering cross-section*, symbolised by the Greek letter *sigma*, σ. In general the incident energy may be scattered in any direction, and not nec-   *Scattering Cross-section* essarily equally in all directions (which would be termed *isotropic*). We therefore must first define the directional scattering cross-section as a function of the angle of observation θ such that:

$$\sigma(\theta) = \frac{\text{Scattered power per unit solid angle into direction } \theta \text{ [W}\Omega^{-1}]}{\text{The intensity of the original incident plane wave/}4\pi \text{ [Wm}^{-2}\Omega^{-1}]} \quad [\text{m}^2]$$
$$(5.14)$$

where Ω is the solid angle, and the 4π in the denominator normalises the plane wave to Ω (since a solid angle of 4π is a full sphere). Note that $\sigma(\theta)$ has dimensions of area (m^2). This is a bit like quantifying the efficiency by asking, "How big a hole in space do would you need to swallow up the same amount of energy that this target subsequently scatters in all directions (isotropically)?" This need not always be equal to the actual physical cross-sectional area of the target, but it would be for an idealised isotropic scatterer.

Total Scattering Cross-section

There are some special cases of the definition given in (5.14) that should be considered. Firstly, we can consider the total amount of power that is scattered. The total scattering cross-section requires integrating Equation (5.14) over all directions surrounding the target and is the ratio of the total scattered power and the incoming intensity:

$$\sigma_{\text{T}} = \frac{\text{Total scattered power [W]}}{\text{The intensity of the original incident plane wave [Wm}^{-2}]} \quad [\text{m}^2].$$
$$(5.15)$$

For a "perfect" scatterer this would equal the actual cross-section, with all the incident energy being redirected, but this may not necessarily be the case. The scattering object may absorb energy for instance[46]. The efficiency of the object to absorb energy is therefore similarly defined as the *absorption cross-section*, as above but with "total absorbed power" in the numerator.

We may also want to consider the energy that is scattered in the same direction as the incident wave (for instance, this would be relevant to the radiative transfer formulation given in the previous section). This would be the *forward scattering* cross-section, which is (5.14) in the forward direction. In the case of microwave radar, we are usually more interested in what is scattered *back* to the instrument — i.e., opposite to the direction of the incident wave. This is the *backscattering* cross-section or radar cross-section (RCS).

5.2.4   Radar Cross-section

In active systems we have control over how much energy is incident upon the target area, and so we want to quantify what proportion of that energy is returned to the sensor. For a radar system we therefore consider only the

---

[46]   This would be the case for objects within a microwave oven, for instance. This is a good example of when the medium is *not* in thermodynamic equilibrium, since it is absorbing more energy that it is losing and consequently gets progressively hotter.

scattered energy from the target that arrives back at the radar system, at range R, so that we can define the *radar (scattering) cross-section* (RCS) as:

$$\sigma = \frac{I_{received}}{I_{incident}} 4\pi R^2 \qquad [m^2]. \qquad (5.16)$$

That is, σ is the target area we would infer, based on the measured intensity $I_{received}$, by assuming area σ intercepted the transmitted beam in the far field and then redirected that power isotropically. Compare this with (5.14), which is slightly re-arranged and there is now an extra factor of R^2 that relates the scattered wave to the received wave at some distance R.

The amount of scattering, and hence the value of σ, depends upon a multitude of parameters of the target: shape, dielectric properties, orientation, roughness, etc. These properties are also likely to vary for different observation angles, frequencies and polarisations. And σ need not have any direct relation to the actual frontal area presented by the target to the radar beam. The cross-section of a target will approach zero, for instance, if the target scatters very little power back towards the antenna. This can occur because the target is small, or absorbing, or transparent, or because it scatters the waves in some other direction away from the antenna.

The cross section σ may also be much larger than the target frontal area in cases where the target scatters much more energy in the direction back to the radar antenna than would an isotropic scatterer — some of these special scatterers are described in a later section. Resonant effects such as Mie scattering from discrete scatterers, or Bragg scattering from surfaces may also result in an apparently large cross-sections.

Only for the very simplest of shapes (such as those used in calibration measurements) can the value of σ be calculated analytically, for example for a perfectly conducting sphere or flat plate, and even in such cases σ depends markedly on frequency (wavelength).

In active microwave systems the primary concern is therefore determining an accurate estimate of the scattering cross-section for the target area. In microwave remote sensing we may deal with scattering from either discrete targets (e.g. water droplets) or distributed targets (e.g. bare ground) or often a combination of the two (a vegetation canopy composed on individual scatterers above a ground surface). Radar data is therefore usually quantified using some form of the radar cross-section. An interesting feature of the RCS is that for an individual discrete target the area over which the measurement is taken has no effect on the calculated RCS (as long as the target is always within the area) since the amount of power returned remains constant. On the other hand, for a distributed target, such as a ground surface, increasing the area of the measurement increases the total backscattered power by the same factor and so the total RCS also changes. Doubling the measurement area doubles the scattered power and therefore doubles your measure of RCS. This is not acceptable in remote sensing since what is required is a description of the target area *irrespec-*

tive of the instrument configuration. For observing natural surfaces we therefore need a convenient and generalised way to quantify all types of scatterers, including distributed targets.

Most active microwave applications observe extended targets — *i.e.* the Earth's surface — rather than individual objects so that σ may refer to the proportion of energy scattered by a distributed area, rather than a discrete object. In the general case this area will correspond to the size of a instrument footprint. In imaging systems it will be related to the area of surface corresponding to an individual resolution cell or image pixel. This could pose a problem if we want to compare measurements from different instruments, or use a combination of non-imaging versus imaging techniques for a given application. We therefore need a normalised measure that is not dependant upon the size of footprint, or pixel. In such a case, one approach is to relate the inferred target area derived from the scattering cross section, σ, to the actual geometrical area A on the ground surface, and define the (*normalised*) *backscatter coefficient*, σ^0, as

Sigma Nought

$$\sigma^0 = \frac{\sigma}{A} \quad [\text{-}]. \tag{5.17}$$

This is the quantity that is most often used in the context of radar remote sensing, and may be referred to as *sigma nought*, the *differential radar cross-section* or the *normalised radar cross-section* (NRCS). Note that it is unitless (m^2/m^2). This gives us a unitless measure of the scattering cross-section per unit area of surface and so is a target property and not a property of the measurement geometry.

Many surface features when illuminated by microwaves will act like a "volume scatterer" — *i.e.* it scatters energy equally in all directions. Forest canopies, snow layers, are two such examples. The value of σ^0 over a volume varies only in proportion to the projected area of the incident energy — *i.e.* as a function of the cosine of the local incidence angle, θ_i, such that:

$$\sigma^0_{\text{volume}} = k \cos \theta_i \tag{5.18}$$

where k is some constant determined by the particular target properties. This is a direct consequence of the geometry of the situation. As the value of θ_i increases, the same incident power is distributed over a larger surface area in proportion to the value of $\cos \theta_i$. The scattered energy per unit surface area is therefore reduced by the corresponding amount simply due to the reduced incident energy. Note that the system itself cannot determine the amount of energy drop-off due to this effect since it is merely comparing incoming to outgoing radiation, so it requires an estimate of θ_i.

Gamma

It is therefore convenient to define a new term *gamma*, γ, such that

$$\gamma = \frac{\sigma^0}{\cos \theta_i} \quad [\text{-}]. \tag{5.19}$$

For an extended volume scattering target, such as a forest, γ will remain approximately constant for all incidence angles, and so is a more convenient measurement parameter to employ than σ^0 when dealing with

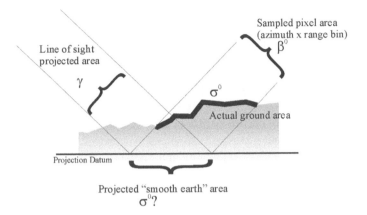

FIGURE 5.4 There are three different ways in which to normalise by area the cross-section values over distributed targets. The normalised radar cross-section, σ^0, should really be normalised over the actual ground area, but knowledge of the local topography across a pixel is required. Radar brightness β^0 is therefore a more appropriate parameter to use as it is normalised only to the projected area on some known reference surface. γ uses the area projected along the line of site, and is suitable for looking at volume scatterers such as forests.

such targets.

Sigma nought is the property we are actually interested in since it quantifies something about the ground surface rather than being instrument specific, but imaging radar systems only directly measure what is known as "radar brightness", β^0. The important distinction is that we do not necessarily know the precise value of the ground area illuminated — the value of A in Equation (5.17). This is because the ground topography will influence the local viewing geometry and therefore the actual area may be different from our predicted area A — without topographic data the area A would be estimated using a flat reference surface and this would be an incorrect representation of σ^0. If we have information on the topography, for instance from an existing digital elevation model, then we can correct for this and derive a more realistic value of σ^0 based on the average slope of the ground, rather than the area on the reference surface. The quality of that estimate will, of course, depend on the coarseness of the topographic data. The two values are related by the following expression:

Beta Nought

$$\sigma^0 = \frac{\beta^0}{\sin \theta_i} \qquad [-], \qquad (5.20)$$

where θ_i is the local incidence angle between the incoming wave and the normal to the local surface.

When topographic data is not available, the radar brightness tends to be used as the standard measure of backscatter in radar imagery. The three values σ^0, β^0 and γ and the different areas of projection they relate to are summarised in Figure 5.4.

5.2.5   Importance of Scattering Theory

Scattering is important in both active and passive microwave remote sensing. In passive sensing, radiant energy from other sources may be scattered into the field of view by particles in the air (such as water droplets) or objects (such as tree canopies) that lie between the sensor and the target of interest. Similarly, microwave energy from the target may be scattered out of the field of view of the sensor. These are all important effects that must be taken into account.

In active sensing, the directionality of the cross-section is all-important. For most cases we are interested only in the energy that is backscattered, but sometimes we may also want to consider the forward scattering when modelling the interaction of microwaves with volumetric targets such as snow, soils or vegetation canopies.

Scattering Regimes   Scientists looking at the optical scattering properties of particles in the atmosphere started studying the physics of scattering long before the advent of remote sensing. This work was principally carried out by scientists between 1870 and 1915, most notably for our purposes, Lord Rayleigh and Gustav Mie. Remember, scattering is governed by relative quantities: it is the size of the target object in proportion to the incident wavelength that counts. The scattering mechanism is the same at any scale so long as the relative size of the object compared to the wavelength is the same. Since we know that microwaves are electromagnetic waves (the same as optical light except at a longer wavelength) then the quantitative theory developed for optical scattering also applies to microwaves.

For simplicity, it is convenient to first consider the scattering from a perfectly smooth conducting (reflecting) sphere, with radius a.

There are two extreme conditions that are easy to imagine. The first is that the target is very much larger than the incident wavelength — by at least ten times. For a sphere, we usually use the ratio of the circumference to the wavelength , so $2\pi a \gg \lambda$. In this case the wave acts as if it were *Optical* incident upon a surface and we can calculate the scattering properties using geometrical optics. This situation is therefore referred to as the *optical region*, or *non-selective scattering* (since the scattering characteristics do not vary with wavelength). In this region the scattering cross-section is directly related to the physical cross-section of the object.

It is the non-selective nature of optical scattering that explains why both clouds and milk are opaque and white. In both cases they are composed of droplets that are microscopic but still much larger than optical wavelengths. In clouds these droplets are composed of liquid water suspended by the buoyancy of the air, whereas in milk they are accumulations of fat. In both cases they scatter light of all wavelengths with equal efficiency, and so appear white (which is the combination of all visible colours, after all).

The second extreme case is when the target is much smaller than the wavelength (again by at least a factor of 10) so that $2\pi a \ll \lambda$ — this is

referred to as *Rayleigh* scattering, which is important when considering   *Rayleigh*
microwave scattering in the atmosphere due to spherical water droplets.
Rayleigh scattering is also distinctive in that forward scattering is dominant. The scattering cross-section in the Rayleigh region drops off very
quickly with increasing wavelength[47] — as a function of $1/\lambda^4$.

As a basic rule of thumb, the Rayleigh region scattering cross-section
of fundamental objects such as spheres or long thing cylinders, is proportional to the square of the *volume* of the object so that the cross-section
drops dramatically with decreasing object size.

The region between these two extremes (about $0.1\lambda < 2\pi a < 10\lambda$) is
known as *Mie* scattering. This is where things are a bit strange because res-   *Mie*
onant effects occur — *i.e.* the scattering becomes very sensitive to small
changes in the target size (or wavelength), but with a variation that is periodic. This resonance is due to coherent scattering from the object since
scattered waves from different parts of the object's surface differ in distance by values close to the wavelength. Recall the interference effects
discussed in Section 4 — if we combine waves that have been shifted in
their path length by values close to the wavelength, we get regions of constructive and destructive interference. A similar effect is happening with
Mie scattering, but in this case the wave is reflected from the object in a
number of different directions, and these scattered waves are combining
constructively and destructively to produce a systematic (but analytically
complex) pattern of increased and decreased scattered wave intensity. A
distinctive feature of Mie scattering is that you can have observed values
of the cross-section very much larger than the physical size of the object
itself — this implies that at some frequencies the object would appear to
much larger than its actual size!

The three scattering regions are indicated in Figure 5.5 which also illustrates the resonant behaviour of Mie scattering. This figure shows the normalised (accounting for the change in apparent size of the target) scattering
cross-section of a spherical target as a function of its circumference measured in wavelengths $(2\pi a/\lambda)$. The magnitude of the cross-section varies
from virtually nothing (in the Rayleigh region where $\sigma \propto 1/\lambda^4$) to πa^2 (the
actual, or optical, cross section) in the optical region. In the Mie region
the cross-section oscillates around the value of the optical cross-section.
The exact calculation of the scattering cross section requires the solution
of Maxwell's equations subject to appropriate boundary conditions on the
surface of the scatterer — a task rather beyond the scope of this text, and
extremely difficult for anything other than simple shapes such as spheres,
plates, cylinders, and the like.

In general the actual scattering cross-section of an object depends on its

---

[47]   This is the reason the sky is blue. The rapid drop in scattering as the wavelength increases means that scattering of solar radiation in the atmosphere is dominated by the short
wavelengths — i.e., the blue end of the spectrum. Similarly, sunsets tend to be dominated
by the red end of the spectrum because the longer wavelengths are scattered much less in the
long path through the atmosphere.

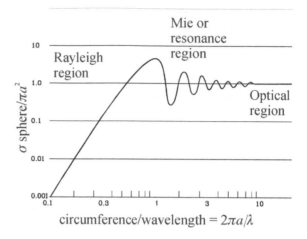

circumference/wavelength = $2\pi a/\lambda$

FIGURE 5.5   The radar cross-section of a sphere of radius a and incident wavelength λ, showing the transition from Rayleigh to optical scattering (after Skolnik, Introduction to Radar Systems, © 1970, reproduced with permission of The McGraw-Hill Companies).

physical size and shape as well as the material it is made of (i.e. its dielectric constant). At the optical extreme the orientation and aspect from which it is viewed or illuminated becomes a dominant influence, as well as the frequency and the polarisations of the incident wave. A rigorous evaluation of cross-section would therefore require a knowledge of the scattering properties of a target for a set of linearly polarised waves incident on the target for every direction with respect to a set of fixed axes. A great deal of work by anyone's standards.

5.3   Scattering and Emission from Volumes

Volume scatterers, or inhomogeneous media, differ from continuous media by being composed of randomly distributed discrete elements with significant cross-sections. Continuous media only absorb and emit radiation while discontinuous volumes have elements that also scatter. The distinction is therefore dependent upon the relative size of wavelength and the scatterers within the volume. Such a medium is referred to as a *sparse* medium when the discrete elements within the medium have a relatively low volume fraction ($< 5\%$), and are separated from each other by more than a few wavelengths[48].

Clouds of hydrometeors (water droplets in the atmosphere) act as volume scatterers/emitters, meaning that they are a collection of discrete targets that are distributed in three dimensions. In the limit of a very high

---

[48]   Another term you might come across when looking at scattering from random volumes is *tenuous*, which refers to volume media where the dielectric properties of the discrete scattering elements do not differ greatly from the background medium.

number of infinitely small targets a volumetric target begins to act like a continuous medium, like the atmosphere. When the scatterers are small compared to the wavelength, their scattering contribution may be minimal compared to the absorption, so that they can be approximated by a continuous media. Snow, ice, dry soils and vegetation all act like volumes under certain conditions. For volumes where the mean distance between scatterers is comparable to the wavelength, the scattering may be difficult to characterise because coherent (near field) scattering effects may dominate the interaction of the microwaves within the medium.

In the context of Huygens' principle, a volume is represented by a truly random distribution of secondary sources — i.e. their positions and orientations are not correlated to each other. This is not always the case — rows of wheat plants in an agricultural field, is a clear example where the volume elements are not randomly located or oriented.

The key thing about a true volume is that it scatters or emits equally in all directions. Each individual scattering element within the volume may have a directional scattering response, but since overall they are randomly oriented and distributed, the combined effects is that the radiation has no preferential direction. From Huygens' approach, we may similarly argue that since there is no correlation in position between scatterers, there is no coherent scattering component and so there cannot be a preferred direction. The lack of regularity in the scatterers, either in individual scatterers or in their relation to each other, would ensure that there is no coherent scattering.

A true volume is therefore difficult to find, since even forest canopies *Key Parameters* are not entirely randomly distributed, even if it is sometimes convenient to assume they are. The nearest thing to volume scattering for microwaves in the natural world tends to be materials composed of randomly located small scatterers suspended in some medium — precipitating clouds, for instance, or deep dry sand, snow or ice. However, for the latter, there is always an upper surface that has some degree of "order" since there is a distinct boundary. For volumes the boundary surface is usually represented by the suspending medium, which will have a low dielectric constant and therefore a much lower contribution to the scattering or emission. When the suspending medium increases in dielectric constant, the surface may play the dominant role, and in such cases the target would no longer act like a volume as most of the interaction would occur at this boundary.

5.3.1   Transmission Through Volumes

In considering the transmission effects of volumes it is possible to apply the same principles as in radiative transfer, except now scattering is considered along with absorption and emission. It is also sometimes appropriate to consider transmission effects that change the polarisation, but we do not cover that here.

The important thing about a wave travelling through some medium, even when it is a volume scatterer, is that it will lose energy, so the same basic terms introduced in Section 5.1 can be used to describe the bulk properties of a volume, such as penetration depth. The other terms of importance for volumes are the *volume-scattering coefficient*, κ_s, and the *volume-absorption coefficient*, κ_a. Note that κ_a is simply relabelling the volume absorption coefficient used in the radiative transfer formulation, κ_ν, so that we can distinguish it from other loss coefficients. The frequency dependence remains, but we no longer use the ν label for simplicity.

The two loss coefficients are defined as

$$\kappa_s = N_V Q_s, \qquad [\text{Np m}^{-1}] \qquad (5.21)$$

and

$$\kappa_a = N_V Q_a, \qquad [\text{Np m}^{-1}] \qquad (5.22)$$

where N_V is the number of scatterers per unit volume, and Q_s and Q_a are the individual scattering and absorption cross-sections respectively. Since these represent loss per unit length, the units are in Np m^{-1}. Note that Q_s is the total scattering cross-section, not the backscattering cross-section since we are interested in the total amount of scattering out of the path.

The *total* cross-section, Q_e, of an individual particle is then given by

$$Q_e = Q_a + Q_s, \qquad [\text{m}^2] \qquad (5.23)$$

and is known as the *extinction* cross-section since it characterises loss by both absorption and scattering.

In the following sections we now apply the radiative transfer approach but using these extinction coefficients, and recognising that $\kappa_e = \kappa_a + \kappa_s$.

5.3.2   Emission

We can formulate an expression for the brightness temperature of a simple, homogenous volume, composed of a sparse collection of small elements, using the same approach that gave us (5.5), but without the background signal. Here we take the simplest case in which the multiple-scattering between particles is small enough to ignore, so that,

$$T_V = \int_0^s \kappa_a(s')T(s')e^{-\tau(s',s)} \, ds'. \qquad (5.24)$$

where the contribution at each point along the path is given by the temperature modified by the absorption coefficient (which performs the function of the emissivity for a surface). Note, however, that the opacity is no longer the integral over the absorption alone, but the total extinction, since we now include scattering as contributing to the loss of energy. If we drop the frequency subscript for simplicity, the opacity is now given by

$$\tau(s',s) = \int_{s'}^s \kappa_e(s'') \, ds''. \qquad (5.25)$$

In fact, in remote sensing, the volumes that we encounter are virtually always a horizontal layer. This is approximately the case for vegetation canopies, snow, soil and even layers of clouds or rain. It is therefore more convenient to re-write (5.24) as an integration over the vertical distance z, but take into account the geometry of the observation so as to include a *zenith angle*, or *incidence angle*, θ, defined as the angle between the line-of-site to the instrument and the normal to the surface (the z-direction). In this case we must also change the integrand, ds'', to be over dz, by recognising that any path length, s, is equivalent to the vertical distance divided by $\cos\theta$, so that $ds'' = \frac{dz}{\cos\theta}$.

Incidence angle

For a layer of thickness, h, the integration is then over the path length from 0 to $h/\cos\theta_i$. Equation (5.24) is then

$$T_V = \frac{1}{\cos\theta} \int_0^h \kappa_a(z) T(z) e^{-\tau(z)} \, dz. \qquad (5.26)$$

Similarly, the opacity is now given by

$$\tau(z,\theta) = \frac{1}{\cos\theta} \int_{z'}^h \kappa_e(z'') \, dz''. \qquad (5.27)$$

For the special case of a homogenous volume, $\kappa_a(z) = \kappa_a$, at uniform temperature, $T(z) = T$, then (5.24) becomes

$$
\begin{aligned}
T_V(\theta) &= \frac{1}{\cos\theta} \int_0^h \kappa_a T e^{-\kappa_e z/\cos\theta} \, dz, & (5.28)\\[1mm]
&= \frac{\cos\theta}{\cos\theta} \frac{\kappa_a}{\kappa_e} T \left[-e^{-\kappa_e z/\cos\theta} \right]_0^h, & (5.29)\\[1mm]
&= \frac{\kappa_a}{\kappa_e} T (1 - e^{-\kappa_e h/\cos\theta}), & (5.30)\\[1mm]
&= \frac{\kappa_a}{\kappa_e} T (1 - \Upsilon_v(\theta)), & (5.31)
\end{aligned}
$$

where the exponential term that represents the transmissivity through the entire volume has been replaced with Υ_v for simplicity, where

$$\Upsilon_v(\theta) = e^{-\tau(h,\theta)} = e^{-\kappa_e h/\cos\theta}, \qquad (5.32)$$

with $\tau(h,\theta)$ the opacity along the total path length (defined by the thickness h and the angle θ), giving a value of Υ_v that ranges from 0 (totally opaque) to 1 (totally transparent).

The first term, κ_a/κ_e, in (5.31) can be rewritten as

$$
\begin{aligned}
\frac{\kappa_a}{\kappa_e} &= \frac{\kappa_e - \kappa_s}{\kappa_e}, & (5.33)\\[1mm]
&= 1 - \frac{\kappa_s}{\kappa_e}, & (5.34)\\[1mm]
&= 1 - \frac{Q_s}{Q_e}. & (5.35)
\end{aligned}
$$

The ratio of the particle scattering cross-section, Q_s, and the total extinction cross-section, Q_e, is often referred to as the *single scattering albedo* of the particle, a, so that our simplest expression for the brightness

Single scattering albedo

temperature of a sparse volume of scatterers at temperature, T, is given by rewriting (5.31) as

$$T_V(\theta) = T(1-a)(1-\Upsilon_v(\theta)). \tag{5.36}$$

The zenith angle has been explicitly left in this expression to stress the importance of the geometry of the observation. When $a = 0$, the wave penetrating the volume experiences no scattering, whereas as it approaches unity the wave would experience multiple scattering between elements before it lost significant energy. For most vegetation canopies we can assume that $a \lesssim 0.2$ so that we need only consider the direct single scattering.

To determine some interesting implications of (5.36) we can consider the two extreme cases of very high and very low opacity in the term for $\Upsilon_v(\theta)$. For very high values of $\tau(h,\theta)$, $\Upsilon_v(\theta) \approx 0$, since $e^{-x} \to 0$, as $x \to \infty$, so that

$$T_V(\theta) = T(1-a), \tag{5.37}$$

for opaque volumes. This can be the case for very thick volumes or for high values of θ. Note that this term is independent of incidence angle (i.e. viewing geometry).

For very transparent volumes, we utilise the simplification that $e^{-x} = 1 - x$, when $x \ll 1$, so that

$$T_V(\theta) = \frac{\kappa_a}{\kappa_e}\frac{T\kappa_e h}{\cos\theta}, \tag{5.38}$$

$$= \frac{\kappa_a h}{\cos\theta}T. \tag{5.39}$$

The extra $\cos\theta$ we find on the denominator is a consequence of viewing along a pathlength through the volume but with no loss due to extinction along that path.

5.3.3   Scattering

To formulate the equivalent expression to (5.36) for backscatter, the source term is now rewritten as $N_V\sigma$: the number density, N_V, times the individual radar cross-section of the scattering particles, σ. Similarly, the opacity is now found from multiplying the number density by the total extinction cross-section, Q_e, that characterises the loss of energy from the path length as a result of the "typical" particles within the volume. The extinction coefficient is therefore, $\kappa_e = N_V Q_e = N_V(Q_a + Q_s)$, and is the combination of absorption and scattering cross-sections of the particles. κ_e then represents the energy lost along the path length and is equivalent to the term in (5.2). Again, for the purposes of this discussion, we consider only the simple case of a homogeneous volume whereby N_V and Q_e are constant, so that the opacity of the layer is given by $\kappa_e h/\cos\theta$, where $h/\cos\theta$ is the total distance travelled through the volume[49].

---

[49]   Of course, the real world is rarely that simple, and readers should at least be aware that vegetation canopies, ice and snow layers and other naturally occuring volumes are rarely as

Whereas above we formulated an expression for the brightness temperature of a volume, for backscatter we want to estimate the radar cross-section, σ_{vol}, of the volume over some area, A. This is given by summing the back-scattering contribution over the length of the path, from 0 to s, and over the area, A. In the interests of simplicity we consider the backscatter from a pencil-beam (a narrow fixed field of view of small solid angle) which has a cross-sectional area, A'. that such that

$$\sigma_{vol}(\theta) = \frac{A'}{\cos\theta} \int_0^h N_V \sigma e^{-2\kappa_e \, z/\cos\theta} dz \qquad (5.40)$$

$$= \frac{A' \cos\theta}{\cos\theta} \frac{N_V \sigma}{2\kappa_e} \left[-e^{-2\kappa_e z/\cos\theta} \right]_0^h \qquad (5.41)$$

$$= A' \frac{N_V \sigma}{2\kappa_e} \left(1 - e^{-2\kappa_e h/\cos\theta} \right) \qquad (5.42)$$

$$= A' \frac{N_V \sigma}{2\kappa_e} (1 - \Upsilon_v^2(\theta)). \qquad (5.43)$$

Note that we have again utilised the term $\Upsilon_v(\theta)$ to characterise the transmissivity of the layer, but there is now an extra factor of 2 (as an exponent) compared to the brightness temperature expression in (5.24). This is due to the "there and back again" aspect of backscatter — the illuminating wave arriving at the edge of the volume loses energy on its way *into* the volume, in addition to the scattered wave losing energy on its way back out of the volume back to the sensor. The passive case only considered the latter.

Note that equation (5.43) is the radar cross-section within a pencil beam of cross-sectional area, A'. To convert this to a normalised RCS, we divide through by the projected area, A, on the surface of the volume (since we are normalising to units of RCS per unit surface area). The projected area is given by $A = A'/\cos\theta$, so that if we assume that κ_e does not vary with height and we divide through by A, then we have

$$\sigma_{vol}^0(\theta) = \frac{\sigma_{vol}}{A} \qquad (5.44)$$

$$= \frac{A' \cos\theta}{A'} \frac{N_V \sigma}{2\kappa_e} (1 - \Upsilon_v^2(\theta)) \qquad (5.45)$$

$$= \cos\theta \frac{N_V \sigma}{2\kappa_e} (1 - \Upsilon_v^2(\theta)). \qquad (5.46)$$

This is the most common form of what is often referred to as the "water cloud" model, which forms the foundation of many more complex models that build on this simplified foundation. *Water cloud model*

We can simplify it even further by taking advantage of the definition of the single scattering albedo used in (5.35) in the previous section, such that

---

simple as this in practice. In such cases, multiple homogeneous layers can be considered, whereby each layer has a different set of scattering properties, or more elaborate models can be considered that also consider the size, shape and orientation distributions of scatterers within each layer.

$a = Q_s/Q_e$, and using the following relationship between the backscattering and total scattering cross-sections for scatterers that exhibit Rayleigh scattering (Ulaby et al 1986):

$$\sigma = \frac{3}{2}Q_s. \tag{5.47}$$

This allows (5.44) to be rewritten with the single scattering albedo, such that

$$\sigma^0_{vol}(\theta) = \cos\theta\frac{3a}{4}(1 - \Upsilon^2_v(\theta)), \tag{5.48}$$

where we have again left the viewing angle as explicit.

As with the volume emission, there are two extreme cases for (5.44): very high or very low opacity in the term for $\Upsilon_v(\theta)$. For very high values of $\tau(h,\theta)$, again, $\Upsilon_v(\theta) \approx 0$, so that

$$\sigma^0_{vol}(\theta) = \frac{3a}{4}\cos\theta, \tag{5.49}$$

for opaque volumes. This can be the case for very thick volumes or for high values of θ, and results in a NRCS that is proportional to the single scattering albedo of the individual scatterers, and the viewing geometry. It is this approximation for opaque volumes that led to the use of Equation (5.18) in Section 2.4, but now $k = \frac{3a}{2}$.

For very transparent volumes, we utilise the simplification that $e^{-x} = 1 - x$, when $x \ll 1$, so that

$$\sigma^0_{vol}(\theta) = \cos\theta\frac{3a}{4}(1 - 1 + 2\kappa_e h/\cos\theta) \tag{5.50}$$

$$= \frac{3a}{2}\kappa_e h, \tag{5.51}$$

or to express it more simply,

$$\sigma^0_{vol}(\theta) = N_V\sigma h. \tag{5.52}$$

In this case we can see that the NRCS of a sparse volume simply becomes the area-normalised sum of the cross-sections over a height, h.

5.4   Reflection and Emission from Smooth Surfaces

In the previous section we dealt with diffraction and scattering as interaction of waves with objects smaller than or similar in scale to the wavelength. At the other extreme, when the wavelength is very much smaller than the object, we can now deal simply with a continuous boundary — a surface, rather than an object, per se. In the first instance we consider reflection.

Reflection   Reflection is a coherent form of scattering that occurs at smooth boundaries, by which we mean that any texture or form to the surface is very much smaller than the size of the wavelength (you should be catching on to this pattern by now). It is the equivalent of the optical scattering re-

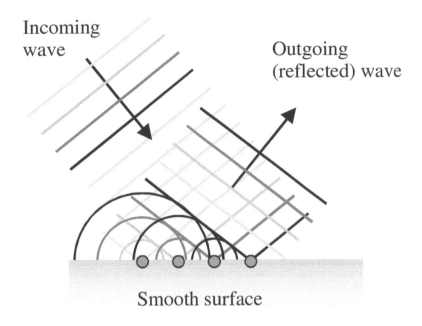

Incoming wave

Outgoing (reflected) wave

Smooth surface

FIGURE 5.6 The scattering from a smooth surface is coherent. Consideration of the scattering at the surface as a line of secondary point scatterers demonstrates how the waves combine constructively in the specular reflection direction only.

gion in Figure 5.5. We can also utilise Huygens' principle here — as a plane wave interacts with a surface, the surface can be considered to be composed of an infinite collection of secondary sources, each re-radiating (scattering) the incident energy and phase of the incoming wave. If these secondary sources are neatly aligned in a straight line, as it would be with a perfectly flat surface, then the scattered field is well defined and ordered, with the distinct peak in the direction that corresponds to that associated with specular reflection. This scenario is shown in Figure 5.6.

We can generalise this idea using Huygens' principle by considering any surface to be composed of many small point scatterers — each one scatters the incident radiation in all directions, acting as individual secondary sources. But we know from the results of Section 4 that many closely spaced wave sources radiating together will produce a wave field the properties of which are governed by the phase differences between the individual scatterers. For emission the phases are random. For reflection of an incoming wave the phase differences are a result of the incoming wave having to travel slightly different distances to each point on the surface. This is the key factor in determining the form of the scattered wave from any surface. If we describe a surface as "rough" on the scale of the wavelength, we mean that if we were to represent the surface by a collection of point sources, they would be displaced from a straight line by a significant portion of a wavelength. We quantify what we mean by this later.

It is also important to realise that any *surface* would never be an *entirely* random placement of points since there is always some degree of correlation between adjacent points (since it is a surface after all, rather than a cloud of particles). As a consequence, there always remains some scattering from the surface that corresponds to coherent scattering (i.e. reflection), even if it is small compared to the dominant incoherent scattering.

A further point of background knowledge is that the speed of an electromagnetic wave within a medium is proportional to the refractive index of the medium — EM waves are slower in materials with higher refractive indices. We have already discussed this in section 5. This is an important consideration when we look at what happens to such a wave when it meets a boundary between two media.

5.4.1   Scattering from Smooth Boundaries

One hypothetical limit of the boundary case is an infinite, perfectly smooth boundary between two media as shown in Figure 5.7. The upper medium, medium 1, has a lower refractive index than the lower one, medium 2 ($n_1 < n_2$). Such a case is a useful approximation to the Earth-atmosphere boundary, or the boundary between two layers in the atmosphere (since lower in atmosphere equates to higher density and therefore higher refractive index). For the solid Earth–atmosphere case the difference in n is very large ($n_{earth} \gg n_{atmosphere}$) so it is often simpler to assume the upper layer, the atmosphere, has the properties of free space.   Note that the boundary between any two media with different dielectric properties will induce some kind of change in the EM waves.

If an incident wave strikes this boundary, at an incidence angle (relative to the normal of the surface) given by θ_i, then part of the wave will be scattered (i.e., redirected), and part is transmitted into medium 2. Since in this case the surface is very regular the incident waves that are scattered at the surface do so in an ordered, coherent manner — each adjacent pair of secondary scatterers on the surface re-scatters a wave with a phase difference exactly the same as the next pair of scatterers, the value of the phase difference being in proportion to the angle of incidence (as was the case in Figure 3.11). The result is that amongst all of the re-scattered waves there is only one direction that exhibits constructive interference, giving a peak of the scattered energy in the direction given by the angle θ_r. This

Specular reflection   direction is known as the *specular* direction, and is given by

$$\theta_r = \theta_i.$$

Such ordered scattering is a direct consequence of the surface being very regular and we call it "reflection". This might seem an elaborate way

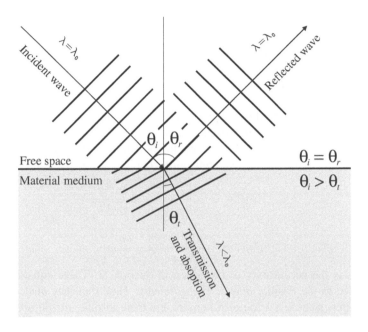

FIGURE 5.7 Summary of the interaction of EM waves with a smooth surface defining the boundary between two media. Note that refractive index for the material is greater than for free space so that the wavelength is shorter for the transmitted wave.

to describe a phenomena that you might already feel very familiar with, but it is important that you have a clear understanding of what reflection actually is in the context of scattering in order to follow what is happening at rough boundaries.

Not all of the wave will be scattered — some of the energy will be transmitted. If the wave arrives at an angle, part of the wavefront that enters medium 2 will enter first and will be slowed down (as a consequence of the higher refractive index). This results in a redirection of the incident wave towards the normal of the boundary or surface (or away from it if the refractive index of the second medium was lower) which is known as *refraction*. The degree of redirection by refraction is proportional to the *Refraction* ratio of the two refractive indices, and is given by *Snellius' Law* (or *Snell's* *Snell's Law* *Law*) which is written as:

$$\frac{\sin\theta_i}{\sin\theta_t} = \frac{v_1}{v_2} = \frac{n_2}{n_1}, \tag{5.53}$$

where the angle of refraction of the transmitted wave has been given as θ_t, as shown in Figure 5.7.

This describes the path of the wave, but what is it that determines the proportion of the wave energy that is reflected as opposed to transmitted? This is described by a reflection coefficient R that relates the reflected *Reflection Coefficients* electric field $\mathbf{E}_r$, to the incident field $\mathbf{E}_i$ such that $\mathbf{E}_r = R\mathbf{E}_i$. When the incident wave is oblique (*i.e.* not straight onto the surface) the interaction of the electric field with the surface is ultimately dependent upon the relative orientation between the electric field vector and the surface — i.e., the reflection coefficient is dependent upon the polarisation of the incident wave relative to the surface.

We therefore need two equations describing the reflection coefficient: one for the vertically polarised waves, and one for the horizontally polarised waves. For an oblique wave the horizontal polarisation is defined as parallel to the surface. The reflection coefficients are given by:

$$R_{VV} = \frac{\varepsilon \cos \theta_1 - \sqrt{\varepsilon - \sin^2 \theta_1}}{\varepsilon \cos \theta_1 + \sqrt{\varepsilon - \sin^2 \theta_1}}, \tag{5.54}$$

$$R_{HH} = \frac{\cos \theta_1 - \sqrt{\varepsilon - \sin^2 \theta_1}}{\cos \theta_1 + \sqrt{\varepsilon - \sin^2 \theta_1}}. \tag{5.55}$$

ε is the permittivity as described in Section 5. These equations are known as the *Fresnel reflection coefficients*. Note that they are complex numbers since ε is a complex number, but their absolute magnitude varies from 0 (no reflection) to 1 (total reflection). When the wave arrives perpendicular to the surface these equations are the same, which should be clear from a symmetry argument since the polarisation should no longer make any difference in that case.

The *reflectivity*, or *power reflection coefficient* , is then defined in terms of the reflection coefficients as:

$$\rho_{HH} = R_{HH} R_{HH}^* = |R_{HH}|^2, \qquad \rho_{VV} = R_{VV} R_{VV}^* = |R_{VV}|^2. \tag{5.56}$$

Similar terms can be used for the backscattered power coefficient.

The equations for reflectivity are summarised in Figure 5.8, which shows the calculated values of ρ_{HH} and ρ_{VV} for a specific value of ε. The most important feature of this figure is that the horizontal and vertical components behave very differently over the range of incidence angles. The horizontal reflectivity varies very simply with θ_i, and peaks at 1.0 for grazing angle. The vertical wave also has a maximum reflectivity at grazing incidence. This seems a strange result, but is easily tested by considering the reflective properties of, say, a sheet of paper — near the grazing angle even a sheet of matt white paper will exhibit almost specular reflection. Try this for yourself — hold up this book and look along the surface of the page towards a bright source, such as a light or window. As you reach the grazing angle, you will see that the paper appears highly reflective, even to the point of almost seeing the mirror image of the light source!

Another key thing to note is that for vertical polarisation, the reflectivity passes through a minimum. This is the *pseudo-Brewster angle* — at this angle, only horizontally polarised waves will be reflected. It is this

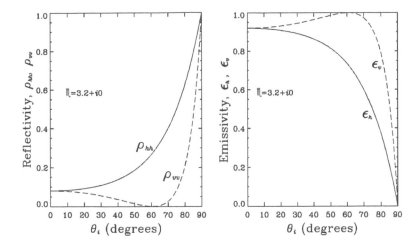

FIGURE 5.8 Reflection coefficients, reflectivity and emissivity as a function of incidence angle for a wave travelling from free space to a medium with $\varepsilon = 3.2$.

principle that allows Polaroid sun-glasses (which only allow through the vertically polarised waves) to cut-out glare from wet road surfaces and ski slopes, since they selectively block out the glare which, being close to the Brewster angle, is predominately horizontally polarised.

Another very important feature of interaction at a boundary is that the phase of the reflected (or scattered) wave will change if $n_1 < n_2$. To be more precise, in Figure 5.7 the horizontal component will undergo a phase change, but the vertical polarisation will not. This may seem strange. It is strange for the same reason that standing in front of mirror results in a peculiar effect — if you point up, your reflection points up. Point down and your reflection points down. But if you point to your left, the reflection points to their right! Why should left-right be switched over but up-down stays the same for you and your mirror image? The answer is that none of these lateral directions change with reflection, in the sense that if you are looking north into the mirror and you point east with your right hand, the reflection also points east (albeit with what they would call their left hand). It is the way you describe the direction that changes, and this is because the mirror image switches front and back. If you point north (into the mirror) the image points south! The result is that reflected waves will experience a change in phase of half a cycle — π radians — to the horizontal component, but not the vertical. The important thing to realise is that this is a consequence of how we describe the wave and its polarisation, rather than anything really mysterious going on at the boundary.

Phase Changes

5.4.2   Emission from Smooth Boundaries

The emissivity of a smooth surface is also shown in Figure 5.8. The emissivity is found by employing Kirchoff's law, expressed as in Equation (3.35), and assuming the transmissivity of the medium is zero, implying that it is completely opaque for the frequencies being observed. The emissivity for the two linear polarisations is then given by

$$\epsilon_V\left(\theta_i\right) \;=\; 1 - \rho_V\left(\theta_i\right), \tag{5.57}$$
$$\epsilon_H\left(\theta_i\right) \;=\; 1 - \rho_H\left(\theta_i\right). \tag{5.58}$$

The apparent microwave brightness temperature of the surface is then given by

$$T_B = \epsilon_p\left(\theta_i\right)T, \tag{5.59}$$

where T_B is expressed as a Rayleigh-Jeans equivalent brightness temperatures and T is the physical temperature of the medium (since brightness is proportional to temperature for the R-J brightness). Note that the brightness temperature is therefore a function of the angle of observation, and the polarisation, since the emissivity is sensitive to these variables.

5.4.3   Summary

To summarise: for a plane wave interacting with a flat boundary, consider Figure 5.7 — the Fresnel coefficients define the magnitude of the reflected and transmitted quantities as a function of incidence angle; Snell's law defines the path of the transmitted wave; emitted radiation in inversely related to the reflectivity; reflected waves may experience a phase shift in the horizontal polarisation; and the imaginary part of the dielectric constant of the medium determines the losses as the transmitted wave passes through the medium.

5.5   Scattering and Emission from Rough Surfaces

We now generalise the boundary interaction shown in Figure 5.7 to include randomly rough surfaces and use the term scattering rather than reflection because the elements that make up the surface features are similar in scale to the wavelength and so the incident waves no longer combine in a regular fashion.

This is the case as long as the surface is *randomly* rough — when there is a periodicity in the surface similar in scale to the wavelength then we can get coherent (resonant) scattering that is known as Bragg scattering (described later in Section 6).

5.5.1   Definition of "Rough"

If we are considering a horizontal boundary (which is usually the case in remote sensing) we can characterise the scale of the roughness of a randomly rough surface using the standard deviation (or root mean square) height deviation h from some mean height $\overline{h}$ of the surface (with both h and $\overline{h}$ defined over an area very much larger than the wavelength).

One important question in this context is, "How large can h be before we should consider the surface rough?" Such a definition must be physically meaningful and be scaled by the size of the incident wavelength, just as we did for the discrete scatterers in Figure 5.5. *What is "Rough"?*

One approach therefore is to ask what the phase difference $\Delta\phi$ would be of two rays if they were reflected from two different levels of a surface separated by one standard height deviation, h, apart. When such a separation starts to be large enough that it is similar in scale to one wavelength, the reflected waves start to interfere in a more unpredictable (incoherent) manner from the surface and the coherent reflection effect for smooth surfaces becomes less significant.

The criterion set by Rayleigh to answer this question states that for a surface to be described as smooth the phase difference due to h has to be less than one quarter of a wavelength, i.e., it must satisfy the condition $\Delta\phi < \pi/2$. An average phase change greater than this would mean the scattered waves would no longer be coherent, on average, and so the scattered wave will become increasingly more diffuse. The term "diffuse" is used here to describe the scattering from a surface that redirects energy equally in all directions, which is the opposite to a specular reflector. In terms of h the Rayleigh criterion gives *Rayleigh criterion*

$$h_{\text{smooth}} < \frac{\lambda}{8\cos\theta_i}, \tag{5.60}$$

where θ_i is the incidence angle, and λ is the wavelength of the incident radiation. You will notice that there is an extra factor of 2 included here (giving 1/8 instead of 1/4) — this is because we have to take into account the two-way path of the wave coming in and the wave that is scattered (incident and scattered path). The incidence angle influence is included since the path difference associated with h will vary with the angle of the incoming wave.

The Rayleigh criterion is useful as a first-order classification of when a surface can be said to be rough. However, for modelling microwave scattering behaviour from natural surfaces, where the wavelength is often of the order of h, a stricter criterion for smoothness is more appropriate. This requires the phase difference to satisfy $\Delta\phi < \pi/8$, before a surface is said to be smooth, giving:

$$h_{\text{smooth}} < \frac{\lambda}{32\cos\theta_i}. \tag{5.61}$$

This limit is referred to as the Fraunhoffer criterion. *Fraunhoffer criterion*

5.5.2   Effects of Roughness

When surfaces are within the limit of Equation (5.61) the surface interaction acts more like a smooth flat surface and exhibits *specular reflection*.

In this case the *coherent* surface scattering dominates and the expressions given in the previous section hold. Such a surface does occasionally arise in remote sensing conditions: a calm water surface, a dry mud flat, a road or an airfield. Some agricultural fields and rough surface water may also fit this criterion if the wavelength is long enough.

The key influence of increasing roughness for both scattering and emission is that the observed radiation in increasingly dominated by the non-coherent component. For most natural surfaces there will be some contribution from the well-ordered coherent scattered/emitted waves, but there will also be radiation that leaves the surface with a random variety of phase contributions and so will therefore combine in a random manner and in an increasing range of directions This chaotic reflection is often termed *diffuse* or *non-coherent* scattering — the rougher the surface, the larger the proportion of noncoherent waves. Note, however, that for a surface there always remains some correlation between adjacent parts of the surface and so there is always *some* contribution from coherent waves.

Non-coherent waves also tend to have no preferred polarisation, so increasing roughness tends to decrease the contrast between horizontal and vertical polarisation for both emission and scattering.

Facet model

One simple model of rough surface scattering entails considering the surface to be composed of very small smooth elements: *facets*. Since on a rough surface these representative facets would be oriented over a range of angles, the resulting radiation is also redistributed into a range of angles.

Roughness effects

We shall see in later chapters that imaging radar measures obliquely at relatively high angles of incidence (compared to most passive sensors). Such instruments are therefore more prone to the influence of incidence angle and roughness. Figure 5.9 illustrates the trend of how an incident plane coherent wave is either specularly reflected by a plane surface, or progressively more diffusely scattered as the surface becomes rougher. For the slightly rough surface the angular radiation pattern consists of two components: first, a reflected component in the specular direction with a magnitude smaller than that for the smooth surface; second, a diffusely scattered component, consisting of power scattered in all directions, the phase coherence deteriorated or even destroyed. The Figure also illustrates how the total scattered energy that is observed from a given observation angle, is dependent upon both the surface roughness *and* the angle of illumination.

5.5.3   Summary

As was the case with discrete scatterers, scattering from a surface is dependent upon the relative dimensions of the surface and the incident wavelength. A given surface may appear rough at optical wavelengths but smooth at microwaves. The degree of roughness must therefore be defined in some manner, often with parameters such as *root-mean-square height* and *correlation length*, or *root-mean-square slope*, or even by fractal dimension. These are all statistical measures used to give a general

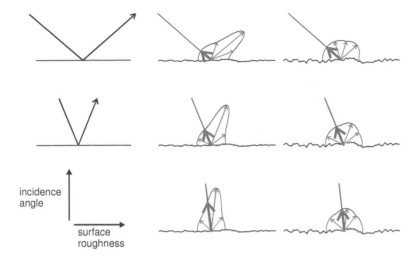

incidence
angle

surface
roughness

FIGURE 5.9 Surface roughness and incidence angle define the scattered field. For smooth surfaces only specular reflection occurs, while for rough surfaces diffuse scattering is also present. For the roughest case, the scattering is almost entirely diffuse, and does not vary significantly with angle.

quantitative measure of roughness, yet are all limited by the fact that they are only usefully applicable to simple flat surfaces, such as agricultural fields, or desert areas.

When discussing scattering, it was mentioned that a key difference in the use of the term scattering was that it was reserved for when the scattering elements are in the same order of size to the wavelength. The same is true for surface scattering, as opposed to reflection at a surface. When the key structural elements of a surface are similar in size to the wavelength, then the wave can be thought of as reflecting off (or emitted from) different elements of the surface, and combining constructively in some directions and destructively in others. For smooth surfaces the redistributed energy is focused into a narrow beam (in the specular direction). For surfaces with effectively a random structure (in particular, no periodic elements) the result is a very wide redistribution of the incident wave into a wide range of directions. The distinction is exactly equivalent to the difference between Figures 3.12 (coherent direction of energy into a narrow range of angles) and 3.13 (incoherent generation of energy into a range of directions).

5.6   Non-Random (Periodic) Surfaces

There is one special case of rough surface where the roughness exhibits a regular and periodic pattern, and the *rms* height variation over a small region is less than $\lambda/8$ (*i.e.* it is smooth with respect to the Rayleigh criterion). It is most commonly employed for the description of sea surfaces

Bragg surface

for radar systems, where the small wind-induced capillary waves or short gravity waves, which have wavelengths in the region of millimetres to centimetres, act as a *Bragg surface*. The important feature of Bragg scattering is that it is a coherent scattering mechanism, but it scatters the waves into a number of different directions, and not just the specular direction. The ordered nature of the surface results in scattered waves that will combine constructively in some directions and destructively in others.

We can use the facet model to describe the Bragg surface. Unlike the randomly rough surface, the facets of the Bragg surface have a regular pattern, so that each facet at a particular angle will occur regularly with the same separation between them. Each wave leaving these facets will combine constructively in some directions, and destructively in others, but since the separation distance can be larger than a wavelength, the scattered pattern is more like Figure 3.8 than Figure 3.12 (the latter equating to normal specular reflection).

From a radar measurement point of view, the important feature is that Bragg surfaces will exhibit a strong sensitivity to observation/incidence angle. We will see in later chapters that this principle is used to measure the size of the wind-induced capillary waves over the oceans using wind-scatterometer instruments.

The principle of Bragg scattering can also be used to model more complex surfaces by assuming that the rough surface can be described by an infinite sum of regular, sinusoidal, waveforms. The mathematics of this is known as Fourier analysis, but here it is sufficient to recognise that if a surface, such as an ocean surface, can be described by adding together a number of regular wave-like surfaces with different amplitude and wavelength, then we can generate a mathematical model of surface scattering based on adding together all the well-defined scattering from each of the component waves. That is, the complex ocean surface would be modelled by a summation of the scattering terms associated with each component wave. The regular wave scattering is relatively straightforward to describe, whereas to do the same for the composite surface as a whole would be very difficult. This is known as the Bragg surface scattering model.

5.7   Scattering and Emission from Natural Surfaces

From Kirchoff's Law and Equation (3.37) we know that emission and scattering (which is dependent upon the reflectivity) are complementary: surfaces that are good scatterers are weak emitters, and vice versa. We can consider this a good rule of thumb in remote sensing, but care must be taken to consider the particular circumstances at hand, especially the viewing direction of the instrument relative to the surface and the surface roughness, as described in the previous sections. Always keep in mind that the *back*scatter measured by a radar is not a measure of the effectiveness of the total scattering, as it may be the case (as it is with a smooth water surface) that a very high scattering efficiency results in almost no signal due the scattered energy being redirected to the specular direction, not the backscatter direction.

For passive sensing of the Earth's surface, it is often the case that the temperature can be relatively uniform across a large area. Over land surfaces, therefore, the key variable that is observed by measuring the microwave brightness is differences in emissivity and not the physical temperature. We therefore use emissivity to characterise natural surfaces. The exception to this is over the ocean, where the emissivity remains relatively constant, but the surface temperature varies. In both case we will characterise the observations in terms of a brightness temperature, T_B.

For active sensing, the normalised radar cross-section, σ^0, is used to characterise surface properties since it is not dependent upon the footprint or spatial resolution of any particular sensor.

The following sections try to summarise the pertinent characteristics of emissivity and sigma-nought for a number of common features on the Earth's surface using the theoretical principles given in the previous sections. The actual measured emission or backscatter response for any natural surface will be a combination of imaging geometry, reflectivity (i.e., dielectric constant) and surface roughness (in proportion to the wavelength).

5.7.1   Oceans and Lakes

As a first approximation, the microwave properties of the ocean should be relatively easy to describe. Indeed, for calm conditions the reflectivity of the ocean is a well-defined function of the water temperature and the salinity. Sea water is about 3.2% salt but much less closer to coastal areas where it is mixed with fresh water runoff from land or freshwater ice. The emissivity of the ocean increases with increasing temperature, and with increasing frequency. Emissivity will also increase with increasing salinity although this is only significant for frequencies lower than about 6GHz.

However, the sea is only rarely so calm as to make things that easy. Variations in surface roughness are a key factor in the microwave properties of water bodies — and all water surfaces are affected by the wind. The friction between air and the water causes small gravity-capillary waves on the scale of mm-cm — if you watch a water surface on a windy day, you will see the gusts of wind roughen the surface as they blow across the water.   These small spatial-scale waves operate on very short time scales and so are closely related to the instantaneous airflow over the surface. Larger water bodies also have longer (∼metres) gravity waves, and in the oceans there are also surface variations due to swell and large-scale currents.   These operate over many metres to kilometres and so operate over much longer time scales. The ocean surface is therefore often modelled as a combination of successive waveforms of differing wavelength and amplitudes (i.e. the Bragg model). This range of waveforms is known as the *wave spectrum*.

Capillary waves

Gravity waves

Even if you are interested in the simplest of model, characterising the ocean surface as a single wave is rarely good enough and it is usually necessary to include at least a two-scale roughness model that describes the small capillary waves superimposed upon the larger gravity waves.

The relationship between the wind and the wave spectrum is not simple, but in general the higher the wind speed the rougher the surface at all scales. Both emissivity and backscatter will change with increasing surface roughness, and so it will also change with the polarisation at off-nadir viewing angles. The rougher the surface, the less contrast there is between H and V polarisations. These trends mean that both active and passive microwave measurements have the potential to offer information on surface roughness, and therefore wind speed. At near nadir angles, it is the gravity-capillary waves that have biggest influence on H-V contrast since they have a preferred orientation (i.e. the polarimetric contrast is related to the orientation of the waves). At high view angles, there is a tendency for a greater H-V contrast due to the incidence angle dependency on the reflectivity coefficients.

In reality, however, the description of a sea surface can be further complicated by the presence of foam on the surface, which reduces the local surface reflectivity, or oil slicks, which will dampen the surface waves by reducing the surface tension. Oil slicks can happen naturally or from commercial shipping.

Sea Foam

Sea foam is a term that is used to cover two types of surface feature: whitecaps, short-lived turbulence at the crest of waves, and foam streaks and patches, which are thinner but longer lasting than whitecaps. By visual appearance white-water may look identical to foam, but note that whatwater be caused by air bubbles beneath the surface that may look white in the visible spectrum, but do not have the same effect as surface foam on the microwave emissivity. On reaching the surface and bursting these bubbles may change the surface roughness and hence the emissivity.

Asymmetry

One further important aspect of ocean scattering to mention here is that the wave spectrum has an orientation relative to the wind (since the wave pattern is approximately in line with the prevailing wind direction). This orientation expresses itself in two aspects of the microwaves. The first is the polarimetry, as mention above, where the measured signal (either emitted or backscattered) will tend to have polarimetric properties aligned with the orientation of the waves. The other point is that emission and backscatter will have an asymmetric response in the azimuth direction — i.e. if measurements are taken at the same incidence angle but from lots of different viewing directions, the response will have a pattern as shown in Figure 5.10.

These waves are extremely small ($\sim$centimetres) but because they are so regular over a large area they act as resonant Bragg scatterers and produce a distinctive relationship between the relative direction of the wind field and the relative scattered power. The magnitude of the scattered power is related to the speed of the wind. The technique is most appro-

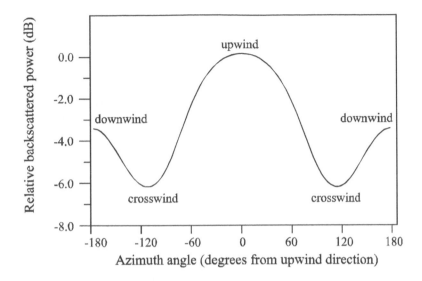

FIGURE 5.10 The theoretical relationship between the relative direction of a radar beam and the wind vector. The magnitude of this curve is defined by the wind speed. Note that exact nature of this pattern is dependent upon frequency, polarisation and inicidence angle.

priate for measuring the small capillary waves that are created when the wind blows across the ocean surface.

Since this condition means that amplitudes of the scattered wave from each individual ripple crest will add coherently, it is possible to have extremely high NRCS (many dB's), even though the individual contributions are very small. For instance, imagine a SAR image pixel of 25m that contains, say, 100 ripples (wavelength of 2.5cm). Each ripple in isolation within this pixel may contribute a NRCS of −45dB, which is not unreasonable for a hundredth of a pixel containing dry sand. If the 100 ripples add incoherently, the NRCS would be about 100×−45dB=20dB−45dB= −25dB. In the incoherent case we can simply add NRCS as it is directly proportional to power. In the coherent case, however, we add amplitudes, so we must consider a hundred times the amplitude of one ripple, and then squaring it, giving a power that is 10000 times as great. This corresponds to 40dB, giving the coherent NRCS when resonant scattering occurs to be in the region of −5dB.

The pattern in Figure 5.10 takes the idealised mathematical form of a function composed of three terms:

$$\sigma^0 = A + B\cos(\phi - \phi_R) + C\cos 2(\phi - \phi_R) \qquad (5.62)$$

where σ^0 is the normalised radar cross-section, ϕ is the azimuth look angle measured from some reference direction (such as North), and ϕ_R the periodic pattern (wind ripples) orientation relative to the same reference direction. A, B and C are coefficients that describe the mean backscatter, the upwind/downwind variation (when present) and the asymmetry due to ripples, respectively (Ulaby et al. 1986).

5.7.2   Hydrometeors

"Hydrometeors" is the term for water particles (liquid or frozen) either suspended in the atmosphere or falling through it. They affect the propagation of high frequency microwaves (>10GHz) within the troposphere and the lower stratosphere by absorbing and scattering radiation. They also have a physical temperature but do not warm or cool significantly (i.e. they are in thermal equilibrium), so they also emit radiation, including microwaves. The absorption, emission and scattering from hydrometeors can cause changes in the apparent microwave brightness temperature and so cloud and precipitation properties might be remotely sensed from appropriate measurements made by passive microwave sensors. In the active context, hydrometeors will attenuate radar pulses, and can therefore be a source of *radar clutter* (unwanted signal) if you are trying to observe through the atmosphere. But the opposite is also true, such that this radar clutter provides information on the hydrometeors, so that radar can therefore also be used as a means of measuring these hydrometeors by detecting the scattering properties at high frequencies. Such sensors are the *rain radars* that are often the source of the "radar image" shown by television weather presenters. In addition to their relevance in remote sensing, hydrometeors can also seriously degrade the performance of terrestrial and satellite radio communication and air traffic control radar.

There is a great deal of variation in the size distribution of hydrometeors depending on the type of cloud and precipitation event. The air temperature, the location and the time of year will also influence the hydrometeor properties. Hydrometeors are generally very sparse, with a fractional volume that is rarely more than about 5×10^{-6}.

The upper limit for the radii of non-precipitating cloud droplets in the Earth's atmosphere is around 0.1mm. For non-precipitating cumulus and stratus clouds, and for fog and haze, radii are typically in the region of 50μm or less. Such particles tend to absorb microwave radiation very effectively, but scatter very little. Such an atmosphere therefore acts much like a hydrometeor-free atmosphere, in that only absorption and emission occur, so that the radiative transfer theory described above in Section 1 can be used.

When cloud droplets coalesce to form rain, on the other hand, the resulting particle sizes approach the wavelength of the higher-frequency microwaves: about 0.1–5mm. Since the size of the droplets (and therefore their backscattering cross-section) usually increases with rain rate, measuring the microwave backscatter from rain cells can be a good indicator of rainfall distribution and intensity.

Frozen hydrometeors can be grouped into two classes: those smaller than 1mm (often needle- or plate-shaped particles) and those ranging from

1-10mm (snow or hail) that consist of a combination of liquid and ice phases, with small air pockets.

For the larger liquid hydrometeors (i.e. precipitation) and all the frozen ones (snow and hail) the scattering can be significant, and increasingly so for higher frequencies. When modelling such scattering, it is usually convenient to assume that the individual scatterers are spherical and uniform, although this assumption is only really appropriate for the smaller hydrometeors. Larger droplets tend to be slightly oblate as they fall, whereas ice particles can be very complex in shape, and these factors are important when analysing the polarimetry of a measured signal.

When measuring properties of the ground surface from aircraft or satellite, therefore, there is quite some advantage in using lower frequencies (longer wavelengths), which are affected much less by the water droplets. Remember that in the Rayleigh region the scattering drops off as the fourth power of the frequency so by halving the frequency, you reduce the scattering by hydrometeors by a factor of sixteen. This extreme sensitivity to wavelength can also be used to advantage, since a dual-frequency measurement can be used to retrieve the size of the hydrometeors directly, giving better estimates of rainfall.

5.7.3   Ice and Snow

In the microwave region, ice has some very interesting properties. Water in its liquid state has an extremely high dielectric constant, allowing virtually no penetration, but when it crystallises into ice, the dielectric constant drops to extremely low values, making it almost transparent to microwaves. You can try this at home if you have a microwave oven. Place an ice-cube next to a cup of cold water and turn to high. The water will boil before the ice cube barely melts (most of the melting due to its proximity to the hot water!) since the ice absorbs less of the microwaves[50].

Conveniently, snow and ice usually occur in relatively simple layers, usually over a surface with a higher dielectric constant, such as the underlying soil or rock, or water. The geometry of the situation is important, because there may be scattering from the upper boundary of the layer, from the layer itself, and from the boundary between the layers (which may be quite numerous if there are different layers of snow, ice, water and soil — e.g., over a snow-covered frozen lake). The scattering at the boundaries is dependent on the roughness of the boundary, and the relative dielectric properties of the materials. In this case, the predominant factor determining the dielectric constants of snow and ice is the *liquid* (or *free*) water content. Wet snow, for instance will have a higher dielectric constant than very "dry" snow. Both snow and ice layers will tend to scatter like a volume scatterer, not unlike a layer of vegetation, and so will tend to scatter

---

[50]   Be careful not to run the oven at "high" with only ice in it as you risk damaging the oven. The defrost function of a microwave oven operates at very low intensities, but for a long period of time.

equally in all directions.

Dry snow

The *liquid* water content of snow is the factor that determines whether it is said to be dry or wet snow. Completely dry snow consists of closely-spaced ice needles where the host medium is simply air. The ice needles are therefore close enough to support each other, and typically make up 10-20% of the volume of the snow when fresh, but up to 40% after it has aged after metamorphism (caused by thawing and refreezing). Although the ice needles are non-spherical they are randomly oriented so that collectively they act as if they are spheres. As a bulk material, fresh dry snow has a low dielectric constant, so microwaves do not interact greatly with the ice needles but will often penetrate through to the underlying surface. The interaction that does take place acts as volume scattering for high frequency microwaves.

Wet snow

Snow is described as wet when there are droplets of liquid water between the ice particles. The average liquid water droplet size is much smaller than ice particle size and so does not make a significant contribution to the total scattering, but it does make a significant contribution to the absorption. The presence of liquid water in a snow medium will tend to increase the absorption, and so reduce the amount of scattering. The water in a wet snow layer can therefore be considered as part of a host medium.

For many of the wavelengths used in remote sensing, snow acts as a volume of Mie scatterers (the ice particles) closely packed and bounded by irregular boundaries.

The permittivity of any particular layer of snow is dependent upon measurement frequency, the snow wetness, and the volume fraction. Variations of the bulk dielectric properties across a snow layer can also be governed by changes in the (ice) grain size and the snow depth. Emissivity and scattering therefore varies with snow thickness since it determines the total snow contribution. In dry snow, the penetration is high so that variations in emissivity and scattering of dry snow on land depends on thickness. This variation can be used as the basis of measuring snow thickness.

The grain size is important since the emissivity and scattering depends on the ratio of the observation wavelength to the characteristic size of the volume elements (the ice particles). This is important as it means that microwave observations have the potential to yield estimates of snow accumulation since low accumulation rates result in larger ice grains near the surface which scatter well and with low emissivities. For 10 GHz, for instance, when accumulation is slow you can expect emissivities of around 0.65, whereas for regions of high accumulation it may be 0.9.

However, the combined influence of snow coverage, accumulation rate, depth, remelting and other physical changes, may disrupt attempts at inferring any one of these parameters.

5.7.4   Freshwater Ice

When water freezes the molecules become bound to each other and their rotational states are no longer available to interact with the microwaves. The result is that pure ice therefore acts as a homogeneous medium with a very low dielectric constant. With very low values of ε'' the penetration depth is very high such that even C-band can have a penetration depth of 10m for pure ice. Any microwave emission or scattering therefore only occurs at boundaries between the ice and whatever lies beneath — i.e. liquid water or ground. The impact of this process can best be described using the example of radar scattering from a shallow lake that gradually freezes-over. Initially, the liquid water provides a good boundary to provide surface scattering so that wind-driven surface roughness is the largest influence on the signal (as discussed above). Normally, as fresh water cools it becomes more dense and sinks to the bottom, but at low temperatures, water just above freezing is actually less dense and so floats at the top until it finally freezes solid (it is for this reason that fresh water freezes over even when the bulk of the water body may not be near freezing). The ice on the surface will tend to have two smooth boundaries — at the top with air, and beneath with liquid water. Since the ice is virtually transparent to microwaves, the scattering all occurs at the ice-water boundary, rather than the ice-air boundary. Depending on how the ice formed, there may also be air bubbles within the ice that act as discrete scatterers, but they will not usually dominate the signal.

The ice-water boundary is usually a very smooth boundary (even for short wavelengths) and so the resulting scattering is dominated by specular reflection. Eventually, if the water continues to freeze, the entire depth of the lake will freeze and the interaction boundary is no longer ice-water, but ice-soil (or ice-rock), which will be a rough boundary governed by the same principles described in Section 5.

5.7.5   Glacial Ice

Glaciers and ice sheets are also fresh water ice but are formed by the compression of many layers of snow. The vertical structure of such ice is therefore subject to many additional factors such as seasonal remelt and metamorphosis and associated seasonal variations. The result is a sequence of horizontal layers that may have quite distinct dielectric, or structural properties, so that each boundary between the layers acts as a surface with which the microwaves will interact. In the accumulation zone, where the glacier gains snow and ice from snowfall and compression, the ice will also be covered with layers of snow, making the boundary between ice and air somewhat less distinct. Additionally, the transformation of snow to ice includes water seeping through the accumulated snow and refreezing to form horizontal ice lenses and vertical glands. These features can act as very effective scatterers. Deeper into the ice the snow will have been compressed sufficiently to produce dense ice. In the ablation zone, where ice is lost through melting and evaporation, it is this older, denser ice that dominates.

The dynamics of ice sheets and glaciers also cause large structural features such as crevassing. Most crevasses occur at the end of the glacier where the surface thins and breaks, and this creates a quite different surface for emission or scattering.

When the ice is dry enough (which usually means very cold Antarctic ice) and sufficiently homogeneous, then wavelengths of L-band or P-band or longer, are able to penetrate through 100m or more, and so can therefore reach the bedrock in areas where the ice is thin enough.

Icebergs

Note that icebergs are the result of *calving* whereby large sections of ice are broken off from an ice sheet or glacier that ends over open water. Individually they have the same vertical properties as described above — quite distinct from sea ice.

5.7.6   Sea Ice

Sea ice is considered separately because the salt content has a big impact on the dielectric properties of the ice and since it is formed from freezing sea water rather than compressed snow, it has different structural properties to glacial ice. In particular, polar ice has imbedded in it a mixture of salt, brine pockets and air bubbles that act as discrete scatterers (although the surrounding ice will tend to have a relatively low dielectric constant of about $\varepsilon'=3$). These scatterers may take up a volume fraction from 5-20% of the sea ice volume and so are not necessarily in far field of each other. Depending on the ice type, which is usually correlated with age, the characteristics of the scattering can change dramatically.

First year ice

Multi-year ice

The vertical structure and variation in dielectric properties is governed mostly by the evolution of the ice, which can be categorised into three main stages: young ice (<0.3m thick), first year ice (between 0.3m and 2m thick) and multiyear ice (typically >2m thick). All have different trends in brightness temperature and scattering with frequency. Young ice and first year ice are salinated, and first year ice may also be covered with snow that has been salinated from spray, which has a high emissivity. New first year ice might not have any snow, and has an intermediate emissivity.

The salt content of the Arctic sea (sufficiently distant from land) is about 3.2%. However, once the sea water is frozen, the salt begins to percolate down through the ice, desalinating the ice from the top and forming pockets of brine (liquid saline water) in the process. First year Arctic ice will have a typical profile that has 0.5–1.6% near the surface (caused by spray), about 0.4–0.5% in the bulk of the ice, and 3% near the ice-water interface at the bottom. Older desalinated ice will have a salinity profile with less than 0.1 % near the surface and only 0.2–0.3% in the main body of the ice, due to the brine pockets draining and being replaced by air.

Saline ice is highly absorptive for microwaves and so is not a good scatterer, but consequently it has a high emission relative to desalinated ice. Desalinated ice also acts more like a volume scatterer since both the brine pockets and the drained air pockets act as dielectric discontinuities within the ice, so that if they are sufficiently numerous they will dominate the scattering. Since they are distributed throughout the ice it will act as a volume scatterer. When the brine pockets have a low volume fraction (5%) then the majority of the scattering is at the surface. If the ice has many distinct layers, or if there are layers of snow on the top, then the emission and scattering properties are much more complicated.

The particular characteristics of the ice will depend on the relative size of the brine pockets or air pockets (approximately 0.24-2mm) to the wavelength being observed, the thickness of each layer and volume fraction. The volume scattering contribution increases with increasing scatterer radius, volume fraction and layer thickness. Surface emission and scattering varies with ice type due both to the desalination but also because older ice tends to have a rougher surface.

Pressure ridges where the ice has packed together increase the roughness in localised areas, and these can occur on any type of ice.

In observation, an important consideration is the contrast between the different ice types and the surrounding ocean, considering both variations with frequency as well as polarisation. Firstly, if we consider passive measurements, the ocean has a high emissivity for both H and V polarisations, both of which increase with increasing frequency, whereas multiyear ice has a high emissivity for low frequencies but steadily decreases with frequency so that it is lower than the ocean above about 20GHz.

First year ice has the highest emissivity for all but the lowest frequencies.

In terms of scattering, the smooth, saline young and first year ice results in very coherent, specular scattering. This will mean that scattering at low incidence angles will be dominated by backscattering whereas at higher incidence angles the forward scattering in the specular direction will dominate. The distinction will become important in later chapters when we consider altimeters, which operate at near nadir, and imaging radar, which operates obliquely (high incidence angles) — the distinction being that altimeters receive a very high return from young ice, whereas in radar images it will appear dark.

Multiyear ice, on the other hand, will have been drained for brine, leaving a low conductivity and so will allow significant penetration into the ice volume. Altimeters will receive a strong response, but less than for young ice, whereas radar images will show multiyear ice as being much brighter.

The emissivity characteristics of sea ice and open water are shown in Figure 5.11.

In the Southern ocean, multiyear ice of the character common in the Arctic is very rare since very little of the ice that experiences summer melt survives into the winter. There is therefore little point in mapping multiyear ice in the Antarctic. However, you do find some sea ice that exhibits similar microwave characteristics to Arctic multiyear ice — the two ice types in the south are therefore referred to as "type A" and "type B", with the terms being intentionally left ambiguous to avoid implying direct correlation with physical ice types.

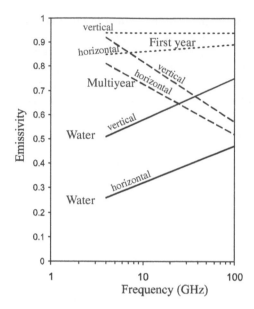

FIGURE 5.11 A schematic diagram based on Svendsen, et al, 1983, showing the general trends of emissivity (at 50° incidence angle) for water, first year and multiyear ice at H and V polarisations.

For open water the mechanism is exclusively surface scattering with a large specular scatter component depending of the surface roughness (i.e., on wind speed).

5.7.7   Bare Rock and Deserts

Hot and cold deserts are discussed before soils, and alongside bare rock, because there is little or no impact from soil moisture variations. Such surfaces are also appropriately grouped together because of their significance to the surface of dry planets such as Venus and Mars. The key factor for bare rock and stony deserts is that scattering and emission will occur at the rock-air boundary so that surface roughness will be the dominant influence on the signal. Note that such surfaces can vary dramatically in roughness, from smooth fluvial deposits to coarse gravel beds. The nature of the elements that make up the rough surface can also vary in character, from smooth eroded pebbles to sharp-edged rock fragments. It is therefore quite difficult to accurately generalise the scattering from rock surfaces other than on the basis of roughness.

A further factor that is sometimes important is the dielectric properties of the dry surface, the classic example being salt flats which have extremely high reflectivities as a consequence of the high salt content.

Sand   The other extreme is observed with hyperarid sand whereby the dielectric constant is so low that microwaves will penetrate many tens or hundreds of wavelengths through the sand. Almost no scattering occurs in such circumstances — the microwaves simply keep going until they meet a boundary, such as bedrock, or their energy is eventually attenuated within the sand. This is a very interested property of microwaves that has allowed long-wavelength surveying in sandy deserts of subsurface features many metres beneath the sand. In the context of environmental archaeology this has allowed mapping of ancient river beds as well as artificial structures such as roads buried beneath many years of wind blown sand. A practical application has been the search for new water sources by finding aquifers beneath the sand — there is a stark contrast in microwave properties between the dry and wet sand.

What scattering does occur from dry sand tends to be volume scattering because the surface does not provide a very distinct boundary. However, there is an exception to this — the Bragg scattering that occurs due to wind ripples. Wind blown sand usually ends up with a rippled surface, much like the surface of the ocean, with the spatial frequency (wavelength) and height of the ripples governed mainly by the size and humidity of the sand as well as the speed of the wind. The wavelength of these ripples can vary quite considerably from cm's to m's, but in a given area are usually dominated by one wavelength. Since these ripples lie across the undulating (and therefore varying local incidence angle) dunes, any observation across a dune field may include many areas where the local incidence angle and ripple wavelength are just right for resonant scattering. Bear in mind, though that unlike over oceans, the ripples will remain even when the wind stops!

5.7.8   Soils

Soils respond quite differently from bare rock because it is composed of a combination of lose grains of soil, air and water. In general, scattering increases and emission decreases when the soil gets wetter. In remote sensing of soils it is the overall bulk properties that are relevant rather than the microstructure (at least directly) since this fine scale structure is usually very much smaller than a wavelength.

Determining the bulk dielectric properties of soil is not straightforward as it is not simply the weighted average of the soil and the water, but is complicated by the fact that the water in the soil can be either free or bound. The free water molecules are able to be excited to their rotational energy states, so that $e' \sim 80$, whereas bound water molecules are absorbed by the surface of the soil particles, so are not so free to rotate. The proportion of free to bound water depends on the soil properties such as surface area, which is dependent upon the amount, shape and size of the soil particles. However, some algorithms do exist for determining the bulk dielectric properties of soil given the physical characteristics and soil moisture content.

As water is added to the soil, the real part of the dielectric constant increases slowly as most of the water molecules bind in the soil. Across the microwave region, we can expect values of $e'=1$ for air, ~ 4 for soil and ~ 80 for water, so that liquid water content is the biggest influence on dielectric properties of soil. For saturated soil, ε' approaches that of liquid water ($\varepsilon'=80$ at $\gtrsim 1\mathrm{GHz}$). Similarly the imaginary part increases with increasing moisture fraction, although at a slower rate. The net effect (at a given wavelength) is that the penetration depth decreases as the soil moisture increases. The reduction in signal penetration with higher free water content also means that microwave measurements of soils are more complicated that first expected, since varying degrees of soil wetness results in the signal being dominated by different depths within the soil.

Very dry soils will act as volume scatters with microwaves emitted or scattered throughout the depth of the soil, as was the case with sand, whereas very wet soils will predominately act like a rough surface. The emission and scattering properties are therefore dependent upon the surface roughness and, strictly speaking, the vertical distribution of soil moisture content.

Therein lies one of the biggest problems in using microwave data to measure soil moisture directly — there is no simple way to distinguish between surface roughness effects and water content effects, and this is currently a serious topic of ongoing research. For this reason there has been a significant amount of research aimed at estimating soil moisture from microwave emission or scattering.

Bulk Density

The bulk dielectric properties of soils are influenced by the bulk density and soil structure. Fortunately the bulk density of soils tends not to vary much over a particular region, so that contrasts within any regional data set is not likely to be influenced by the bulk density. The soil structure is based on the texture variations due to varying proportions of sand and clay, as well as the aggregation (i.e., the "clumping").

Aggregation

Aggregation may change depending on local rainfall conditions which means it is a very important consideration in the case of estimating soil moisture for two reasons. Firstly, the variability of the soil moisture and moisture profile may influence the measurements such that short wavelength sensors may be more sensitive to the superficial variations in the very top of the soil, whereas longer wavelengths (L-band) should be more appropriate for examining the long term moisture conditions. Secondly, if the depth of sensitivity depends on wavelength, then there is the potential for multi-frequency sensors to determine information on the soil moisture profile.

The soil roughness is important. At high incidence angles the contrast between H and V polarisations, for both emission and scattering, reduces since the orientation information is diminished by increasing the roughness. At the high frequencies used in some passive sensing ($>80\mathrm{Hz}$) the micro-structure becomes more important and the soil looks permanently rough. The H-V contrast reduces and the emissivity is no longer sensitive

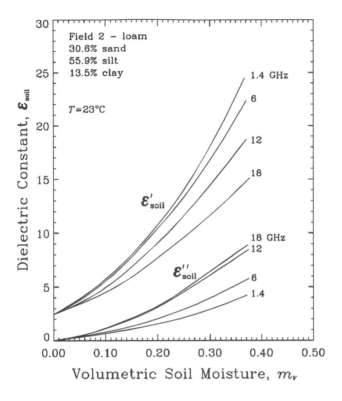

FIGURE 5.12 Relationship between soil moisture content and dielectric constant. (Reproduced by permission from Fawaz T. Ulaby, et al, *Microwave Remote Sensing: Active and Passive, Vol. III: From theory to applications*, Norwood, MA: Artech House, Inc., 1986. © 1986 Artech House, Inc.).

to roughness.

Emitted radiation from soil is a combination of upwelling radiation from all depths, although the contribution diminishes with depth. A significant contribution comes from the layers within the top few 10ths of a wavelength, the actual depth depending on the specific dielectric properties of the soil for the particular wavelength being considered. The response from soil can be greatly complicated by heterogeneities within the soil (air pockets, large stones) and vegetation cover, which can be so great as to totally obscure the soil from observation.

Another important feature that is appropriate to discuss here is that frozen water molecules are also not free to rotate so that frozen soils have very low dielectric constants compared to the value when thawed — ε' can be different by a factor of ten or more. The change is dramatic, so that spatial or temporal boundaries between frozen and thawed soil have an extremely high contrast. As we saw above, the extremely low dielectric constant of hyperarid desert sand is also a results of it containing virtually

Groundfrost

no liquid water. As a rule of thumb, it is the liquid water content that determines the dielectric properties of the soil, and in particular the penetration depth. Measurements of wet soils will therefore express the characteristics of the very top layer of soil (much less than a wavelength deep) and will be very sensitive to the surface roughness. Observations of progressively dryer soils become increasingly sensitive to the properties of the deeper layers and the soil will begin to act like a volume. The vertical soil moisture profile is therefore an important parameter to consider rather than just a straightforward average soil moisture.

5.7.9   Vegetation

The interaction of microwaves with vegetation canopies is a rather complicated affair since in general the canopy forms a heterogeneous volume with structural components of varying sizes and number densities. The scattering elements are leaves (flat ellipse-like, or needle-like, scatterers) and branches (long thin cylinders) and in a large forest you will find variations in the size of these elements with diameters ranging from millimetres (twigs) to metres (trunks), with number densities tending to increase for the smaller elements. It can also be said that in general the larger structural elements tend to be nearer the ground while the smaller elements are more numerous higher in the canopy.

Since the wavelengths of the microwaves are of the same order as these structural elements of a forest the physics of the scattering can be quite complex, and resonant Mie scattering may complicate the theory even further (depending on the particular microwave frequency being considered). It is worthwhile at this stage to refer back to Figure 5.5, as the types of scattering within a forest can encompass the full range of scattering types from Rayleigh to optical. We can use this figure to make some generalisations (remembering that the figure plots the cross-section normalised to the physical cross-section, so that even in the optical region the actual cross-section increases with increasing scatterer size). The greater the average diameter of the elements, the greater the backscatter and the greater the extinction by the canopy. The depth of the canopy will determine the total extinction through the vegetation. The smaller wavelength microwaves such as X- and C-band therefore tend not to penetrate a dense canopy, since the canopy elements are comparatively large for these smaller wavelengths.

For low frequencies the canopy components appear smaller compared to the wavelength so that microwaves in L-band and P-band can penetrate further into the canopy. P-band for instance, has backscatter over forest that is mostly determined by the size of the trunks and larger branches, rather than the smaller crown elements. At VHF with wavelengths $\gg 1$ m the signal is virtually all from the underlying ground and largest tree trunks.

Forests   The detailed description of emission and scattering properties of forest canopies require a thorough description of the individual scattering elements including their location, number densities, shapes, sizes, dielectric properties and orientation angles. Conveniently, forest canopies tend to be so complex that it is often appropriate to model it as a random volume. The assumption is that there are so many shapes and sizes at different orientations that the cumulative effect for either scattering or emission is for the canopy to *appear* like a volume composed of idealised identical objects, with the properties of that object being in some way representative of the canopy objects "on average". These objects are randomly oriented to represent the wide range of actual orientation angles.

At the level of an individual cylindrical element of the tree structure, we would expect the backscatter to increase in proportion to its volume squared when in the Rayleigh region, or its physical cross-section when in the optical region[51]. This sensitivity to the size of the elements is one reason why radar backscatter at long wavelengths shows a sensitivity to above ground vegetation biomass density. The last decade has seen an increasing number of studies looking at empirical relationships between radar backscatter and biophysical forest properties such as above ground biomass[52], stem volume or basal area. However, due to the increasing attenuation of a forest canopy with increasing frequency (decreasing wavelength) any practical retrieval of such parameters requires long wavelength sensors (L- and P-band or longer). For the radar case, observed trends usually display a positive correlation between biomass density and backscatter up to some "saturation" level whereby the sensitivity to biomass is lost (and occasional appears to be slightly negatively correlated), as illustrated in Figure 5.13. The usual explanation for this observed pattern is that at high biomass the forest eventually becomes virtually opaque such that the backscattered signal returns from only the upper layers of the forest canopy after saturation.

The correlation of biomass and backscatter is usually best for longer wavelengths and for cross-polarised measurements (since the cross-polarised response is dominated by the canopy contribution, rather than the surface component).

The depolarising effect of vegetation is observed in both backscatter and brightness temperature, whereby the H-V contrast reduces as the amount of vegetation increases (as opposed to a bare ground which has a high H-V contrast).

Scattering properties scale in proportion to the wavelength, so that, say, the X-band response to a miniature bonsai tree should be similar to the P-band response of a full sized equivalent. However, the structure of short vegetation tends to be more ordered than the effectively random canopy

Short Vegetation

---

[51] The boundary of the regions is defined as for the sphere radius in the original discussion, but this time with the cylinder radius. This is an appropriate approximation as long as the cylinder is "long and thin" (i.e. the length is greater than ten times the radius).

[52] For simplicity in writing, I also slip into using the incomplete terms "biomass" instead of the more accurate "biomass density". The reader should simply be aware that the sensitivity

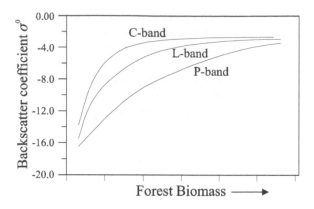

FIGURE 5.13 A schematic illustrating the general trends expected for backscatter against forest aboveground biomass density.

of mature forest, and their shallow depth means that the contribution of the ground will be significant for long wavelengths. Agricultural crops such as wheat have elements that are predominately vertical, for instance, whereas potatoes have predominately flat, horizontal leaves. The difference between these two structures is manifest in the polarimetry, whereby the vertical wheat stalks will tend to increase any vertical emission or scattering, whereas large flat leaves will have less contrast between H and V.

The other important features of agricultural vegetation is that it changes on very short (monthly) timescales, as opposed to forests which change only slowly. The combined observation of soil moisture and vegetation properties are also much more important for agriculture, where the aim is to predict crop growth (and subsequently crop yield). Agricultural observations are further complicated by the fact that crops tend to be grown in rows, and the underlying soil surface also exhibits periodic features. Such patterns can create coherent effects akin to Bragg scattering.

5.8   Special Scatterers

5.8.1   Corner Reflectors

One special kind of object that is worth introducing here is the corner reflector. These are based on a very simple principle — that when a wave is incident on the inside corner of two flat plates that are joined at 90° it is reflected directly back along the path of the incident wave. This is true for a wide range of incident directions, as illustrated in Figure 5.14, as long as the incident wave is perpendicular to the line where the plates join. This is

---

is to the biomass per unit area, and not the biomass of individual trees.

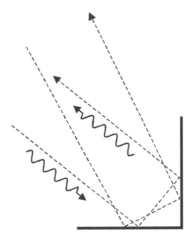

FIGURE 5.14 This figure illustrates the principle of a dihedral corner reflector — i.e. two highly reflective plates at right angles to each other. For a wide range of directions in this plane, the incoming wave is scattered back in the exact opposite direction. In the case of a radar instrument, the signal is directed back towards the sensor, resulting in an extremely high radar cross-section.

known as a "dihedral" corner reflector. This can be extended to three dimensions by adding a third plate to make the equivalent of the corner of a cube (all three plates meet at right angles), so it is now a "trihedral" corner reflector. In this case the reflected wave can, in theory, arrive from any direction "within" the opening of the corner and still be returned back towards the source of the wave. This has important implications in radar, of course, where the transmitter is usually located at the same place as the detector. Such targets therefore provide a very large return signal and are often used as a method of calibration or as a ground control point in radar imagery. The radar cross-sections for a trihedral corner reflector and a number of other well-defined targets is given in Table 5.1.

Dihedral

Trihedral

Trihedral corner reflectors can be seen in a number of other situations. Some coastal features that may be of danger to shipping, for instance, may have collections of corner reflectors to provide a strong radar echo. Similarly, small sailing boats may also carry a set of small corner reflectors to heighten their radar visibility[53]. The optical equivalent of a trihedral corner reflector is also what allows highway "cat's eyes" and "reflectors" (on

[53] A more unusual example of a corner reflector in the public domain is the official explanation for part of the strange metallic objects found in Roswell in 1947 that at the time were claimed to be remnants of a crashed alien spacecraft. Part of this object was a lightweight trihedral corner reflector constructed from metallic foil, designed to hang below a balloon so that it could be tracked by ground-based radar. The balloon was carrying sensitive audio-detecting equipment designed to measure the sound waves caused by Soviet nuclear bomb tests, hence the panic and apparent cover-up at the time. Unfortunately, many people still believe the original explanation that it was a crashed UFO.

Reflector	Characteristic dimension	Max. RCS	θ_{3dB}
Square Plate	Length of one side, a	$4\pi a^4/\lambda^2$	$0.44\lambda/a$
Sphere	Radius, r	πr^2	$360\ (2\pi)$
Triangular trihedral	Length of open edge, a	$4\pi a^4/3\lambda^2$	$\sim40\ (\pi/4)$
Square trihedral	Length of square edge, a	$12\pi a^4/\lambda^2$	$\sim40\ (\pi/4)$

TABLE 5.1   The theoretical maximum RCS for four idealised targets. (Curlander and McDonough, 1991).

cars, or bikes) to reflect the light from car headlights directly back to the driver[54].

If you recall in Section 4.1 we mentioned that the horizontally polarised reflected wave would incur a phase shift of half a wavelength (π radians). Each single reflection will produce this phase shift of π radians so that an *odd* number of reflections will result in a π phase difference, whereas an *even* number of reflections will result in zero phase difference between the horizontal and vertical waves. In Chapter 4 we will see how this can be exploited for interpreting the signal from a target, but the key relevance here is that a trihedral corner reflector will exhibit the same characteristics as a flat plate scatterer, so will have a phase difference similar to its surroundings (assuming it is located in an area, such as a bare agricultural field, where it is surrounded by single-scattering-type surfaces). It is therefore not easy to detect when the amplitude of the scattered wave is relatively low compared to other targets in the vicinity. A dihedral would be a better choice since it has such a distinctive phase difference between horizontal and vertically scattered waves that a detector that is sensitive to this phase difference could more easily identify which part of the image corresponds to the corner reflector. The problem here is that a dihedral has such a narrow range of azimuth angles from which it will function effectively. One solution to this is the "top hat" corner reflector. As the name suggests, is a cylinder with a large rim around the bottom and so looks like a top hat. This provides a dihedral phase difference response but forms a fairly good dihedral corner reflector from a complete 360° range of azimuth angles.

Top hat

5.8.2   Moving Targets

One topic that should be mentioned here is the question of what happens when an EM wave strikes a moving target. Clearly this is an important topic for many applications such as air traffic control, ship detection, surface wave measurements, and rain-radar. More than that, it is an intrinsic feature of any radar system flown on an aircraft or satellite since the instrument itself is moving. In fact, we have effectively already discussed this

---

[54]   Cat's eyes were invented in 1934 by Percy Shaw. Apparently he came up with the idea one night after shining a torch into the eyes of an approaching cat. As Ken Dodd once pointed out, had the cat been facing the other way he would have invented the pencil sharpener.

topic in section 5.2 when we considered moving sources and the Doppler *Doppler effect*
effect. If we consider the target as scattering a wave from many Huygen
secondary sources, then we can expect the scattered waves from targets
that are moving relative to the source to exhibit a Doppler shift. Note that
for echolocating systems that transmit a signal and measure the backscattered waves there is a double Doppler effect — the scattered wave is
Doppler shifted due to the relative velocity, and the scattered wave appears
to be Doppler shifted again by the moving detector.

The important thing to remember about moving scatterers is that it is
the *relative* velocity in the direction between source and target that counts,
not the absolute speed of the target or instrument.

5.8.3 Mixed Targets

Until now the descriptions of natural targets have focused on particular
elements within a landscape, but the surface of the Earth is composed of
many elements in various combinations. Snow layers lie over a rough
surface and soils are covered by a vegetation canopy, to name two common
cases. The heterogeneity of the Earth's surface is horizontal as well as
vertical, so that within any given pixel there may also be spatial patterns
across the surface, as well as differing properties with height.

In the first instance, let us consider the impact of horizontal variability. *Horizontal Mixing*
This may not seem significant if you are dealing with pixel sizes of 1m or
so, but is clearly an important consideration at the other extreme of pixels
(or footprints[55]) many tens of kilometres in size. The important question
to ask is whether there is any interaction between the different features.
This may be a scene covering sea ice and open water, or patches of forest
interspersed with bare agricultural fields, and so as a first approximation
it is appropriate to assume that the features are independent and do not
interact in any special way — this is the *incoherent* approach. In this case
we can simply add the scattering or emission terms from each feature in
proportion to the fractional area they cover, so that if C_i corresponds to the
fractional cover of surface type i, then

$$\sum_{i=0}^{N} C_i = 1, \qquad (5.63)$$

for N different surface types. The apparent brightness temperature,
T_B, of a scene composed of N elements with emissivity ϵ_i and physical

---

[55] We might argue whether it is appropriate to talk about pixels for non-imaging devices
with large instantaneous fields of view, but for now it makes the discussion a little more
straightforward.

temperature T_i is then given by

$$T_B = C_1\epsilon_1 T_1 + C_2\epsilon_2 T_2 + ... + C_N\epsilon_N T_N \qquad (5.64)$$

$$= \sum_{i=0}^{N} C_i\epsilon_i T_i. \qquad (5.65)$$

Similarly, the NRCS of a scene composed of N elements with NRCS σ_i^0 is given by

$$\sigma^0 = C_1\sigma_1^0 + C_2\sigma_2^0 + ... + C_N\sigma_N^0 \qquad (5.66)$$

$$= \sum_{i=0}^{N} C_i\sigma_i^0. \qquad (5.67)$$

As a second order approximation, however, it may be appropriate to consider, say, the edges of the ice or the forest. This is particularly significant for radar systems because these edges can provide conditions that result in the kind of "double bounce" interaction that was described for dihedral corner reflectors in the previous section. While it might be easy to see how artificial features can have edges that contain well-defined 90° angles between two scattering surfaces, such as a brick wall and a road surface, it might not be so obvious why such features would occur in a natural environment. In fact gravity ensures that they occur far more often than you might have thought. Geotropism ensures plants grow vertically, so that for a forest growing on flat ground every tree trunk forms a corner reflector with the surface. In forests that seasonally flood, this effect is exaggerated by the specular scattering of the very flat water[56] and the near vertical trunks (both orientations being driven by gravity). In sea ice, the cleaving of large ice regions tends to produce vertical edges next to horizontal water, and many rock formations may also cleave at right angles.

Vertical Mixing

The situation of strong double-bounce conditions under a forest canopy leads us to the other form of mixed targets, whereby the different target types are arranged in the vertical direction. We can generalise such feature types by classifying them as "layered media", so that we can consider a vegetation canopy or a snow covered area to be a single layer (either vegetation or snow) above a ground surface. Sea ice is a layer of ice above sea water. Glaciers are compacted ice above a ground surface.

Layered media

This simplified radiative transfer approach of Section 3 can be extended to consider multiple layers. A forest is then described by layers of consecutive leaves, branches and trunks, all above a ground surface. Sea ice can be a layer of snow, then ice, then water. Glacial ice can be described by different layers of ice compactness, again, with a layer of snow on the top. It is quite usual, then, in the practical application or microwave remote sensing, to find that we are never measuring the emitted or scattered signal from a single type of idealised target, but a combined signal originating from a number of different types of targets, as well as signals resulting from the

---

[56]   The water is unlikely to be significantly roughened by wind, since the vegetation canopy acts as a wind shield.

interaction between different targets. Any measurement is therefore likely to contain varying degrees of contributions from the atmosphere, from the underlying surface and from the different layers of material in-between, depending on the frequency of the radiation being measured.

Creating quantitative descriptions of this process is not easy — a great deal of current research is devoted to trying to develop models that will describe the interaction of microwaves with such complex targets, and some simplified versions of these are described in later chapters (in the context of measurement techniques). These will include many other properties of microwave interaction than are described here, including the very important topic of polarimetry, since the polarimetric response can be altered dramatically by the interaction of the different scattering elements within a scene. The very simplest model that is used in understanding measurements of the Earth's surface, is the *water cloud model*, or "random volume over a surface", which is the approach described in Section 3. Although simple, such models are important because they allow some very general trends to be observed and to support the interpretation of both passive and active measurements. How this approach can be used when dealing with volumes above a surface are considered later within the context of different measurement techniques.

5.9   Further Reading

Ulaby, Moore and Fung's three volumes of "Microwave Remote Sensing: Active and Passive" (1981, 1982, 1986) are clearly a fundamental source of further details on microwave interactions. For specific details on modelling approaches, you will find Fung's "Microwave Scattering and Emission Models and Their Applications" (1994) a detailed source of information on how to formulate both passive and active models. A more general, and concise, approach is given in Schanda's "Physical Fundamentals of Remote Sensing" (1986). For discussion on snow and ice, see Wadhams (1994). For the atmosphere and hydrometeors I highly recommend Janssen (1993).

6
DETECTING MICROWAVES

"Exactitude is not truth".
— Henri Matisse (1869–1954)

Chapter 3 described how any object with a physical temperature will radiate microwaves with a brightness determined by its temperature and its radiative properties. That section also considered how to generate microwaves artificially. These microwaves could then be used to illuminate a target area, and the scattered energy may be sufficient to measure and subsequently infer something about the physical properties of the target area. Chapter 5 then gave descriptions of microwave emission and scattering and how they are dependent upon physical attributes of natural targets (including the atmosphere).

This chapter now focuses on how to detect and quantify the microwaves arriving from the target, whether they originate from a natural source, or have been scattered off a target that has been illuminated by an artificial source.

Even if you are only interested in applying the final retrieved product from an instrument, it is always important to know how measurements are made in order to understand the relationship between data and target, and to become aware of the sources of error or uncertainty in the measurements.

6.1 General Approach

In basic terms the first objective of a microwave remote sensing system is to detect and quantify the electromagnetic radiation arriving at a detector. Of principle interest is the average intensity (energy) of the radiation at a predetermined frequency, but it may also be of interest to measure other properties of the radiation such as polarisation and phase. This chapter must therefore address two main topics: the first is how to collect the radiation in the first place, and the second is how to measure and quantify, with a useful degree of certainty, the properties of the collected radiation.

Most microwave systems consist of two basic elements: an antenna (and, when necessary, an associated mechanism for pointing the antenna) which collects incoming radiation primarily from a narrow range of look angles; and a receiver, which will amplify and detect the collected radiation within a specified range of frequencies. A third element that is essential but is specific to the measurement technique, is a data handling system that will perform all the digitising, formatting, calibration and will record other housekeeping data such as instrument orientation or pointing direction (Elachi, 1987). For simple systems this element will also handle the processing of the data, but more elaborate instruments such as imaging radars have such a huge data rate that the bulk of the processing has to be carried out by a separate system within the ground segment (on-board processing being limited by space, power and time).

A further element is also required for active (radar) systems since they must incorporate a pulse generator and a transmitter. Although the transmitter could use a different antenna to direct the microwaves, it is usually more efficient for the same antenna to both transmit a narrow beam, as well as receiving the returned signal.

Collector

The wavelength of microwaves is up to six orders of magnitude greater than those within the optical range, so that a focusing mechanism analogous to optical systems is not always feasible. The main constraint is size since the key parameter that determines the directivity of the collector is the ratio of the wavelength to the size of the aperture (refer back to Section 4.6). When size is not an issue, it *is* possible to construct an optical-type focusing system for microwaves, such as those used for radio telescopes (which use a large parabolic reflector to focus the incident waves). When remote sensing from aircraft or satellite, the size and weight are extremely important considerations as they equate directly with expense. This means that similar reflectors are impractical for low frequency microwaves (<10GHz) and a different mechanism must be employed that will normally be based on an array of transmitting and/or receiving units. Microwave sounders and most passive imagers utilise the higher frequency microwaves (>10GHz) and can therefore be focused using some kind of reflecting antenna.

One important difference between microwave and optical systems is that microwave antennas are not as directional — microwave systems will collect energy from a comparatively wide range of directions, even though it may be most sensitive to signals within a small range of angles.

Receiver

The detection of EM radiation involves the inverse process from their generation; *i.e.* rather than moving an electrical charge to initiate radiation, we want to measure a movement of electric charge that is induced by the electromagnetic wave. A detector will therefore provide the raw unit of measurement as an electrical signal, usually a voltage V. The relationship between this voltage and the incoming signal is determined by the specific detector type — it may be proportional to the amplitude of the electric field vector, E_0, or for *square law detectors,* it is proportional to the energy (or power, which is energy per unit time) of the wave, which is proportional to E_0^2.

6.2 Conceptual Approach to Microwave Systems

Visual metaphors dominate the wider remote sensing literature (at least those texts directed mainly towards remote sensing of the land) and it is quite common for introductory text books and lecture courses to reinforce the visual context by introducing the eye (or a lens camera) as a simple example of a remote sensing instrument. Within remote sensing, visual metaphors are common— brightness, (false) colour, shadow, etc.— even when referring to non-optical systems such as radar or sonar (e.g. multiple *looks*). Unfortunately our familiarity with visual methods — graphs, maps, photographs, television, and so forth — naturally leads us to conceptualise any 2-D information within the framework of visualisation. This can be a problem for those readers who subsequently go on to look at microwave remote sensing, where eyes and cameras bare little resemblance to the workings of many microwave systems. In some cases, the visual context may even be detrimental to gaining a clear understanding of what is going on — ironically this is perhaps most important within the context of *imaging* radar.

As an alternative, this chapter gives some of the analogous features of microwave systems to the ear, and how humans detect and analyse sound. Making comparisons between radar and acoustic systems is not new (see Griffin 1958 for a good example) and much of the technical overlap between radar and sonar is still evident in the literature. The important thing is that for the rest of this book you stop making comparisons between microwave remote sensing systems and eyes, and instead think more about the context of how you (or in the case of radar systems, bats) use sound to perceive our surroundings.

A non-visual conceptual framework for microwave sensing is taken for granted by many engineers and physicists who work on the technology of microwaves (probably as a result of the common history with telecommunications). The conceptual differences between optical and microwave remote sensing are certainly apparent in some of the comparative phraseology . For instance, those working with optical systems usually refer to wavelengths, while microwave researchers commonly use frequencies. Other common radar terms such as chirp, echo, amplification, decibel, also imply audio metaphors. Indeed, the use of the "deci-Bel" (dB) (named after Alexander Graham Bell, the Scottish-Canadian inventor of the telephone) originates from early research into auditory perception.

There are two principle differences between visual and auditory sens-

Terminology

ing. The first is that vision measures intensity (energy) as a function of direction, or look angle, to great precision, whereas hearing is far superior at mapping intensity as a function of time or frequency. The second, and perhaps more important for some systems, is that sound is a *coherent* wave phenomena (i.e. both the amplitude and the phase are important, not just the average energy). These fundamental differences are also the key distinctions between optical and (most forms of) microwave remote sensing.

The great value of using audio analogies is that it introduces some of the more difficult concepts of microwave remote sensing within a context with which we are all familiar from everyday experience.

6.2.1   A Word of Warning

It should be noted that the intention of introducing this audio context is merely for comparing analogous mechanisms of measurement, and definitely not about making direct comparisons between electromagnetic waves and sound waves. They are both forms of wave, and exhibit wave phenomena such as inteference, but it is important to bear in mind that the audio analogy is limited since sound waves are longitudinal pressure waves, rather than transverse waves, and so do not exhibit polarisation.

6.3   Basic Microwave Radiometer

A basic radiometer is intended to produce an output voltage proportional to the amount of radiation collected at an antenna. A typical microwave radiometer now used for Earth observation purposes is a *total power radiometer* using a *heterodyne receiver* (although these are not the only types). Such radiometers use a square law detector which gives an output voltage proportional to the received power. A schematic of a simple total power radiometer is shown in Figure 6.1 and illustrates features common to most microwave radiometers. The following sections describe in more detail the basic principles of a total power radiometer, and throughout these sections reference will be made to Figure 6.2, which compares a schematic of a simple microwave radiometer with a description of the human ear. In a general way, the individual components of the ear perform analogous functions to each of the steps within a microwave radiometer system (or the receiving part of an active system).

6.4   The Antenna

Perhaps the most obvious similarity between the ear and a microwave system is that the outer ear (the pinna or auricle) is essentially an acoustic antenna, collecting the sound energy and providing some degree of direc-

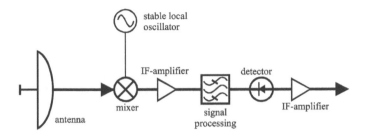

FIGURE 6.1 The typical components of a heterodyne radiometer system.

tivity[57]. Since the wavelength of sound is about six orders of magnitude greater than that of light, an acoustical focusing mechanism with similar resolving power to the eye would require ears that were about a kilometre across! Clearly not feasible and so nature has ensured that a different mechanism has been employed.

Remote sensing systems have the same problem — the wavelengths commonly used for radar imaging range from $1mm$ to $1m$ (P-band), again almost six orders of magnitude larger than those wavelengths employed for visible and IR imaging (0.3–$10\mu m$). This is the ultimate challenge in microwave remote sensing: achieving the kinds of resolving power that is standard for optical sensors, but using technologies that are appropriate for the longer wavelengths.

6.4.1 Parabolic Antennas

A parabolic antenna is a reflecting device that focuses the incident energy from a localised direction onto the detector (or into a waveguide which then carries the signal to the detector) much like the primary mirror system on a reflecting telescope or optical scanner. The reflector is a curved

---

[57] The following discussion will sometimes interchangeably talk about transmitting or receiving waves using an antenna. Conveniently we can rely on the reciprocity principle introduced in Section 4.4, which means that for every path a microwave may take out of an antenna, it would take the same path back into the antenna (i.e., along the same path, but in opposite directions). This means that in general an antenna will work in much the same way as a receiver as it does as a transmitter. A simple way to remember this principle is to note that the word "radar" is a palindrome, the literary equivalent of reciprocity — i.e. it reads the same backwards as it does forwards. That ears and antennas can act as both good receivers as well as transmitters is nicely illustrated by case I once read about of a horse which appeared to have sound coming out of its ears. Due to some unusual brain activity the nerves to the inner ear where working in reverse, so that they were stimulating the sensitive hairs that lie within the cochlea of the ear. The vibrations from these hairs was being amplified by the cochlea, causing vibration of the ear drum, which in turn was being amplified by the inner ear and radiated *outwards* by the outer ear!

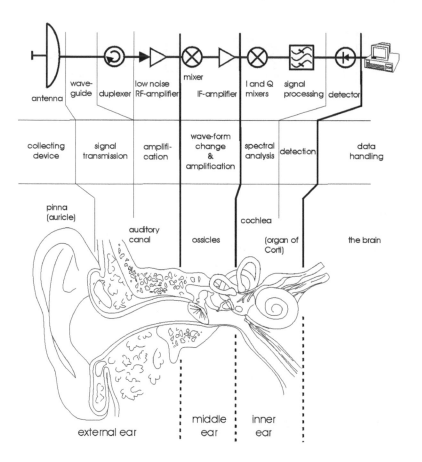

FIGURE 6.2 A schematic diagram illustrating the comparitive elements of the human ear and a basic microwave radiometer.

surface made of a material that is highly reflective at the relevant frequencies. At short wavelengths this could be a solid metal antenna, but at longer wavelengths it can be composed of a wire mesh (as long as the holes are much smaller than the wavelength). The choice is purely a practical one, since a mesh is both lighter and has less air resistance.

The pinnae of the ear act as a reflecting antenna for our auditory system, although they are significantly more complex than most radio antenna: the various folds, cavities and ridges in our outer ears result in a frequency response that is directionally dependent. This frequency dependence (and additional frequency dependent diffraction by the head and torso) provides the brain with cues as to the location of the sound relative to the head.

Simpler acoustic reflectors are more evident in bats and cats, which have relatively large, cupped ears which focus the sound towards the opening of the middle ear. If you ever watch a resting cat, you will notice that they can "steer" their ears to point in the direction of an interesting sound source. Humans too, when trying to hear a faint sound, will cock their heads slightly – this motion points the direction of highest sensitivity of one ear towards the sound source (the peak direction being slightly off-centre).

Figure 6.3 illustrates the principle of some reflecting antennas. The first diagram shows a basic geometry of a parabolic-reflector and illustrates how an incident plane wave is brought to focus at a point by the parabola. Note also that the reciprocal process means that a plane wave wavefront can be generated from a point source at the focus of the parabola (an important consideration for active sensors). The detector itself might be placed at the focus, but such an arrangement would block the incoming waves. An alternative would be to have the focus offset, as in Figure 6.3 (b), the key advantage being that the detector no longer lies within the outgoing/incoming waves — this is a typical arrangement for satellite television "dishes". A Cassegrain arrangement takes a different approach and instead of placing a large detector at the prime focus within the incoming wave it has a smaller secondary reflector that focuses the waves behind the main antenna where a large detector can be placed without much intrusion. Finally, (d) shows a Newtonian arrangement, which is similar to (c) but the secondary reflector focuses the waves to some location off to one side. The main advantage of the Newtonian is that if the antenna needs to be mechanically steered to provide a scanning motion, then the secondary reflector can be located along the axis of rotation, allowing the detector to remain fixed while the antenna assembly is moved.

When it is necessary to change the direction of the antenna sensitivity, this need not always be achieved by mechanically scanning the whole antenna assembly, since some limited scanning potential can be found by scanning the much smaller secondary reflector alone.

6.4.2   The Dipole Antenna

The simple dipole antenna forms the most basic microwave transmitting and receiving device. In fact a dipole is only barely an antenna at all since it has very little directivity when used as a single device. In section 6.3 it was stated that any accelerating charge would produce EM radiation. In the case of a current oscillating back and forth down a wire, we would generate EM radiation in proportion to the frequency of oscillation. Such a device would be known as a simple dipole antenna (its a dipole since at the maximum of one oscillation, one end of the wire is positive, the other negative). When used as a transmitter it will generate EM waves when an oscillating current is passed through it and, likewise, when used as a receiver, it will generate an oscillating current when in the presence of an

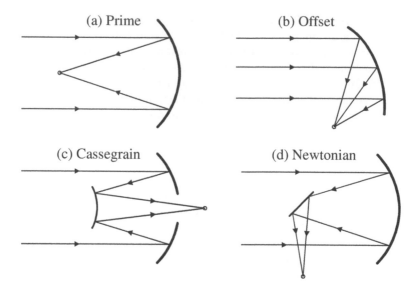

FIGURE 6.3 This diagram summarises the main type of reflecting antenna arrangements used for microwave remote sensing: (a) focuses the incoming waves to the prime focus of the parabola, but has a major disadvantage in that the detector would block incoming waves; (b) an offset focus gets around this problem by locating the detector out of the way of the incoming waves; (c) a Cassegrain arrangement replaces a large detector within the incoming field by a smaller secondary reflector that focuses the waves behind the main antenna where a large detector can be placed; and (d) is a Newtonian arrangement, which is similar to (c) but the secondary reflector focuses the waves to some location off to one side.

EM wave.

On its own it has very little directivity, since it will transmit and receive in all directions with a maximum in a plane perpendicular to the long axis of the dipole. The oscillation is also only in the direction of the long axis of the dipole which means that in principle the directivity drops to zero in the two "polar" directions (along the axis, since from these perspectives the current can only oscillate longitudinally "back and forth", and not transversely). This means that it can only transmit and receive EM waves that are polarised in the direction of the axis of the dipole. This is potentially very useful, since it means it is relatively easy to construct a measuring device that is sensitive to polarisation — in fact, for microwaves it is difficult *not* to make the instrument sensitive to the polarisation.

The properties of a dipole make it ideal for a number of applications, such as radios or mobile telephones — in such cases directivity is definitely not what you want (since you want transmission and reception in all directions) and it is also relatively simple to have all appliances with their antennas in an "upright" position to match the permanent transmitters. Dipoles are also useful for finding the direction of a source since the reception will always be a minimum when the source is in the direction of the dipole axis (where it will be least responsive).

In remote sensing, however, the isolated dipole is insufficient since we what to achieve some level of directionality of transmission and/or reception. This is achievable if we use a collection of dipoles arranged in some kind of array. Section 4.6 considered the coherent addition of a number of closely spaced sources in an array, and demonstrated that if the signal at the different locations could be generated or received in a coherent manner, we could generate a significant peak in the sensitivity towards one specific direction. This principle was used by the first radar systems (such as the Chain Home system in the UK), and the earliest astronomical radio telescopes, and are described as "array antennas".

6.4.3 Array Antennas

An array antenna has a number of microwave transmitters (and/or receivers) arranged across a panel (the panel being some rigid structure). In so called "phased array" antennas, these transmitters are electronic elements that *Phased array* can each transmit microwaves of a specified frequency with a well determined phase. The important thing about a phased array antenna in the context of radar is that the ability to specify the phase (and amplitude) of each individual transmitter allows control over the direction of the peak in the antenna sensitivity pattern. This is termed *electronic scanning* or *steering* of the antenna pattern. This can be explained by referring back to Section 4.6 where we saw that you could generate a change in the direction of the peak power from an array of transmitters by changing the relative phase differences of the individual transmitters. In receive mode, the receivers must record the exact phase, as well as the amplitude, of the waves it measures. Only those parts of the received waves that are in-phase (or a specified phase difference apart) are the ones that are used in the processing.

This same principle has been used since the early development of microwaves and radio but through the more brute force approach of the "slotted waveguide" array antennas. In these systems, rather than individual *Slotted waveguide* transmitters/receivers, the antenna panel is composed of outlets from an array of waveguides. Waveguides are the microwave equivalent of fibre optics, being narrow metal pipes (of rectangular cross-section) that allow channelling and control of microwaves. For sound waves, a tube composed of an acoustically reflective material (such as metal) will act as a simple waveguide, as does the auditory canal which transmits sound waves from the outer to the middle ear.

The use of waveguides means that for transmission it is only necessary to have one central source that produces microwaves that are directed by a number of different waveguides to the antenna array. In this case the microwaves leaving each outlet on the array will be coherent (as they come from the same source) and as long as each waveguide is the same length, the emitted waves will all be of the same phase. We then have a situation similar to Figure 3.12 in Section 4.5. Note, however, that unlike the phased array, it is not practical to instantaneously shift the phase of each emitted wave in the array (as this would require a physical change in the length of the waveguides). It is the advance of microwave technology that has meant it is now possible to have a number of independent small receivers or transmitters that can act coherently in an array.

In receive mode, after the incident EM radiation is collected by the slotted waveguide antenna, it is transferred via the same collection of waveguides before being combined coherently. Since the waves that originate from within the main antenna beam direction are in-phase, while those outside the main lobe are, for the main part, out of phase, the signal that goes to the detector is primarily representative of the main lobe signal.

Due to the requirement for most imaging radar systems to have a relatively large antenna (i.e., more than a few metres) the phased array is a common choice. The panel array can generally be made proportionally larger without a significant increase in cost, and they are also lightweight, a significant design constraint for spaceborne systems in particular.

It is not a requirement to use an array antenna for imaging radar, indeed the radar imager onboard the Magellan space probe to Venus used a 2.3m circular parabolic dish, but it is often the most efficient way to construct a large antenna.

6.4.4   Antenna Properties

The purpose of an antenna in microwave remote sensing is to provide a means of selecting the angular distribution of the received electromagnetic energy radiated by the scene under observation (and for directing the transmitted waves in active systems). The total amount of power that enters the system is determined by the antenna's sensitivity to direction, which is defined by the *antenna gain pattern*.

To achieve selectivity in direction, a practical gain pattern will consist of a well-defined main beam (the central peak) with smaller maxima of much lower gain away from the beam axis — these are the *sidelobes* discussed in Section 4.5. An example of an antenna pattern for a rectangular
Deci-Bels   aperture is shown in Figure 6.4. Note that the figure is in *deciBels* (dB), a dimensionless quantity that defines a ratio as a logarithm, such that,

$$\text{Quantity of interest in } \textit{deciBels} \text{ (dB)} = 10 \log_{10} \frac{\text{Quantity of interest}}{\text{Reference quantity}}.$$
$$(6.1)$$

By choosing unity as the reference quantity, you can express any value in dBs. The use of dBs is very common when working with microwaves, and is a throw-back to the very early research into radar systems since the electrical engineers of the time came from a culture of communication engineering (the *Bel* being called after Alexander Graham Bell). Logarithmic quantities are useful in communication engineering because our ears'

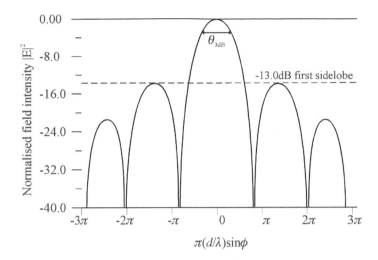

FIGURE 6.4 The sensitivity pattern of a rectangular antenna. The diagram has been generalised by using d/λ so that the diagram is appropriate for any size of antenna and wavelength
of radiation. Note that the first sidelobe for a rectangular antenna is at about -13bB.

sensitivity to loudness has a logarithmic relationship with power so that
logarithmic relationships had to be accommodated for in electrical components such as loudspeakers.

Aperture An antenna can be considered as an *aperture*, in that it forms an effective "hole" between the "outside" and "inside" of the instrument. This
is a rather simplistic way of thinking of it, but it helps to get a picture of
what is going on by relating this back to the discussion on the diffraction
caused by an aperture (Section 3). The diffractive distortion is what essentially defines the antenna gain pattern. The same physical processes are
involved when we consider the resolving power of an optical instrument.
"Diffraction limited" implies that the aperture of the instrument is the limiting factor of the directional sensitivity of the whole instrument. At optical
and IR wavelengths, even quite a small aperture still provides a very narrow gain function — usually referred to as the instantaneous field of view
(IFOV) or angular resolution (i.e., an angular measure of how sensitive the
instrument is to direction).

Boresight The direction of the peak of the antenna pattern is called the *boresight*,
and not surprisingly, this is usually taken as the reference direction when
displaying a sensitivity pattern.

Gain The antenna gain G is a dimensionless quantity of the antenna that
represents the combination of the directivity, $\mathcal{D}$, which defines the relative
sensitivity to direction, and the antenna efficiency ρ, which characterises
the magnitude of the gain pattern, so that

$$G = \rho\mathcal{D} = \rho\frac{4\pi A}{\lambda^2} \qquad [\text{-}] \qquad\qquad (6.2)$$

Effective area

where A is the physical area of the antenna and λ is the wavelength. For spaceborne systems we might typically expect gains in the region of 30dB or more. A_e is the *effective area* of the antenna and is given by $A_e = \rho A$. The effective area can be thought of as a cross-section of the antenna as it defines the equivalent "area" of energy that the antenna intercepts when receiving.

In all systems the size of the antenna is constrained merely by practical limitations of size and weight, traded off with requirements for sufficient sensitivity to make the measurements worthwhile. Imaging radar systems have an additional constraint that is governed by the need for certain power transmission, pulse repetition frequencies and beam sizes. At this stage it is sufficient to be aware that there are such constraints without going into the details of what they are, since many are beyond the scope of this book.

HPBW

The angular resolution for microwave antennas is often characterised by the *half-power beamwidth (HPBW)* — the diameter of the gain pattern where its value is half that of the central peak[58]. Since in deciBels a factor of 2 or $\frac{1}{2}$ equates to approximately $+$ or $-$ 3dB, respectively, you may also see this referred to as the "3dB beamwidth", θ_{3dB}. Note that this is narrower than the beamwidth discussed in Section 4.6 where the width between the first minimum on either side of the main lobe was considered.

A uniformly illuminated circular aperture of diameter D, for example, will have a minimum-to-minimum of $2.44\lambda/D$ but has a HPBW of $1.02\lambda/D$, for radiation of wavelength λ. However, in practice, a more stringent limiting factor than the beamwidth for active systems is more likely to be the gain level of the first sidelobes since it effectively broadens the sensitivity to almost three times the angular diameter (bearing in mind that microwave sensors need to be extremely precise). For a uniform illuminated circular antenna, the intensity of the first sidelobe is around 18dB below the main beam (and about 13dB for a rectangular antenna). This may seem high enough not to worry about — factors of 63 and 20 respectively — but must be put into the context of variations in, say, radar backscatter between land and sea being in the order of 20dB. It is therefore common for practical systems to be designed to sacrifice beamwidth for a corresponding reduction in the sidelobes of the antenna pattern. It is far better to work with a larger beamwidth in a known direction, than to have to accommodate spurious signals from large sidelobes. In radar systems this is done by modulating the transmitted signals with a smooth function across the width of the aperture, usually with a higher weighting in the middle than the edges. For both transmission and reception, this can be easily controlled on phased array antennas by adjusting the properties of the individual array elements. For instance, if a cosine-squared weighting is applied across a rectangular antenna the sidelobe ratio drops to 32dB, but with a beamwidth that is now $1.45\lambda/D$.

A better rule of thumb, therefore, for describing the beamwidths of

---

[58] This is the same term as the *Full Width at Half-Maximum*, or *FWHM* described in Section 4.6.

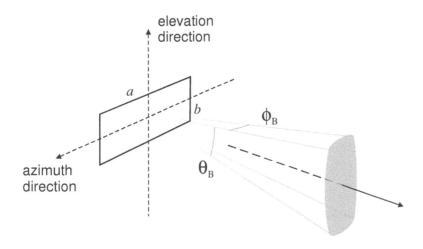

FIGURE 6.5 The definition of the angles within the beam of a rectangular antenna. Note that the thinner of the two angular beamwidths, ϕ_B, corresponds to the larger dimension of the antenna.

most practical microwave antennas to an accuracy of about 10% is therefore

$$\theta_{HPBW} = \theta_{3dB} = 1.5\lambda/D. \tag{6.3}$$

θ_{3dB} is directly proportional to wavelength so that for longer wavelengths the physical size of microwave antennas must be significantly larger to achieve comparable resolution to those of infra-red or optical systems[59]. Antenna size is therefore an important concern for many microwave applications and can be a limiting factor in the design of satellite and airborne systems.

One important feature to note is that for a rectangular antenna, the narrower beamwidth corresponds to the wider axis of the antenna. This is illustrated in Figure 6.5.   *Rectangular antennas*

6.5   The Receiver

In the most simple case of the dipole antenna, a voltmeter attached to end of a wire is sufficient to act as a very crude receiver. Indeed, so crude that the signal would be hardly detectable, contain power from a wide range of frequencies and be composed of mostly noise, so it would not be particularly useful. A practical detector system therefore has to be able to amplify the signal, measure only those signals that lie within a narrow bandwidth and be able to minimise the effect of noise — thus giving an accurate estimate of the radiation being collected by the antenna. For coherent systems,

---

[59]   Note that in all this discussion we have been assuming that for small angles in units of radians, $\sin\theta \approx \theta$.

it is also a requirement to measure the phase of the wave, and not simply the average intensity.

Narrow Bandwidth

The first two problems would seem relatively straightforward because all we need do is make sure we can amplify and filter the signal, preferably without introducing any more noise. After the antenna has intercepted the incident waves it could direct it to some kind of low noise amplifier in order to make the signal more easily detected. This is similar to the process in the ear, whereby the auditory canal acts both as a waveguide to transmit the sound energy from the pinna to the tympanic membrane (the ear drum) but also as a resonant amplifier, selectively promoting frequencies corresponding to the peak sensitivity of the whole ear (2–5.5kHz).

Heterodyne receiver

The historic problem with microwave receivers is that signals at microwave frequencies are very difficult to deal with directly – low noise amplifiers and effective filters at not easy to build for microwaves. They are, on the other hand, much easier to build for lower frequency signals so many microwave detectors employ a *heterodyne receiver* system whereby the received signal (the *radio-frequency*, or RF, signal) is not directly amplified or detected, but is converted to a different, and usually lower, frequency (the *intermediate*, or IF, signal). It is this signal that is further amplified, filtered and detected since this down-converted IF frequency can be handled much more effectively.

Local oscillator

The down conversion is carried out by a non-linear circuit element, known as a *mixer*, in which the RF signal is combined with a constant-frequency signal, ω_{LO}, generated by the *local oscillator* (LO). This process is related to the effect of "beats" whereby two waves of nearly the same frequency are added together, resulting in a modulated signal that has a much lower frequency. Mathematically it can be expressed as the sum of two sinusoids, such that if we have the original signal $s(t)$ and the LO signal $s_{LO}(t)$ described by

$$
\begin{aligned}
s_{RF}(t) &= A_{RF}\cos(\omega_{RF}t), \\
s_{LO}(t) &= A_{LO}\cos(\omega_{LO}t),
\end{aligned}
\tag{6.4}
$$

then their combined signal is

$$
s_{RF}(t) + s_{LO}(t) = 2A_{RF}A_{LO}\cos\frac{t}{2}(\omega_{RF}+\omega_{LO})\cos\frac{t}{2}(\omega_{RF}-\omega_{LO}).
\tag{6.5}
$$

This combined wave has a high frequency component equal to the average of the two original frequencies, and a low frequency component equal to half the difference[60]. It is the latter of these two that is of interest to us since it means we can control this low frequency component by choosing an appropriate frequency for the LO. Conveniently the signal that comes out of the mixer whose frequency is the difference between the RF and LO frequencies has a power proportional to the power of the original RF

---

[60] This effect can be used to tune two strings on a musical instrument, even if you are tone deaf! When the strings are only a small tonal (frequency) difference apart it is possible to hear the modulated signal as a change in the volume if the two strings are played simultaneously.

signal, which is the value we actually want to detect.

6.5.1 Detector

After down-conversion, amplification and filtering, the signal can then be detected by some electrical component (such as a diode) which is designed to convert the microwave energy into an electrical signal. The most common kind of detector is a square-law detector which produces an output voltage proportional to the power of the wave. It is called a square-law detector since the instantaneous voltage of the wave itself is $E(t)$ (the amplitude) but the output voltage of the detector is proportional to $E^2(t)$, the power of the wave.

6.6 Coherent Systems

A total power radiometer only measures the *total* energy incident on the antenna. However, there is more information to be gained from the observation by designing a system that measures both the amplitude and the phase of the incident electromagnetic radiation — these are *coherent* measuring systems.

Coherent passive radiometers are becoming progressively more important, but coherent measurement systems are fundamental to modern imaging radar systems. The trick used in coherent systems is to make two power measurements: one after the signal has been mixed with a reference signal as above, then again after the same signal has been mixed with a *second* reference signal that is $\pi/2$ (90°) out of phase with the first.

The complete process therefore starts with the incoming measured signals being passed through a low-noise amplifier, and then mixed with a reference frequency, this time provided by a *stable* local oscillator (SLO) that has control over the phase, down to some intermediate frequency (IF). The signal can then be further amplified at the IF stage.

The next step is to mix the signal down to some carrier frequency with the output of another oscillator but this time at intermediate frequencies, the phase of which is locked with the SLO. It is at this stage that, the signal is converted to *in-phase* (I) (the *real* part of the wave) and *quadrature* (Q) (the *imaginary* part of the wave) format by mixing it with two signals, one $\pi/2$ out of phase with the other. This effectively converts the input signal from a sinusoidal format to the complex exponential format. For example, let us say that the received signal can be described by

$$s(t) = A\cos(\phi_T + \omega t). \tag{6.6}$$

This is the in-phase part of the signal, denoted by $s_I(t)$. The quadrature part of the signal is obtained by applying a phase shift of $\pi/2$ to $s(t)$, i.e.:

$$s_Q(t) = A\cos(\phi_T + \omega t + \frac{\pi}{2}) = A\sin(\phi_T + \omega t). \tag{6.7}$$

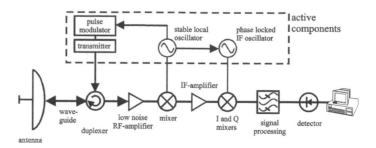

FIGURE 6.6 The typical components of a coherent active microwave system, with the active components highlighted.

This gives the signal in its complex form, since:

$$S(t) = s_I(t) + is_Q(t) \tag{6.8}$$
$$= A\left[\cos(\phi_T + \omega t) + i\sin(\phi_T + \omega t)\right] \tag{6.9}$$
$$= A\exp[i(\phi_T + \omega t)]. \tag{6.10}$$

The I and Q channel information is the *raw data* of a coherent system. This allows recording of both amplitude *and* phase information, which is fundamental to interferometric radiometry and aperture synthesis, which we will deal with in the later chapters.

6.7   Active Systems

For an active system we also need to transmit microwave signals. For convenience, this is usually done using the same antenna system, in which case the set-up is said to be *monostatic*. This need not always be the case, and there are sometimes certain advantages to having your transmitter and receiver in different locations — the *bistatic* case.

Figure 6.6 shows a simple block diagram of the essential features of a radar system. Because radar usually employs coherent processing it is necessary to use a coherent transmitter system. The stable local oscillator (SLO) is used to ensure that each transmitted pulse originated by the pulse modulator has exactly the same phase, ϕ_T. The transmitter then provides high-power amplification of this pulse.

The duplexer is a two-way switch that makes sure the transmitted pulse is radiated out of the antenna while the received signals are directed into the receiver network. When the system is polarimetric, a further element switches the pulse signal between the two orthogonal polarisation antenna elements.

6.8   System Performance

As in any measurement system, the signal-to-noise ratio is the important measure of the instrument's performance. In this respect, for an active system, the transmitted power will also have an influence on the final performance of the system.

6.8.1   Noise and Sensitivity

The noise level, and subsequent sensitivity, of a microwave system also has analogous processes in the ear. In eyes and optical remote sensing systems, it is primarily the detecting medium that introduces noise or random errors into the final measurements (although systematic errors may arise due to imperfections in other parts of the instrument). In both microwave systems and ears, however, noise may arise from all parts of the system.

The antenna or pinna, for instance, may collect "background" noise originating from sources far removed from the direction of interest (because of the many sidelobes of the antenna pattern). The inner ear will also detect signals originating from a number of other sources — the basilar membrane, for instance, will pick up vibrations in the skull from such sources as teeth clicking, or stomach gurgling, for example. Fortunately, the low frequency cut-off of the cochlea sensitivity is just high enough not to hear the continuous sounds generated by the various other organs of the body. In a similar way the net IF power reaching the detector of a microwave system will include internally generated noise power arising from electrical leakage or even the physical temperature of the instrumentation.

At all levels it is the signal-to-noise ratio that ultimately defines the usefulness of an instrument (or individual component of it). Its definition is given by:

$$SNR = \frac{\text{received signal}}{\text{system noise}} \quad [\text{-}]. \tag{6.11}$$

Or more generally,

$$SNR = \frac{\text{signal of interest}}{\text{unwanted signal}} \quad [\text{-}]. \tag{6.12}$$

If the SNR is less than one, then the system is generating more noise than the signal you are receiving. Clearly the higher the SNR the better the performance of the measurement system. Although unitless, the SNR will often be quoted in dB for microwave systems.

6.8.2   Sensitivity Considerations for Receivers

In the microwave region the radiance emitted by an object in thermodynamic equilibrium is approximately proportional to its physical temperature T. It is therefore convenient to measure the radiance in units of tem-

Antenna temperature

perature by defining a *radiometric brightness temperature*[61]. Similarly, a radiometric *antenna temperature* T_a can be defined that is related to the radiance received by the radiometer antenna. These temperatures may not relate to actual physical temperatures, they are merely a convenient way of expressing the energy in the signal.

T_a includes contributions from the target being observed as well as from emitted and scattered radiation from other sources, but not from internal elements. This usage of temperature as a convenient unit of radiation power can be used for all parts of the system, including active systems, especially for discussions related to system noise. Noise throughout a system is additive and usually independent for each individual component so that noise powers, or equivalent noise temperatures, can simply be added numerically.

Radiometer Sensitivity

The performance of a microwave detector is characterised by the *radiometric sensitivity* (or *radiometric resolution*) ΔT which represents the precision with which T_a can be estimated. For most radiometers this can be shown to be

$$\Delta T = \frac{k_r T_s}{\sqrt{B_n t}} \qquad \text{[K]}, \qquad (6.13)$$

where B_n is the bandwidth of the noise power through the system (i.e., the range of frequencies), t is the integration time of the individual measurement and k_r is some constant determined by the particular type of radiometer (ranging from about 1 to 3). T_s is the system temperature of the instrument, defined as the equivalent noise temperature for a one-Hertz bandwidth over a one-second integration time. Since noise is quoted as an equivalent blackbody temperature it is often referred to as an NEΔT or

Noise-equivalent ΔT

noise-equivalent ΔT.

The noise contributions at most stages of the measurement process are produced from thermal radiation processes and so can be treated as having a Gaussian distribution with white-noise characteristics (meaning that it is evenly distributed in frequency).

Note that Equation (6.13) tells use about the ability to accurately determine the signal coming into the system (i.e. at the antenna) and not the signal at the target source. For radiometry, the calibration is in terms of radiometric antenna temperature which is then used to represent the intensity of radiation incident upon the antenna. The antenna pattern must be well known by using appropriate calibration techniques (discussed in Section 9.1) in order to convert the *antenna* temperature to a *target* brightness temperature.

Noise-equivalent σ^0

In radar systems the aim of the measurement is to estimate the normalised radar cross-section, σ^0, rather than a radiant temperature so that it is more usual to quote noise contributions as a *noise-equivalent sigma nought*. This is determined by estimating the target radar cross-section that would give a backscatter response with an SNR= 1, which is the radar

---

[61]   This is the *Rayleigh-Jeans equivalent brightness temperature*, which is discussed in more detail in Section 1.1, and is distinct from the frequently used thermodynamic definition.

equivalent of the NEΔT above.

6.8.3 Other Sources of Uncertainty

The sensitivity of a measurement is also constrained by a number of other factors that are not discussed in detail here. What do you do with the output voltage from the detector, for instance? The manner in which this is recorded or stored also introduces a degradation or noise into the final measurement. Even highly precise digitising equipment puts some noise into the finally recorded measurement, through digitisation errors (rounding errors). In an optimised system, such rounding errors will be small compared to the other noise in the system.

6.9 Calibration

Calibration is the quantitative characterisation of the performance of an instrument. Most instruments record a voltage as the raw output data (or voltage as a function of time) but we are really interested in the physical properties of some distant object — whether that be a brightness temperature, or a backscattered power — not the raw electrical signal at the detector. For any given instrument, therefore, the relationship between the recorded voltage and the physical property of interest has to be determined. Additionally, we also want to now how well this relationship is known, given that all systems contribute noise to the final detected signal, but that in practice this noise signal is difficult to determine theoretically before the instrument is deployed.

Determining this relationship and its noise-related uncertainty for a particular instrument is referred to as *calibration*. The aim is ideally to achieve *absolute* calibration whereby the direct relationship is known quantitatively for that particular instrument. Sometimes that is not possible, in which case the next best thing is *relative* calibration, such that the scaling or the trend in sensitivity is known so that quantitative estimates can be given on proportional differences within the data.

The engineering specifications of an instrument will often characterise the calibration in terms of the performance of particular subsystems or even individual components (e.g., the antenna). If you are interested in using the data from the instrument, however, you are only really interested in the "end-to-end" system performance, not all the details of each individual component in-between. However, even then it is important to know some of the general principles of calibration in order to understand the quality of any data you might use.

Since we cannot determine noise theoretically, the basic approach to *Basic Principle* all types of calibration is to feed some signal with known properties into various stages of the sequence of instrument components. For instance, it should be possible to input some artificial signal directly into the RF amplifier in Figure 6.1 instead of the antenna signal. The resulting response further down the chain of components can then be quantified for this known signal. Such an approach can then isolate the noise contribution from each part of the chain.

Note that this process is not the same as "testing" the instrument, which is a term usually referring only to something you do before or after the operation of the instrument to make sure it is working properly. Calibration quantifies the manner in which it is working, not whether it is working or not, and it needs to be carried out on a regular basis while the instrument is actually operating. The reason for this is most apparent on spaceborne satellites which, for a number of reasons, are particularly difficult to calibrate. The launch of the satellite, for instance, can result in extreme acceleration and vibration that may have an impact on the final performance of the instrument. Once in orbit the instrument will operate in a microgravity environment, but it will have been tested while under the gravitational acceleration of the Earth. And throughout each orbit, thermal conditions will vary from a few tens to thousands of degrees K, depending on whether it is illuminated by the sun or not. Finally, as with any piece of machinery, the characteristics of an instrument will alter with time due to degradation of individual components, which is an important consideration for an instrument that may have a useful lifetime of as long as ten years with no opportunity of physically checking or servicing the satellite. The term for the associated changes in sensitivity over time is known as instrument "drift".

The calibration signal may be a target (with known properties) either inside or outside of the instrument itself, or may be an electrically generated signal. When this reference signal is originates within the instrument, it is called *internal* calibration. Internal calibration relates what is measured with what is collected and redirected by the antenna. Such a process normally bypasses the antenna and so can never characterise the end-to-end performance, but it does allow long term monitoring of the detector system. Active systems usually have internal calibration through feeding a pulse directly from the transmitter to a calibrator that generates an equivalent (but much lower power) RF pulse, which is fed directly to the RF amplifier, and an IF pulse which is sent to the IF amplifier.

Internal Calibration

In passive systems, a common internal calibration method is to use hot and cold calibration "targets", which are often well specified objects within the radiometer itself with brightness temperatures above and below the usual range of measurements. Although the word "target" is used, it should be stressed that these need not be features observed outside the system, and so the calibration does not include the antenna. Hot targets are usually made from material such as iron, designed to act as closely like a blackbody as possible, so that by maintaining it at a predetermined temperature, we can be confident about knowing the intensity of the emitted radiation.

Hot-Cold Calibration

A "cold" target may simply be no signal at all (i.e., so you can measure the background noise signal generated by the instrument itself). A common cold target is the darkness of outer space — the cosmic background radiation discussed in Chapter 2 which has a radiant temperature of about 3K, conveniently well-below signals that would normally be measured in Earth observation. Observations of deep space usually bypass the antenna (since space is a large distributed target so direction is not so important) and since the intention is to compare it to an internal hot target the measurement is often made simply through a hole in the side of the instrument. In that case the process should be strictly classed as *external calibration.*   *External Calibration*

External calibration uses a known signal from outside of the instrument and so can include the antenna response. The primary aim of external calibration is to include the influence of the antenna on the relationship between the measured signal and the target.   One strategy for external calibration therefore involves pointing the antenna towards some external source that has well-defined and stable properties (the equivalent of a colour "test card" in photography). Passive radiometers may use natural targets such as the moon for this approach, whereas active systems may utilise artificial point targets or natural distributed targets.

The ideal way in which to calibrate the end-to-end performance of an active system is to have a point target of known cross-section placed within   *Point Targets* the region being observed. The easiest way to do this is to use microwave corner reflectors, mentioned already in Section 8 as one of the special scatters that have well-defined cross-sections. Importantly, corner reflectors provide an extremely intense backscatter which means that they are not only easily detected in the radar data, but they are also strong enough to generate a signal in the antenna sidelobes. If a point target is brought through the antenna beam pattern the response will therefore give calibration information that includes the antenna pattern as well as the rest of the processing chain, including the manner in which the data has been processed. This last point is particularly pertinent to imaging radar where the particular software used to generate the image may also influence the final result.

Another approach to external calibration in active systems is to use   *Distributed Targets* the signatures of *distributed* targets of known radar cross section (RCS), such as extensive forests or, more rarely, agricultural fields. The assumption is that the area in question is uniform, and that its average RCS for the particular radar configuration and time of year is known (from previously calibrated data). As well as providing a calibration methodology for large footprint sensors (such as windscatterometers) this approach has been useful in determining relative calibration errors within higher resolution imagery, and has also been proposed for absolute calibration of SAR images.

Of the potential calibration sites, tropical rain forests such as the Amazon and the Congo forests show remarkably stable RCS over a large area and extended time periods. However, these regions do exhibit some degree of spatial and temporal variability, which if unaccounted for can lead to biases and increased uncertainty in the calibration of the radar instrument. Seasonal variability has been included in some calibration studies but only as a seasonally fixed response without adjusting for any particular ground parameter variation. Characterising such fluctuations as a function of measurable ground parameters could be of great benefit to the calibration of active microwave instruments, allowing for distinction between the seasonal and inter-annual trends in the forest and slow variations of the system gain.

Corner reflectors and distributed targets passively provide strong well-defined targets and so are well-suited for end-to-end calibration of radar systems. However, they are not able to distinguish between the calibration parameters of the system in transmit versus receive mode. To this end, "active" targets are required — i.e., a transponder that transmits a signal that is compatible with the radar system so that it is recorded by the instrument[62]. This allows analysis of only the receiving part of the system (and by inference, the transmitting part) and is therefore an important tool in calibrating the specific properties of the antenna assembly.

6.9.1   Antenna Calibration

In any microwave system, the antenna is perhaps the major source of calibration error. Since it is the key component that links the inside and outside of an instrument, only external calibration techniques can be used. This poses a particular problem for spaceborne systems since they operate at an altitude of many hundreds of kilometres — most definitely farfield conditions. But how do you replicate farfield conditions before launch to provide an accurate calibration of the antenna? For a typical spaceborne radar antenna (such as the one on the Envisat satellite) you require to be a distance of at least 4km from the antenna to be in strict farfield conditions! A further complication is that while in orbit the antenna is influenced only by the satellite platform itself, and nothing else, whereas on the ground (or in a laboratory) there is the platform, plus the support, the ground, the various instruments and the laboratory itself.

One solution is to use a suitably designed anechoic chamber that minimise scattering and reflection from everywhere but the platform and antenna, and sensors (transmitters or receivers) that can replicate farfield conditions. This is then followed up by an extensive calibration campaign using external targets and sources that takes place following a successful launch (as described in the previous section). Such "cal-val" (calibration and validation) campaigns can be both laborious and expensive but they are a fundamental requirement to properly characterise the end-to-end performance.

---

[62]   A transponder is a ground-based instrument that detects an incoming radar pulse then re-transmits a very strong, well-characterised signal that the radar system will have no trouble detecting. By transmitting such signals well before and after an instrument over-flight, it is also possible to determine information about the whole antenna receive gain pattern. Transponders used for polarimetric systems are known as PARCs (Polarimetric Active Radar Calibrators).

Another significant problem with antennas when it comes to calibration is that many of their intrinsic properties can change while in orbit. The microgravity environment (as opposed to the one-*g* environment it will have been tested under) and changes due to temperature fluctuations that occur during an orbit, will cause distortions in the structure of the antenna. In order to maintain any degree of consistency, the rigidity of an antenna should remain with $\lambda/8$. Such considerations have to be traded off with the fact that a larger antenna may mean greater gain or higher resolving power. At C-band, this can mean that a 10×2m antenna requires to remain rigid to within 7mm!

For phased array radar antennas there is a further problem associated with the transmit and receiver modules since each one has its own characteristics and potential flaws. The Envisat ASAR system, for example, contains 320 Transmit/Receive Modules (TRM). Instead of a single point of possible failure it has 320, and it is reasonable to assume that not all modules will survive getting into orbit and a lifetime of 3 or 4 years continuous operation. ESA estimate that about 20 or so modules will fail by end of lifetime, each having an impact on the overall calibration of the antenna.

6.9.2   Verification and Validation

Two other terms are worth mentioning here, if only to stress that they are not the same as calibration. *Verification* is the comparison of measured values between two (or more) independent instruments. If the measured quantities are the same (within the uncertainty for the calibration performed) this offers some confidence in the calibration process and it also provides for a continuity of results between the different instruments used[63]. It does not necessarily mean that both measurement are correct, merely that there is consistency between them. The ability of an measurement technique to repeatedly give consistent results is referred to as precision. Precision is different from accuracy, which tells you how close your measurement is to a "true" property of the target. Assessing the accuracy is known as *validation*, since you are determining whether or not your very precise measurements are actually valid.

Both of these tasks are extremely important for the operation of a sensor, but neither of them are calibration. However, they will help to determine whether the calibration is reliable.

6.9.3   Types of Calibration

---

[63]   This is particularly important when trying to measure variables related to changes in the Earth's environment that may require a sequence of satellites to measure the impact of long term trends. Each instrument carried by a different satellite in a series must then at least be calibrated to each other so that long-term assessments may be made.

Different instruments are dedicated to measuring different properties of the incoming microwave radiation, and if you are using the acquired data, you may only be interested in one aspect of the measured radiation. The performance of the instrument may vary, depending on the particular parameter, so that a scatterometer, for instance, may be focused on making very accurate measurements of power, but at the expense of making inaccurate measurements of phase. There are therefore different end-to-end calibration characteristics that are both instrument and wave-property dependent.

Radiometric calibration

Most microwave instruments rely on well calibrated measures of the intensity, or power (which is related to the amplitude), of the incoming wave. Characterising an instruments ability to do this is called *radiometric calibration*. Because this is the essence of useful information from most passive radiometers, a great deal of effort is put into radiometric calibration of such instruments. The internal hot-cold calibration described above, for instance, allows very accurate radiometric calibration, especially if the viewing of hot and cold targets is made on a very frequent basis (i.e., intermittent with normal operation).

It is also valuable to have good absolute radiometric calibration in a radar device. Scatterometers are non-imaging radar system that are designed specifically to make very accurate measurements of radar backscatter and so require very accurate radiometric calibration.

Phase calibration

Coherent microwave instruments are distinct in that they record the phase of the measured signal as well as its intensity. Aperture synthesis, for instance (which is discussed in Chapter 10) requires very accurate phase measurements. Phase estimates are also required for polarimetry and interferometry, and so for all these techniques accurate phase calibration is required.

Polarimetric channel imbalance

Systems that measure signal polarisation have some additional calibration criteria that need to be considered. Usually such systems measure an H and V component separately, using a different detector, so it is first necessary to make sure that the response of both these channels is the same. This is called as *Polarimetric Channel Imbalance*.

Polarimetric phase calibration

For full polarimetry it is also necessary to calibrate the phase measurements between the H and V channels. Most importantly, it is vital that the phase difference is well known — for full polarimetry the absolute phase is not the issue, but the phase difference is vital. For radar polarimeters, corner reflectors or PARCs are valuable tools here since the former have characteristic phase differences, and the latter can generate well-defined phase differences.

Cross-talk

Another problem with having two sets of measurements is being able to isolate the two signals so that they do not interfere with each other. A completely vertically polarised wave, for instance, should not register a signal in the H channel. However, as a consequence of antenna alignments and the limitations of the engineering, there is always some *cross-polarised leakage* or "cross-talk". The cross-polarised leakage is therefore a measure of the isolation of the two polarisation channels.

6.9.4 Strategies for Calibrating Receivers

There are two approaches to maintaining consistent and frequent calibration for passive systems: Dicke and total power radiometers. Both of them rely on making measurements relative to some known quantities in order to characterise the system complete with noise, since noise is not determinable theoretically.

In a total power radiometer the voltage output from the square-law detector is proportional to the total power through the system. The net IF power reaching the square-law detector can then be considered to consist of two components: that part proportional to the antenna temperature T_a and the internally generated noise power, or *receiver temperature T_r*, which is often predominantly electrical noise arising from individual components such as the mixer. The system temperature T_s is then defined as the sum of these components. In this case we expect the relationship between detected signal temperature and the actual brightness temperature to be linear, so that by using hot and cold targets with temperatures that cover the observed extremes, we assume that we can derive a linear relationship between the two.

A total power radiometer can then be calibrated by observing black-body emitters at two different temperatures. Assuming the radiometer is linear, which is usually the case to within practical limits, the antenna temperature for an unknown target is then determined from

Total power radiometer

$$T_a = C(V - V_0) \qquad (6.14)$$

where V_0 is the voltage offset that acts as a reference. The radiometer calibration constant C can be found from the measured voltages for the two blackbody calibration targets and fitting a straight line through these points, since C corresponds to the slope of the straight line as shown in Figure 6.7, such that

$$C = \frac{T_{hot} - T_{cold}}{V_{hot} - V_{cold}}.$$

Estimating the *radiance* of the target from T_a then also requires that the antenna has been calibrated. To allow for changes in the instrument, this calibration procedure can be executed at frequent intervals to maintain high precision on the actual measurements.

Total power radiometers have a significant advantage as they are very sensitive — the value of k_r in Equation (6.13) is unity. But such systems are not always well-suited for a given application since they are difficult to stabilise and the calibration procedure is not necessarily as straightforward as the above discussion might imply. An alternative approach, therefore, is the Dicke radiometer. The principle behind this type of system is to make a comparative measurement between a stable target that has an radiative temperature similar to the incoming antenna temperature T_s. A fast switch

Dicke radiometer

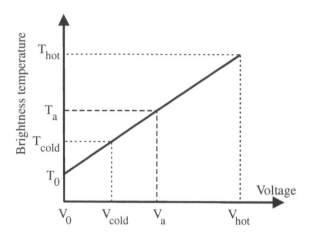

FIGURE 6.7 A typical calibration method for a total power radiometer is based upon observing hot and cold targets with known brightness temperatures. The voltage recorded for each target is used to plot a linear relationship between voltage (power) and brightness temperature.

is used to alternate between the antenna and the reference signal and the detector outputs a voltage in proportion to the *difference* between the two, rather than the total input signal. Since half of the time available is spent observing the reference source the sensitivity is halved since now $k_r = 2$ in Equation (6.13), but the calibration is improved.

6.10   Final Remarks on Calibration

A fundamental property of microwave systems is that it is possible to achieve well-calibrated quantitative results. In passive systems a radiometric brightness temperature is required to accuracies of a fraction of a Kelvin to be useful in atmospheric or land-based applications. In radar systems we require quantitative measurements of sigma nought to within a fraction of a dB, and phase measurements that are no worse than a handful of degrees. This is not always achieved, of course, but a great deal of time and money and effort is expended to aim for the best calibration possible.

Good calibration is vital because it allows you to say something quantitative about the physical properties of the object being observed, rather than simply something about what happens at the instrument. This allows quantitative limits to be placed on the sensitivity of the instrument (so that you can gauge its suitability for a particular purpose). It allows for inter-comparison between different measurements and instruments so that ground-based, lab-based, airborne and spaceborne data can all be compared or used together. And it allows for direct comparison between microwave observations and physical models, either for testing our understanding of the measurement process through the use of models, or by providing the basis for quantitative parameter retrieval, rather than simply attractive looking pictures. It is vitally important therefore, that whenever you are dealing with a data set, you have some idea of the manner in which it has been calibrated.

6.11 Further Reading

As always, Ulaby, Moore and Fung's three volumes are the ultimate source. John Curlander and Robert McDonough's "Synthetic Aperture Radar: Systems and Signal Processing" (1991) gives an extensive summary of the hardware and processing of SAR. I also recommend Skolnik (2002) for radar systems and Skou (1989) for radiometers. Tony Freeman has written widely in the topic of calibration, including useful (but hard to find) notes from a CGI course from a decade ago and Freeman (1992) in the open literature.

For readers interested in learning more about the ear, I can recommend von Békésy (1957) for a general description of the workings of the human ear, or Allen and Neely (1992) for an in-depth description of the mechanics of the cochlea.

7
ATMOSPHERIC SOUNDING

We may mount from this dull Earth, and viewing it from
on high, consider whether nature has laid out all her cost
and finery upon this small speck of Dirt. So, like Travellers
into other distant countries, we shall be better able to judge
of what's done at home, know how to make a true estimate
of, and set its own value upon every thing.
— C. Huygens, *The Celestial Worlds Discovered* , c.
1690.

90% of the Earth's atmosphere lies within the troposphere, a thin layer
about 15km deep — the equivalent of the thickness of a piece of paper
on a football-sized classroom globe. Yet despite its apparent geometrical insignificance the atmosphere is vital to life on Earth. Over the last
few decades it has become increasingly evident that human activity also
contributes to the delicate balance of processes in the atmosphere and that
changes in the physical properties of the atmosphere such as temperature
can have dramatic effects on the human population. This has heightened
awareness of the importance of the atmosphere to life on Earth and has fuelled major programmes aimed at developing an understanding of global
atmospheric processes through the essential monitoring of basic data such
as temperature, pressure and wind, as well as the distribution of water
vapour, ozone and other constituents. Such data enable testing of existing
models of ozone depletion, the hydrological cycle, global climate change
and other aspects of the atmosphere that are of vital importance for understanding and predicting the Earth system.

Tropospheric measurements are particularly important for monitoring
pollution and for identifying changes that influence the Earth's radiation
budget, water vapour in particular being a major contributor to the so-
called "greenhouse effect". In the stratosphere, the greenhouse gases contribute negatively (since the radiation budget is defined at the top of the

troposphere so that greenhouse gases above this layer help to cool the troposphere) but are therefore equally important to measure. The importance of stratospheric ozone is now widely known for the protection from harmful UV radiation that it affords life on the surface of the Earth.

This chapter looks at the way in which microwaves can be used to measure the atmosphere, including hydrometeors, from remote instruments in general but with a clear focus on satellite systems.

7.1   Atmospheric Sounding

7.1.1   The Need for Measurements

The reason that microwaves are of interest for atmospheric sounding is that at frequencies between about 10 and 300 GHz a number of important atmospheric constituents have rotational and vibrational spectral lines. Molecular oxygen (O_2), ozone (O_3), chlorine monoxide (ClO), sulphur dioxide (SO_2) and water vapour (H_2O) have absorption/emission lines within these frequencies, as do BrO, OH, HCl. These chemicals are important in long term climate change and in the upper atmosphere and many of them are important in the creation/destruction of ozone. In addition to individual spectral lines, with increasing frequency from about 10 GHz the absorption and emission from the atmosphere generally increases, largely as a consequence of the water vapour continuum that acts as a continuous background signal.

Microwave measurements can make an important contribution for a number of reasons (Waters, 1993). Firstly, it is relatively straightforward to make measurements of the relevant physical parameters, such as temperature and pressure, as well as molecular concentrations required for monitoring the global atmospheric system; e.g. temperature, pressure, liquid water and water vapour, rainfall, and a collection of molecular concentrations, including O_2, O_3, ClO, BrO, OH, HCl, HO_2, CO, HCN, N_2O, HNO_3, BrO and volcanically-injected SO_2. Measurements of gas concentrations are made by measuring the emission from spectral lines and since it is pressure broadening that dominates the shape of such lines in the troposphere and stratosphere, these lines are relatively insensitive to temperature so that highly accurate temperature measurements are not required to determine concentrations to a useful level of accuracy. Importantly, many of these measurements can be reliably made even in the presence of tropospheric aerosols and cirrus or polar stratospheric clouds, the scattering from which would normally degrade measurements made at shorter wavelengths such as ultraviolet, visible and infrared. Clouds and hydrometeors are generally not that important across the region considered for microwave sounding, and indeed their effect is almost non-existent below 20 GHz.

An additional property of microwave sounding is that passive microwave systems measure emitted radiation, so there is no requirement for some external source. Measurements can therefore be made at all times of day and night, so that satellite instruments can achieve global coverage on a daily basis.

In the microwave region, it is also possible to resolve the actual shape of the emission lines at all altitudes, which allows measurements of even very weak lines in the presence of nearby strong ones. It is then possible to make useful measurements of chemical species with very low atmospheric abundances (down to parts per billion, ppb). These measurements are supported by the availability of very accurate spectral line data based on laboratory experiments.

From a practical perspective, microwave instruments can now be developed for orbiting satellites that have long (many years) operational lifetimes with accurate and reliable long-term in-flight calibration. Unlike most IR instruments, microwave instrumentation can achieve the required sensitivity without the necessity for elaborate, fixed lifetime cooling systems[64]. The gradual improvement of microwave technology to allow measurements into the Tera Hertz (THz) region also means that practical sized antennas can provide improved angular resolution.

7.1.2   The Earth's Atmosphere

To understand the context of measuring properties of the atmosphere, it is first worthwhile to summarise the key physical attributes of the Earth's atmosphere.

The Earth's atmosphere below about 100km can be described by three regions: the lower, middle and upper atmospheres, the boundaries of which are defined by the distinct changes in the atmospheric temperature profile. Figure 7.1 shows a typical temperature profile through the atmosphere as a function of pressure and approximate altitude (the temperature follows pressure more closely than it does altitude). *Lower Atmosphere*

The "lower atmosphere" refers to the troposphere, the region which begins at the ground and reaches to an altitude of about 15 km. This is the region in which most of the weather takes place and is of most concern in terms of global climate change. 90% of the mass of the atmosphere resides within the troposphere. The vertical heat transfer in the troposphere consists mostly of the thermal radiation emitted from the Earth's surface and convection from direct heating at the lower boundary of the troposphere. The temperature is therefore highest at the ground and steadily decrease to the top of the troposphere, the *tropopause*[65].

The "middle-atmosphere" refers to the *stratosphere* and *mesosphere* *Middle Atmosphere*

---

[64]   The cooling systems only have a fixed lifetime because they rely on a supply of coolant, such as liquid nitrogen.

[65]   The suffix *-pause* comes from *pausis*, the Greek for "ceasing", so that the tropopause is the *end of* the troposphere.

which span the altitude range 15–85km. The exact lower and upper boundaries are defined by the temperature minimum at the tropopause at a pressure level of about 300-hPa ($\sim$15km) and an even colder *mesopause* minimum at 0.001-hPa ($\sim$85km [66]). The boundary between the stratosphere and the mesosphere is defined by a temperature maximum at the *stratopause* ($\sim$1-hPa or $\sim$50 km). The temperature inversion within the stratosphere is driven by direct solar heating, largely due to absorption of solar ultraviolet radiation by ozone (O_3) which has a maximum concentration at around 100-hPa ($\sim$20 km). The decline of atmospheric temperature in the mesosphere is a result of decreased O_3 and increased cooling rates from CO_2 infrared radiation, which radiates more easily to space above the stratosphere.

Satellite remote sensing techniques are of particular value in monitoring the middle atmosphere, a region which is little understood and has limited measurement coverage from the ground. Although less than 10% of the atmospheric mass lies above $\sim$15 km, the middle atmosphere plays a fundamental role in maintaining the global habitability of the Earth. In particular, long-term climate changes and the stability of the ozone layer shielding the Earth from solar ultraviolet radiation are primarily middle-atmosphere problems, and although there has been a great deal of progress in the last few decades, detailed understanding of these processes is still limited.

Upper Atmosphere   The "upper atmosphere" goes from the *thermosphere*, defined at the mesopause, into the local interplanetary space. The temperature increases again into the thermosphere due to the interaction of high energy radiation from the sun.

Over the land areas of the world much of the information about the atmosphere comes from a high density of surface observations together with measurements from a large number of radiosondes[67] and rocketsondes, but such methods are clearly unable to provide the continuous and detailed observations of the global atmospheric state required by modern numerical models. It is for this reason that remote sensing from satellite platforms has become increasingly important as the only way to obtain the spatial and temporal coverage needed to understand global atmospheric processes.

7.1.3   Water Vapour and Oxygen

By far the most significant contributors to the microwave properties of the atmosphere are water (either as vapour or in liquid form as hydrometeors) and oxygen. Water vapour has strong spectral lines at 22 and 183 GHz and molecular oxygen has a sequence of lines between 50 and 70 GHz

---

[66]   The units of *hecto-Pascals* (hPa) are approximately equivalent to millibars. In order to have a linear altitude scale related to pressure, p is often replaced by $z = -\log_{10} p$.
[67]   These are simple instruments carried by balloons which measure pressure, temperature and humidity.

and at 118 GHz. Oxygen is well-mixed and does not vary significantly with time, so measurements of oxygen are important as an indicator of temperature and pressure of the atmosphere, rather than being an important gas to measure in itself.

In contrast, water vapour is a very dynamic variable in the atmosphere and knowledge of the transport of water vapour in the troposphere is important for understanding the hydrological cycle, as well as monitoring aspects of climate change, since water vapour is the major gas that contributes to the global greenhouse effect.

Measurements of H_2O in the middle atmosphere are also important as it plays a major role in middle atmospheric ozone photochemistry (as it is the major source of HO_x radicals) as well as having a significant impact on stratospheric radiative heating, troposphere-stratosphere energy exchange and ultimately tropospheric climate change.

A further important reason for measuring column water vapour is in support of altimeter measurements that will be discussed in Section 3. Water vapour is one of the major variable factors that contribute to uncertainties in measurements made by microwave altimeters and so if the total column content can determined the altimeter measurement can be corrected to a much greater level of accuracy.

7.1.4   Clouds and Precipitation

Clouds and precipitation are measurable in microwave regions either because they scatter the upwelling radiation from the underlying surface and therefore look cooler than the surrounding surface, or they appear as heightened emission. The details of how to make such observations are given at the end of this chapter.

Since the presence of clouds or precipitation can hinder accurate retrievals of temperature or composition, if they lie within the field of view, it is usually necessary to have at least a rudimentary method of detecting them in order to account for their impact on the sounding measurements.

7.1.5   Ozone

The discovery of the "ozone hole" over the southern pole by ground based measurements made by the British Antarctic Survey in 1985 confirmed some scientists predictions of excess ozone depletion. Ozone is created and destroyed by photochemical processes initiated by high energy UV solar radiation. The balance of creation versus destruction tends to result in a net production near the equator and a net loss near the poles. The ozone hole is not a direct consequence of this loss, but is a particular anomaly caused by the formation of extremely high clouds within the polar vortex that forms during the Antarctic winter.

Ozone is important because it absorbs high energy UV. Ozone is also a greenhouse gas and contributes to the radiation budget — in a positive sense in the troposphere but a negative sense in the stratosphere (since the tropopause defines the boundary of the radiation budget). In the troposphere, ozone is also a potent pollutant.

7.1.6   Chlorine Monoxide

ClO measurements on a global scale are important for understanding and monitoring the destruction of ozone by chlorine since ClO is the rate-limiting molecule in the catalytic chlorine cycle that destroys ozone. Its abundance is therefore a measure of the rate at which chlorine compounds destroy ozone. Such processes are of special concern to modern society since the major source of chlorine in the stratosphere is industrial chlorofluorocarbons (CFCs). Despite the Montreal Protocol in 1987 that implemented the controlled phasing-out of CFCs, they remain in the stratosphere for many decades. It is predicted that it will be the middle of the 21st century before Antarctic ozone reaches pre-1980 levels. Indeed, recent assessments confirm the need for continued monitoring as they have shown greater ozone loss in all seasons at mid- and high latitudes than previously thought as well as associated increases in ClO abundance, as well as identifying coupling between lower-stratospheric ozone chemistry and potential climate change.

7.1.7   Other Relevant Measurements

Many other atmospheric molecules and atoms have been shown to play a part in the complex chemical processes occurring in the middle atmosphere and it is clear that continued monitoring and study is essential to developing an understanding of the global atmospheric system and our influence on it. The lack of understanding of middle atmosphere processes is due mainly to a lack of sufficient data coverage and satellites such as UARS and Aura have been invaluable in filling this gap by providing daily global coverage of many of the relevant trace species in the stratosphere and mesosphere.

7.2   Principles of Measurement

Atmospheric sounding from satellites was initially developed using observations of emission lines in the infrared region. The principle of microwave sounding is the same, namely measuring the brightness temperature of the atmosphere at frequencies that correspond to the emission lines of key constituents. The apparent brightness at these frequencies will depend upon the physical temperature of the atmosphere and the concentration of the emitted molecules, so that satellite measurements of T_B should allow you to infer the temperature and/or the concentration. That T_B varies in response to changes in both of these parameters is the first problem. For

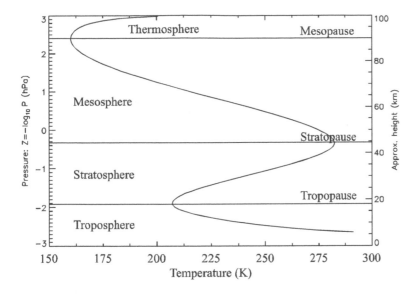

FIGURE 7.1 Cross section of the atmosphere showing the main atmospheric layers. These are defined by the shape of the temperature profile.

temperature sounding this is overcome by looking at emission lines corresponding to uniformly mixed gases such as oxygen and carbon dioxide that have concentrations that vary by only small amounts within the atmosphere, both spatially and temporally. This makes the oxygen emission at around 60GHz extremely useful for determining atmospheric temperature since it means that we can assume changes in the measured radiance at this frequency are related only to temperature and pressure variations, and not to changes in concentration.

To make measurements of concentration of, say, water vapour, then requires two measurements: a temperature measurement from the oxygen line, and another measurement at a water vapour line at, say, 183GHz. The two measurements together are necessary to account for variations in both temperature and concentration. The ideal scenario is to measure a number of different narrow frequency bands around a collection of emission lines so that the complete shape of the line can be observed in addition to simply its peak brightness. This allows the retrieval of temperature, pressure *and* concentration.

The theoretical considerations for sounding the atmosphere are covered in more detail in the following section where we will also consider different viewing geometries. Many microwave sounders use a geometry whereby the sensor is directed straight down, towards the nadir. This nadir viewing geometry has the ability to sound deep into the atmosphere and has an optimum horizontal resolution, but is poor at resolving differences in the vertical profile through the atmosphere. An alternative geometry is

Limb sounding

to look sideways, perpendicular to the nadir view, so that the sensor looks across the atmosphere's edge, or *limb*. This is known as *limb sounding*, and improves the ability to resolve the vertical profile, but with a corresponding reduction in the horizontal resolution.

One further practical consideration for microwave sounders is that they must be very sensitive so as to detect a sufficient amount of radiation. There are two approaches to making sure a radiometer measures sufficient energy to detect. The first is to have a long integration time, meaning that the time spent detecting radiation from a specific direction is maximised giving a better SNR. For instruments on aircraft or satellites, however, the flight speed places a strict limit on available dwell time. The alternative approach is to make measurements over a very large beamwidth. The larger the collecting area the higher the signal compared to the instrument noise.

Microwave radiometers will therefore tend to have relatively poor angular resolution. The other factor that influences the angular resolution is the size of the antenna. In Section 4.4 we saw that the angular resolution of an antenna is directly proportional to the wavelength, so that as the measurement frequency gets lower, the angular resolution gets proportionately larger. For atmospheric sounding, this is not a drastic problem — a footprint in the order of tens of kilometres or more in diameter is quite appropriate for mapping variations in atmospheric properties such as temperature and chemical composition since these properties change slowly over horizontal scales. For atmospheric studies it is usually more important to achieve high measurement accuracies, with high temporal frequency (< 1 day) and with a high *vertical* resolution for profile measurements (ideally down to kilometres).

7.3   Theoretical Basis of Sounding

7.3.1   The Forward Model

Any remotely sensed measurement is only as good as our understanding of the measurement process — not just the physics of the observation, but the complete measurement chain. Quantitatively, this understanding is expressed in the form of the *forward*, or *direct*, model, F. The forward model should be an accurate characterisation of the relationship between a remote physical attribute and a local set of digital numbers, which is ultimately the nature of the final product of any modern remote sensing system. The forward model is a simulation of the measurements given the physical variables of the atmosphere. In practice a forward model will be characterised not in terms of the raw data (digital numbers) but a calibrated measurement quantity such as an observed brightness temperature, the explicit assumption being that the instrument is sufficiently well calibrated that the link between the measured voltage and T_B is reliable.

In atmospheric sounding the forward model is a fundamental part of the measurement process — it allows us to go beyond simple statistical correlations between measurements and *in situ* data and instead to interpret the physical meaning of the observed quantities. Indeed, the principle of the forward model can apply equally well to all forms of remote sensing, although practically it is not always possible to describe the forward model sufficiently accurately to take such an approach.

The model should include the physics of the emission/scattering from the target area, and the radiative properties of the atmosphere (even if only to ignore it for longer wavelength microwaves). It therefore should include our knowledge of the EM radiation being used or measured (Chapter 3), including the polarisation (Chapter 4), as well as our understanding of how such waves interact with natural objects (Chapter 5). However, it will also include the behaviour of the instrument (Chapter 6) in order to characterise the measurements and to be able to describe how they depend on the state of the target object or material.

Formally we can represent the forward model using the vector $\mathbf{y}$ to represent a set of measurement values, and vector $\mathbf{x}$ to represent the values of the physical attributes of the atmosphere and underlying surface. The forward model F can then be expressed mathematically as

$$\mathbf{y} = F(\mathbf{x}, \mathbf{b}_M, \mathbf{b}_I) + \boldsymbol{\epsilon}_y \tag{7.1}$$

where the vector $\mathbf{b}_M$ represents known physical parameters relevant to the measurement (such as spectral line data), and the vector $\mathbf{b}_I$ represents instrument parameters such as antenna patterns, filter responses, imaging geometry, and any other factor that may influence the measured signal. $\boldsymbol{\epsilon}_y$ is a vector representing the noise or random error in the set of measurements, $\mathbf{y}$. By way of example, for the Microwave Limb Sounder (MLS) 183-GHz radiometer, the vector $\mathbf{y}$ represented radiance measurements made over 15 spectral channels and 26 view angles (i.e. a total of 390 values) and $\mathbf{x}$ was a state vector which included the unknown vertical profiles of temperature, pressure, H_2O, O_3 and ClO.

In remote sensing the function $F(...)$ often takes the form of an integral equation,

$$y_i = \int_a^b K_i(x) f(x) \, dx + \epsilon_i \tag{7.2}$$

where $f(x)$ is the function that describes the physical properties of the system that is to be determined, y_i are the set of observations of the system and ϵ_i is the error, or noise, associated with each measurement y_i.

This is known as a Fredholm integral (since the integral limits are fixed) and of the first kind ($f(x)$ only appears in the integrand). The function $K_i(x)$ is generally known as the *kernel* or *kernel function* and is determined by the particular observational setup being modelled. The kernel is a function that transforms the input variables into a set of output values (the observations). For example, in imaging the kernel corresponds to an end-to-end impulse response function that determines how the true image is altered (by blurring, for instance) by the specific imaging device.

Often the forward problem is nonlinear, meaning that Equation (7.2) is only an approximate, linearised version of the problem, although for most application in remote sensing this is sufficient. The main source of non-linearity in microwave sounding models is likely to be the dependence of transmittance on absorber concentration. Two further potential sources of nonlinearities are the temperature dependence of the atmospheric trans-mittance and the frequency dependence of the Planck function. These ef-fects can be minimised by choosing temperature insensitive spectral lines, and by using microwave measurements so that the Rayleigh-Jeans limit is valid, respectively.

If the forward model characterised the measurement system exactly we would expect the differences between simulated observations (if we knew $\mathbf{x}$ exactly) and actual observed values (for the actual state $\mathbf{x}$) to lie within the limits of the measurement error. However, this is often not the case since knowledge of the quantities within $\mathbf{b}_M$ or $\mathbf{b}_I$ usually have some additional uncertainty associated with them beyond the known noise, and these must be accommodated within ϵ_y. The differences between the modelled and real observations are known as *residuals*.

7.3.2   Simple Formulation of the Forward Model

A simple forward model for atmospheric sounding can be formulated as follows.

The measurement made by a satellite radiometer is the top of the at-mosphere (TOA) brightness, which may contain varying contributions from different layers in the atmosphere, or contributions from sources other than the atmosphere such as the underlying surface. With reference to Figure 7.2 and Equation (5.10) we can then formulate a simple description of the measurement as follows:

$$T_{\text{TOA}} = \Upsilon T_{\text{BG}} + T_{\text{UP}} \tag{7.3}$$

where T_{UP} is the upwelling radiation from the atmosphere and T_{BG} is the brightness temperature of the background signal, which, for the partic-ular geometry shown in the figure, is the underlying surface beneath the atmosphere. Since the background term depends on the viewing geometry, we will discuss this in more detail in a later section.

Υ is the transmissivity of the entire atmosphere and attenuates the sig-nals originating from behind the atmosphere. Note that this expression is a version of the radiative transfer equation given in (5.6), with the exiting term now T_{TOA} and T_{BG} corresponding to the background brightness, T_{B0}. The integral component of (5.6) is now generalised to T_{UP}, the contribu-tion from the atmosphere which is actually composed of emission from all the layers of the atmosphere, each attenuated by the layers above.

Ideal conditions for measuring the atmosphere are therefore at frequen-cies where the atmospheric transmissivity is low, so that a large part of the measured signal originates from within the atmosphere (usually above about 10GHz) combined with a relatively low brightness temperature for the surface.

We can rephrase the transmissivity of the entire atmosphere in manner that is specific to a remote sensing geometry, by assuming that the atmosphere is horizontally homogeneous and that the viewing path length is at an angle θ from the vertical. The more general term for the transmissivity of the path length is then given by

$$\Upsilon = e^{-\tau/\cos\theta}$$

where τ is the vertical opacity of the entire atmosphere[68]. Similarly, we can write the upwelling signal from the atmosphere using the vertical dimension, z (rather than path distance, s), so that

$$T_{\mathrm{UP}} = \frac{1}{\cos\theta} \int_0^{\mathrm{TOA}} \kappa_\nu(z)T(z)e^{-\tau(z)/\cos\theta}dz. \qquad (7.4)$$

Note the similarity of this expression with the generalised expression for the forward model given in Equation (7.2). The additional cosine term at the beginning of this expression accounts for the increased atmospheric contribution now that we are taking into account the potential for off-nadir viewing. The integral is then over the vertical distance, z (the height), rather than along an arbitrary path length.

7.3.3   The Inverse Model

The *inverse* problem in atmospheric sounding is the process of trying to figure out the state of the atmosphere given a set of satellite measurements. In atmospheric sounding the process is also often referred to as *retrieval* or simply *inversion*.

While instrument performance and platform technology have progressed significantly since the onset of satellite remote sensing, retrieval methods have changed very little. The approach to the inverse model for atmospheric sounding is now well-established and despite occasional attempts to improve the details of the methodology, the basic approach has remained the same for more than three decades. The review paper by Rodgers in 1976 still covers almost everything that is relevant for retrieving atmospheric parameters from modern instruments.

The retrieval of atmospheric temperature and composition from satellite measurements of emitted thermal radiation requires finding the inverse solution of the full radiative transfer equation, (5.3). An interesting question to ask is under what conditions we think we might be able to invert this equation and what information such an approach might provide. To illustrate the point, it is useful to consider the very basic case of an isothermal[69] atmosphere, viewed at nadir, so that we can rewrite Equation (7.3)

Ill-Posed Problems

---

[68]   In some versions of this relationship the cosine terms is replaced by $\sec\theta = 1/\cos\theta$.
[69]   An isothermal atmosphere has a uniform temperature.

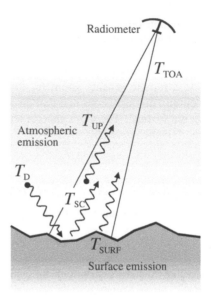

FIGURE 7.2 A diagram summarising the main components of the forward model for passive (nadir) sounding. Atmospheric emission contributes both an upwelling (T_{UP}) and a downwelling term (T_{D}), the latter being scattered by the surface. The surface also contributes by emission (T_{SURF}).

as

$$T_{TOA} = T \int_0^{\text{TOA}} e^{-\tau} d\tau, \tag{7.5}$$

where we have assumed no background signal, and have changed the integration term so that it is over $d\tau$ instead of ds (the advantage being that the absorption coefficient is now embedded into the integral[70]). Calculation of the integral (as for Equation (5.10)) gives

$$T_{TOA} = T(1 - e^{-\tau_a}) \tag{7.6}$$

where τ_a is the total optical depth through the entire atmosphere (from 0 to TOA). We can consider two extreme cases of this expression. The first is for frequencies where the opacity is high, so that $\tau_a \gg 1$, in which case $T_{\text{TOA}} = T$. The brightness temperature measurement therefore acts like a thermometer — the observation gives the temperature of the isothermal atmosphere directly.

The second case considers frequencies where the opacity is low (but not entirely transparent), so that $\tau_a \ll 1$, giving

$$T_{\text{TOA}} = T\tau_a, \tag{7.7}$$

where we have used the approximation $e^{-x} \approx 1 - x$, for $x \ll 1$. In this case, if we know the temperature we can estimate the opacity, which for

---

[70]   This is found from noticing that $d\tau = \kappa ds$, from Equation (5.2).

a given frequency will be directly related to the concentration of a particular atmospheric constituent. For example, measuring the opacity at the
183 GHz water vapour line, *and* knowing the temperature, allows an estimate of the water vapour concentration. Similarly, if the concentration is
known, the temperature can be estimated.

The important thing to realise here is that it is the choice of observational frequency that determines the effectiveness of the atmospheric retrieval. A frequency within an true atmospheric wall is no good, since it
has high opacity — the atmospheric chemistry cannot be inferred from
such a measurement (although the temperature can). However, a totally
transparent atmosphere (zero opacity) is also not ideal, since in practice
it will be the background signal (not shown in the above equations) that
would dominate the measurements.

This is the basic premise by which any atmospheric sounding measurement is made, but as you might guess, it is not quite as simple as that.
A real atmosphere is not isothermal, for instance, so we must somehow
accommodate the variability of temperature with height. And not all constituents in the atmosphere are well-mixed — many gases, such as water
vapour and ozone, have significant variations both vertically and horizontally. Any practical retrieval procedure must try to accommodate these
factors, making it very difficult to analytically solve the inverse model. Indeed, it is so difficult in fact, that it is impossible. We devote the next
section, therefore, to explaining how to overcome what would seem to be
an intractable problem.

7.3.4   Solving the Inverse Problem

Rather than dealing with the complete radiative transfer equation when
trying to solve the inverse problem, it is more convenient (i.e. faster and
less computationally intensive) to use the matrix equivalent of Equation
(7.2). The radiometric measurements made by the satellite radiometer are
then represented by a kernel *matrix*, $\mathbf{K}$, a discrete matrix version of the
forward model. This dependence is usually assumed to be linear so that
we use the simplified form

$$\mathbf{y} = \mathbf{Kx} + \epsilon_y, \tag{7.8}$$

where the elements of the vector ϵ_y represent the error or noise associated
with each measurement y_i. Note that this assumption of linearity is only
really valid if $\mathbf{K}$ is calculated for an atmosphere that is already very near
the solution.

In mathematics, this kind of problem is referred to as *ill-posed* or
under-constrained, meaning that there are an infinite number of possible
solutions (states of the atmosphere, $\mathbf{x}$) that give the same set of observations values, $\mathbf{y}$, from Equation (7.8) — at least to within the expected
measurement error, *i.e.* $\pm \epsilon_y$. The problem is often made worse by the fact
that there are not as many measurements being made as there are influen-

tial parameters in **x**, or that the problem is simply too non-linear to provide a unique and stable solution.

A simple analogy is to ask you to solve the equation $3x_1 + 2x_2 = 12$. There is not enough information in that one equation, since there are an infinite number of x_1 and x_2's would satisfy this equation (you might be able to work out a handful in your head). Introduce a second equation with the same unknown variables, however, such as $x_1 + 4x_2 = 11$, and it is now possible to find a solution: $x_1 = 3$ and $x_2 = 2$. Two equations were required to solve for two unknowns. If we consider these two "simultaneous equations" in geometric terms they describe two straight lines. The solution is then the intersection of the two lines, since this is the point that satisfies both equations. However, even this is not straightforward when we introduce random measurement error onto these equations. If the lines have uncertainty associated with them, then the solution is no longer a discrete point but a *region* of possible solutions.

The fundamental difficulty in inversion problems is therefore not one of finding *a* solution but of deciding which of the infinite number of *possible* solutions that match the data is the most appropriate to the particular problem at hand. These problems can be differentiated by describing the former as the *inverse* problem and the latter as the *estimation* problem.

The main question that estimation theory addresses is that of the uniqueness of the solution. Since the original observations **y** are non-unique, due to the presence of measurement noise or error, there is a corresponding non-uniqueness in the solution. Since there are components of **x** which make little or no contribution to the observations (and could therefore be extremely large or even infinite) even small changes in **y** can lead to large changes in the solution. Practical inversion methods must therefore attempt to avoid such instabilities and provide some criteria for choosing a "best" solution through the application of constraints.

In order to solve the atmospheric inverse problem we have to introduce new information. Most atmospheric retrieval methods incorporate additional information by relying on constraining the solution in some manner, often making use of some *ad hoc* "smoothness" constraint. In practice there are a limited number of useful constraints that can be applied to the solution, and therefore a limit to the different retrieval methods.

Empirical Approach   The simplest method of constraining the solution is to use the statistics of a large set of *in situ* measurements. This empirical approach uses regression analysis for a number of experimental observations whereby many **y**'s are measured by the instrument while simultaneous *in situ* measurements of **x** are made. The two are then investigated for a statistical relationship which is used to characterise **K**. The method assumes no knowledge of **K** from physics and is purely statistical. It is a very good approach for identifying those variables to which the measurements are sensitive, or in cases where the physics is not known with sufficient certainty as to be able to describe an accurate forward model. However, this approach has many problems. Firstly, the empirical **K** tends to be specific to the experiment, so it may be valid only over a small range of conditions as provided by the *in situ* data. Secondly, it cannot be used to predict, or simulate, other situations that are beyond the *in situ* conditions, such as extreme cases of atmospheric state, or new instruments configurations, etc. To tackle these issues you need to apply more rigorous methods based on physical modelling.

A physical modelling approach is very different in that we apply our available knowledge of the physics and engineering of the measurement process to characterise $\mathbf{K}$. Chapters 3 and 5 summarised the type of physical understanding required to provide the values of $\mathbf{b}_M$. The type of calibration characteristics described in Chapter 6, on the other hand, would provide the relevant parameters to describe the instrumentation, $\mathbf{b}_I$. The solution is then found through an analytical solution of the inverse problem. *Physical Modelling*

The inverse problem is encountered in many branches of physics and engineering, and much literature has been devoted to analytical solutions. However, for most physical problems, including atmospheric remote sounding, where the number of unknown parameters is larger than the number of independent measurements, the relationships characterised by $\mathbf{K}$ cannot simply be inverted as you might with a set of simultaneous equations. Instead the problem is overcome by imposing on the solution an additional *a priori* condition or criterion (independent of the measurements) which enables a solution to be found. This criterion or *penalty function* is selected arbitrarily, and is not a consequence of the measurements, but it is obviously advantageous to choose it with regard to the physical problem at hand. The most simple example of such an approach is to assume that the solution is smooth — i.e., there are no discrete jumps in the solution. This may be a good assumption for, say, a vertical profile of atmospheric water vapour, but may be ineffective when trying to detect clouds. However, in the latter case, an alternative constraint might be applied that limits the retrieved boundary to a specific region based on time of year and temperature. *Inverse Modelling*

Some authors still use the term "smoothing" to describe this application of a penalty function on the solution, but, while in common usage, is really too restrictive a terminology as there need not be any correlation between adjacent elements of $\mathbf{x}$. In an attempt to clarify this some authors distinguish between "local smoothness" and "general smoothness".

The most common approach is to constrain the solution to one that is statistically "most likely" given our understanding of the atmosphere. We therefore start with a prior solution — a first guess — based on physical models and prior in situ measurements, which will be specific to a latitude and time of year. The idea is not to fix this as the solution (which would mean we had no extra information from the satellite) but to prioritise one solution from those indicated by the measurements by finding the solution most like the answer we would have expected to observe for a given set of conditions.

There is one major problem with the use of such *a priori* constraints, which is that they are, by definition, governed by your current knowledge of the atmosphere. The presence of new or unexpected observations can therefore go unnoticed. In the folklore of atmospheric sounding there is a story of exactly such a case. It concerned the Antarctic ozone hole, which was discovered by researchers from the British Antarctic Survey using ground-based instruments, despite the Total Ozone Mapping Satellite having been making global measurements of stratospheric ozone for many months. On announcement of the ground-based findings, the TOMS team reprocessed their data and found that the instrument had indeed observed the extreme reduction in ozone over the Antarctic during the Southern hemisphere spring but the retrieval algorithm was programmed to consider such low values of ozone to be outwith normal expectations. The measurements implied values so low that they had never before been observed, so the software assumed these measurements to be erroneous. The prior expectation, while necessary for the retrieval algorithm to work properly, had been caught out by an anomalous phenomenon.

7.3.5   The Influence Functions

In many cases it is sufficient to be able to determine the total amount of an atmospheric constituent within an entire vertical column of atmosphere. It is the total amount of ozone, for instance, that is important when considering the potential impact from increased UV radiation from the sun reaching the surface. In fact, the unit used to measure the total columnar ozone is the Dobson unit, measured in centimetres, which corresponds to the thickness of the ozone were it all to be combined in a single layer at sea level.

However, it is also important to be able to study the vertical distribution of concentration or temperature. Measuring vertical profiles is possible because for any given frequency the signal is dominated by a particular layer in the atmosphere due to the exponential decrease in atmospheric density with height. The general shape of the vertical sensitivity pattern for nadir viewing instruments is such that the observations are very insensitive to layers in the lower atmosphere or the upper atmosphere, with a peak of sensitivity at some region in-between. This shape is due to two opposing trends. Firstly, the exponential drop in atmospheric pressure with altitude means that the total amount of radiation emitted drops off exponentially as you get higher into the atmosphere. Radiation from the lower part of the atmosphere, on the other hand, has the entire column of atmosphere to pass through, so if the atmosphere is sufficiently opaque, much of the radiation from low in the atmosphere will be absorbed before it reaches the instrument. Between these two extremes there will be a level of the atmosphere where most of the measured radiation originates from and so will dominate the signal. This level corresponds to the peak in the sensitivity function that we call the *weighting function*.

Weighting function

These weighting functions are actually expressed through the kernel, **K**, since it is this matrix that describes the sensitivity of the observations to the different layers in the atmosphere. Figure 7.3 shows an example of some weighting functions for the first eight spectral channels of the MLS. The peak of the weighting function indicates the region of the atmosphere where most of the measured signal is originating, and its shape dictates the vertical resolution of the instrument. Measurement frequencies and/or viewing geometry can be chosen so as to obtain a set of weighting functions that peak at different altitudes. For a typical atmosphere, **K** is a function of the absorption coefficient and therefore depends on the atmospheric temperature, pressure and composition. The curves shown in Figure 7.3 are therefore for a specific atmospheric profile.

The assumption that the atmosphere is "sufficiently opaque" is usually the case because any atmospheric sounding instrument will be, by definition, using a part of the EM spectrum that does not have too high a transmissivity. Choosing an atmospheric window would mean that there would be no atmospheric signal to measure, whereas choosing an atmospheric wall would mean that the measurements would only be of the top of the atmosphere. Choosing a variety of frequencies, each with a different atmospheric transmissivity will therefore result in a collection of weighting functions with peaks in the vertical sensitivity pattern at differing heights in the atmosphere. An appropriate choice of frequencies will therefore give sufficient information to retrieve a profile measurement, rather than simply a columnar total.

Note that the shape of the influence functions also indicates that it becomes increasingly difficult to measure deeper in the atmosphere because the Earth's surface provides an ever increasing background signal. Making detailed measurements into the troposphere therefore requires a good knowledge of the background signal (which in practice could be estimated using simultaneous measurements in a different channel).

7.4   Viewing Geometries

7.4.1   Nadir Sounding

A vertical nadir-viewing geometry can be used effectively to make regular global measurements of the atmosphere. For the nadir case the instrument looks directly (or nearly directly) towards the nadir point below the satellite (as suggested in Figure 7.2). The background term is therefore the underlying surface of the Earth, including both the emitted radiation from the surface, T_{SURF}, as well as the downwelling radiation emitted by the atmosphere which has been scattered by the surface, T_{SC}. The forward model for nadir sounding can therefore be written as

$$T_{TOA} = \Upsilon \left(T_{SURF} + T_{SC} \right) + T_{UP}. \qquad (7.9)$$

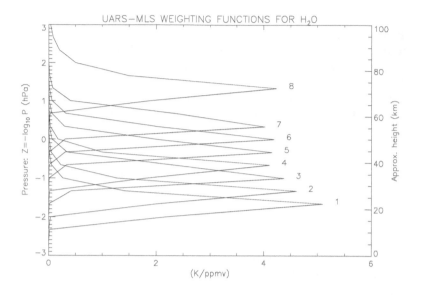

FIGURE 7.3 The weighting functions for the first 8 measurement bands of the Microwave Limb Sounder on UARS. Note that for each frequency the peak sensitivity is at a different layer in the atmosphere. The limb viewing geometry helps to keep the narrow shape.

The scattered brightness temperature, T_{SC}, is also a function of the downward emitted radiation from the atmosphere, T_D, modified by the surface reflectivity, Γ, which can be rewritten as $(1 - \epsilon)$, so that the term T_{SC} can be expressed as

$$
\begin{aligned}
T_{SC} &= \Gamma T_D \\
&= (1 - \epsilon)\, T_D.
\end{aligned}
$$

Similarly, we can rewrite T_{SURF} to in terms of the surface emissivity times the physical temperature of the surface,

$$T_{SURF} = \epsilon T_s,$$

where we are maintaining the convention of upper case subscripts meaning that T is a brightness temperature, and lower case subscripts to mean that it is a physical temperature.

Note that ϵ will actually be polarisation dependent for most systems if the viewing angle is off-nadir (i.e., when θ not equal to zero). The final expression for a near-nadir-viewing geometry is then:

$$T_{TOA}(v, \theta) = \Upsilon\left[\epsilon T_s + (1 - \epsilon)\, T_D\right] + \frac{1}{\cos\theta}\int_0^{\text{TOA}} \kappa_\nu(z) T(z) e^{-\tau/\cos\theta} dz.$$

$$(7.10)$$

Compare this expression with Equation (5.5), where the term $T_B(0)$ is now a term that includes the emissivity and temperature of the Earth's surface — for any given measurement this may be land, sea, ice or cloud. It is therefore important that a good estimate of this parameter can be made before inferring the atmospheric properties. How much influence the surface will have is determined by the weighting function, so that the higher the peak of the weighting function the less influence we might expect from the ground. It is for this reason that sounding gets progressively more difficult the lower you go into the atmosphere. One solution is to purposely make measurements at frequencies of very high transmissivity so that the measurement is dominated by the surface brightness. Models of emissivity across different frequencies are then used to predict the background signal at the other observed frequencies.

Temperature Sounding

For nadir looking instruments the majority of the measured signal originates from a localised layer in the atmosphere such that the thickness of the weighting functions are about 5-10km for most nadir viewing systems.

Nadir sounding makes optimum use of the IFOV to achieve good horizontal spatial resolution, but has a poor vertical resolution because the weighting functions are so broad. Nadir sounding is also not very effective at measuring middle-atmosphere properties because the background signal always dominates — i.e. it is difficult to find a frequency that gives a weighting function that peaks in the middle atmosphere.

For instance, the ERS satellites carried a Microwave Sounder, a nadir-viewing passive radiometer with two channels at 23.8 and 36.5 GHz. It provided measurements of the total water content of the atmosphere within a 20km footprint. The main purpose of this microwave radiometer was the measurement of the tropospheric path delay for the altimeter through the measurements of the atmospheric integrated water vapour content and the estimate of the attenuation of the altimetric signal by the liquid water content of the clouds.

7.4.2   Limb Sounding

An alternative viewing geometry that addresses some of the limitation of nadir sounding is to view the edge or *limb*[71] of the atmosphere and is known as *limb sounding*. This technique measures atmospheric thermal emission spectra as the instrument field of view (FOV) is scanned down through the atmospheric limb. The viewing geometry is illustrated in Figure 7.4.

This technique offers several advantages over downward-looking systems for study of the upper and middle atmosphere.

Although it has a poorer horizontal resolution compared to nadir sounding (hundreds of kilometres rather than tens) it has a number of advantages that make it particularly appropriate for middle and upper atmosphere measurements. The first advantage is that measurements of atmospheric radiation are made with cold space as a background rather than the relatively bright (and fluctuating) signal from the Earth's surface. For limb sounding,

---

[71]   The word "limb" comes from the Latin *limbus* meaning border.

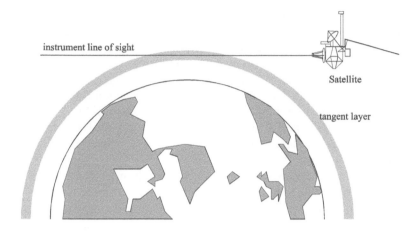

FIGURE 7.4 This figure shows the configuration for limb sounding, whereby the instrument line of sight goes through a tangent layer of the atmosphere.

therefore, the forward model can be written as

$$T_{\text{TOA}} = \Upsilon_{\text{limb}} T_{3\text{K}} + T_{\text{UP}}. \tag{7.11}$$

The background signal, $T_{3\text{K}}$, is deep space, but as we saw in Chapter 2, deep space acts like a blackbody with a temperature of 3K. This is a small signal and is easily characterised, so that the low emission from the middle and upper atmosphere can more easily be observed (in the nadir viewing case the deeper atmosphere will tend to dominate the signal).

The T_{UP} term is now integrated over a long horizontal path length at the tangent layer. This is another advantage for measuring the high atmospheric layers where density is low and emission weak. This increases the ability of a microwave instrument to measure many atmospheric trace gases. Note also that the total transmissivity term, Υ_{limb}, is for the tangent path through the atmosphere, not the vertical transmissivity.

The final advantage is that no part of the signal received at the instrument originates below the tangent point (assuming an infinitesimally narrow field of view), so that the weighting function is effectively "cut-off" at the lower tangent point of the field of view. This allows vertical resolutions on the order of about 1.5km, a significant improvement on nadir sounding.

Limb sounding techniques from satellites have been used successfully at infrared wavelengths and the MLS is one of three limb-viewing instruments that flew onboard The Upper Atmosphere Research Satellite (UARS) from the Space Shuttle Discovery in 1991. UARS is of interest because it carried the first microwave sounding instruments ever to have flown into space — the Microwave Limb Sounder (MLS). The MLS was one of the ten instruments carried by UARS and was dedicated to measuring vertical distributions of ozone (O_3), chlorine monoxide (ClO) and

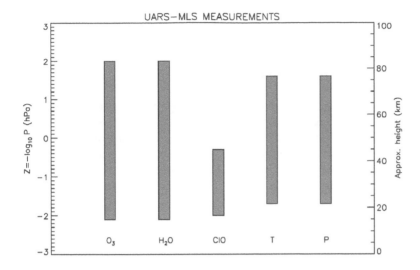

FIGURE 7.5   As an example of the coverage of limb sounding instruments, this diagram
illustrates the measurement regions of the Microwave Limb Sounder.

water vapour (H_2O). Figure 7.5 shows the vertical measurement range.

The instrument is described in detail by Barath et al. (1993). It used
three ambient-temperature heterodyne radiometers operating near 63 GHz,
183 GHz and 205 GHz to determine levels of stratospheric ClO, O_3 and
H_2O. The 63GHz radiometer measured a molecular oxygen line to determine atmospheric pressure and temperature at the tangent point of the observation path to provide a vertical reference for the other measurements.

The MLS measurements of H_2O and O_3 were to higher altitudes than
previously explored on a global basis and provided new information on
chemistry in the mesosphere as well as providing measurements of stratospheric ozone distribution. The MLS provided a number of key measurements that continue to improve our understanding of the complex cou-
plings in the atmosphere, and so improve our ability to predict climate
variations. The follow-on to the UARS MLS was the EOS MLS, an improved version of the UARS MLS that was launched on the Aura satellite
in 2004. Improvements in microwave technology mean that the radiometers can now measure at 118, 240 and 640 GHz, with a new Tera Hertz
radiometer at 2.5 THz, allowing new stratospheric measurements to be
made, as well as more measurements further down into the troposphere.
A further adaptation of the EOS MLS is that the limb measurements are
made in the orbital plane (i.e. in the direction along the flight path) rather
than perpendicular to the flight path so that sequential measurements can
be combined to improve the retrieval process.

7.5 Passive Rainfall Mapping

7.5.1 The Need for Measurements

Precipitation is a key component of the global hydrological cycle and knowledge of regional rainfall is essential to improve weather and climate predictions. Rainfall is important for heat transfer between atmosphere and the surface, especially for ocean-atmosphere interaction. It is estimated that almost three quarters of the energy that drives weather comes from the two heat transfer processes associated with rain: evaporation and condensation.

Satellite measurements are invaluable because they over the spatial and temporal coverage required to monitor the variable and discontinuous patterns of rainfall over space and time. Accurate global coverage of rainfall is not possible from ground based meteorological stations since rain gauge data are available primarily on land and mainly in densely populated areas, with little offshore information. This should be considered within the context of the fact that 78% of all precipitation falls on the ocean compared to 22% on land.

At a regional level, rainfall mapping is vital for drought or flood warning, as well as being an important input to hydrological models, and crop yield prediction. Long term, the threat of changes in global precipitation patterns and the resulting impact on water resources is likely to become the most important environmental issue on the planet.

7.5.2 Principles of Measurement

Visible and IR estimates of precipitation are based on observing the properties of cloud tops and correlating these with ground-based rainfall measurements. The advantage of microwave systems is that they are able to measure deeper into the cloud to measure properties of the precipitation directly. It is the microwave region below about 100GHz that is best placed to measure rainfall since at these frequencies that atmosphere is almost transparent, but the hydrometeors are not. The disadvantage is that patterns of precipitation vary greatly depending on cloud type and the nature of the precipitation event. Additionally, the poor spatial resolution of passive microwave sensors means that measurements tend to be averages of rain and rain-free areas within the field of view.

At high frequencies greater than 50GHz propagation of microwave radiation through the atmosphere is affected by scattering from ice particles, whereas at lower frequencies it is the thermal emission from liquid water droplets that dominates the atmospheric effects. The emission and scattering characteristics depends on a number of properties of the cloud particles, such as size, shape and vertical distribution. It is this distinction

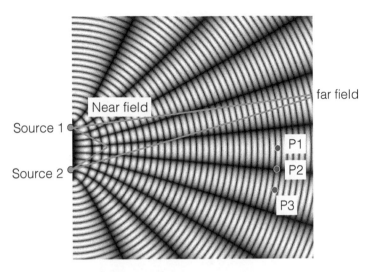

COLOR FIGURE 3.8 If two nearby coherent sources generate the same wave signal, an interference pattern is generated. With electromagnetic radiation, this pattern exists in the space in front of the sources rather than as something you actually see. If a sensor is placed at P_1 or P_3, a signal maximum is detected, whereas at a P_2 signal minimum is detected.

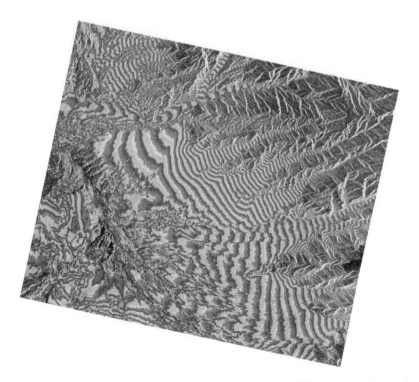

COLOR FIGURE 11.7 A colour interferogram using hue (phase difference), saturation (coherence magnitude) and intensity, instead of RGB. In this way, the hue cycle corresponds to the cyclic phase angle and the saturation represents the "meaningfulness" of the phase, such that the phase angle becomes less visible when the saturation gets low (since the phase will be noisy).

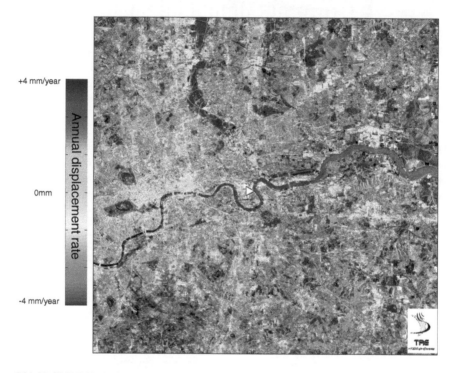

COLOR FIGURE 11.13 This remarkable image shows the results of PSInSAR for central London over a period of 1995–2000. Note the red line to the left of the image, just below the river. This is subsidence due to the tunnelling of a new extension to the London underground. (Image is courtesy of TRE. ESA-ERS data were used for PS processing).

across the range of microwave frequencies that determines the two methods used to estimate cloud structure and rainfall rates.

7.5.3 Emission Method

The first method operates over oceans at frequencies below 50 GHz. At these frequencies the emission from the ocean surface is relatively low ($\lesssim 0.5$) so that the emission from precipitating clouds can be a significant contribution to the measured signal at the top of the atmosphere (TOA). The contrast in brightness temperature between the ocean signal and the more emissive precipitating regions can be as much as 50K or more. The apparent brightness temperature will also tend to increase with increasing rainfall rate (assuming the underlying sea surface remains the same) but increasing cloud thickness will also results in increased brightness temperature. This method has limited use over land surfaces due to the high emissivity (typically around 0.85–0.95) meaning that there is not sufficient contrast to measure rain rates effectively.

At high rain rates (e.g. at about 10-20 mmhr^{-1} for 19.35 GHz (Ulaby et al, 1986)) the sensitivity to rainfall rate decreases until their is no sensitivity at all, which corresponds to a particular optical thickness (which is dependent upon both rain rate and cloud thickness) whereby the rainfall deeper down is obscured by the rainfall at higher altitude. In some cases, particularly at the higher frequencies, the brightness temperature even drops due to scattering. It is this scattering that forms the basis of the second method.

The Special Sensor Microwave Imager (SSM/I) instrument being flown   *Water Vapour*
as part of the Defense Meteorological Space Program (DMSP) from 1987 is able to estimate rain rate. The SSMI water vapour estimate is generated from a combination of the following channels: 19 GHz vertical polarisation, 22 GHz vertical polarisation and 37 GHz vertical polarisation. The output is in units of millimetres (mm) of total perceptible water (TPW) in the column observed by the satellite. The resolution of this product is 40×60 km. The accuracy has been found to be about 3 mm, when compared to sounding derived TPW. The SSM/I TPW product is useful detecting areas of the atmosphere that are either high or low in moisture content. The TPW product output is only produced over water surfaces. Over land a surface moisture product is produced although is too crude for most applications.

7.5.4 Scattering Method

For frequencies higher than 50GHz it is no longer emission that is dominant but scattering from the millimetre-sized ice particles that form above the rain layer. Upwelling radiation from the Earth's surface is therefore scattered out of the field of view of a satellite instrument resulting in an observed decrease in TOA brightness temperature. This is not as direct a technique as the emission method since it measures an associated property of the rainfall, rather than the rainfall itself. But it does have the advantage of being applicable over land surfaces as well as the ocean.

These ice particles at the top of rain-bearing clouds are efficient scatterers but with very little absorption or emission. It is therefore quite straightforward to characterise the TOA brightness temperature as a function of the background surface radiation, T_{BG} (where $T_{BG} = \epsilon T_s + (1 - \epsilon) T_D$) and the opacity of the ice layer, τ_{ice}, such that

$$T_B(v, \theta) = T_{BG} e^{-\tau_{ice}/\cos \theta}. \tag{7.12}$$

The opacity is found by integrating the total scattering cross-sections, Q_s, over the size density distribution, $N(r)$, for all possible radii, r, across a cloud of height h, such that

$$\tau_{ice} = h \int_0^{\infty} Q_s N(r) dr. \tag{7.13}$$

Identifying the presence of scattering due to precipitation can then be done using a scattering index that quantifies the difference between the T_B and T_{BG} for a given frequency. For example, for the SSM/I vertically polarised 85 GHz channel, the scattering index $SI(85V)$ is given by

$$SI(85V) = T_{BG} - T_B(85V). \tag{7.14}$$

Values of $SI(85V)$ greater than 10K would signal the presence of precipitation. In order to determine this for a particular measurement of $T_B(85V)$, however, requires an estimate of T_{BG}. Ideally the instantaneous T_{BG} would be known, since the emissivity of the surface can change rapidly (with soil moisture variations, for instance). One method requires generating a database of cloud free T_B measurements for a number of different frequencies including low frequency measurements that are not effected by scattering. In the case of SSM/I this would be the 19 and 22GHz channels that provide information on the non-scattering component. The brightness temperature is not the same across all these frequencies, but if measurements are made simultaneously at all frequencies the database can be used to estimate a value for T_{BG} at 85GHz given measurements at 19 and 22 GHz. The database over land would be generated to cover a wide range of vegetation cover types, soil moisture variations and surface temperature, while over oceans it would cover different ocean surface temperatures, column water vapour and cloud liquid water content.

7.6   Further Reading

The classic publication summarising retrieval theory is Rodgers (1976), with Twomey (1977) being perhaps a clearer introduction to the main elements of the theory. Harder to find, but worth the effort, is Houghton and Rodgers (1984), which has an excellent discussion on atmospheric retrieval theory. For a general overview of all microwave techniques for measuring the atmosphere, Janssen (1993) is highly recommended, which includes an interesting chapter on measuring the properties of atmospheres on other planets.

8
PASSIVE IMAGING

Sitting quietly, doing nothing,
Spring comes, and the grass grows by itself.
— a Zenrin poem[72].

The microwave region of the EM spectrum contains both atmospheric walls and atmospheric windows. In this chapter we consider in more detail the application of passive measurements in the microwave window region to measure properties of the Earth's surface.

Passive microwave sensing offers a unique view of the Earth's surface since the measurements are sensitive to changes in temperature, roughness, salinity and moisture content. The latter three of these are of particular interest because they are difficult to measure with sensors utilising other parts of the EM spectrum. Active radar sensors are also sensitive to such parameters, but active instruments are typically much larger, require larger antennas, use an order of magnitude more power, and produce six orders of magnitude more data. Satellite passive microwave sensors are therefore good value for money since size, mass and power are the expensive factors when launching a satellite into orbit. The minimal impact of cloud cover and aerosols on the middle microwave region also affords it an extra advantage over thermal IR systems (its nearest neighbour for measuring properties such as sea surface temperature).

The challenge for land applications, though, is that passive microwave sensors have poor spatial resolution when compared to other platforms looking at the surface of the Earth, since the wavelengths are large compared to the size of the antenna. Footprints of the order of tens of kilometres are typical for spaceborne systems so that they are limited to regional to global scale applications. Localised studies can be done from airborne platforms, but these suffer from the usual constraints of limited coverage.

---

72   Quoted in Alan W. Watts' "The Way of Zen".

In this chapter, because the use of microwave imagers divides neatly into three categories — oceans, sea ice and land surfaces — the generic measurement principles are described first, followed by a description of the need for measurements and the details of the methodologies used for each topic.

8.1   Principles of Measurement

8.1.1   Background

The terrestrial blackbody curve actually peaks in the infrared, not the microwave region, but there is still sufficient emitted energy in the microwave region to allow useful measurements to be made from aircraft and satellites. Across the oceans, the emissivity is fairly constant, so that variations in the brightness temperature of the ocean surface corresponds to changes in the physical temperature of the water. Over land and ice, however, it is differences in emissivity that dominate the horizontal variation in brightness temperature and similarly, the dynamics of the surface over time is governed by changes in emissivity. In both cases, the goal is an accurate estimate of the surface brightness temperature from which the physical attributes of the surface can be estimated using a method appropriate for the application at hand.

Frequencies less than about 40GHz are considered appropriate for measuring properties of the surface since the attenuation of the atmosphere is still low. Clouds and rainfall can add a contribution to the measured signal down to about 5GHz, but as long as these can be identified within the data, then they can be accounted for (or indeed, used to measure properties of the clouds as described in Section 5). What cannot be accounted for is the large instantaneous field of view from the microwave antenna — with an antenna in the region of 1m across, the instantaneous footprint on the ground is still more than 10km for a centre frequency of about 85GHz. However, the unique data afforded by passive microwaves and the fact that most meteorological and climate models work on units much larger than 10km, means that such measurements are still extremely valuable when acquired globally and on a regular basis. The fact that they are not very power hungry and are relatively lightweight, also helps their case.

8.1.2   Practical Radiometers

Practical microwave radiometers are required to be very sensitive in order to detect a sufficient amount of the weak emission from the surface. There are two approaches to making sure a radiometer measures sufficient energy to detect. The first is to have a long integration time, meaning that the time spent detecting radiation from a specific direction is maximised giving a better SNR. For instruments on aircraft or satellites, however, the flight speed (in the order of a few kms^{-1}) places a strict limit on available dwell time. The alternative approach is to make measurements over a very large beamwidth. The larger the area the higher the signal compared to the internally generated instrument noise. Fortunately this should not be a problem since we are already limited by the fact that microwave radiometers will tend to have a relatively poor angular resolution anyway.

In terms of instrumentation, the SSM/I instrument provides an appropriate example. It used an offset reflecting antenna with dimensions of 61×66 cm, which focused the microwave radiation onto the receivers. The radiometer was a total-power type, using deep space and a warm reference load as calibration targets every 1.9s, giving an estimated absolute accuracy of less than ± 3K.

8.1.3   Viewing Geometries

The geometry of a radiometer is governed by two major requirements: maximum coverage based on a wide swath, and consistency of incidence angle across the measurements. A simple linear, cross-track scanning approach can be applied when the sensitivity of the emission to viewing angle can be accommodated for by an appropriate model, say, of the ocean emission and scattering. However, polarimetric measurements are more effectively exploited when a constant incidence angle is maintained. This is achieved by using a conical scanning system — in this case, the scan pattern approximates a circle on the ground, giving a wide swath, but with constant incidence angle over all observed locations across the swath.

A further reason for desiring a large incidence angle (as opposed to a near-nadir scanner) is that polarimetric information becomes more meaningful. Near nadir there is no difference in the horizontal and vertical emissions for all surface types (other than for oriented surfaces such as ocean surfaces), whereas at increasing incidence angles the ratio of one polarisation to the other changes quite dramatically depending upon the surface properties (in particular, the polarimetric contrast reduces as the surface gets rougher).

8.1.4   The Generic Forward Model

In this discussion we generalise the actual nature of the surface, so that the same description applies for ocean, sea ice and land surfaces. In each case we characterise the properties of the target through its brightness temperature, T_B, which we relate to geophysical variables depending on the context of the measurement. Over ocean surfaces we assume that the emissivity remains constant so that variations in T_B are due to changes in the physical temperature of the ocean surface. For land and sea ice, the largest influence on the T_B will not be the physical temperature, but differences in emissivity, which in turn is related to surface roughness or variations in the dielectric constant due to changes in liquid water content or salinity.

The measurement made by a satellite radiometer is the top of the atmosphere (TOA) brightness, which is both directly and indirectly related to the brightness temperature, T_B, of the surface of the Earth. The direct component is the radiation emitted by the surface — this is the signal we would really like to determine as it is this quantity that links our models of surface emissivity to the physical properties of the surface. The indirect signal arises from microwaves originating from elsewhere but are scattered by the surface into the field of view of the radiometer. In addition, there may also be a contribution from the atmosphere above the surface, which has a physical temperature and so emits microwave radiation (and the shorter the wavelength of the microwaves being measured, the more signal the atmosphere will contribute). A significant contribution in the TOA signal might actually originates from atmospheric emission that is reflected in the ocean surface.

With reference to Figure 8.1(a) we can then formulate a description of the forward model for passive sensors as follows:

$$T_{\text{TOA}} = \Upsilon \left(T_{\text{SURF}} + T_{\text{SC}} \right) + T_{\text{UP}}. \tag{8.1}$$

where T_{UP} is the upwelling radiation from the atmosphere, T_{SC} is the result of downwelling radiation that is scattered by the surface and Υ is the transmissivity of the entire atmospheric path. Since most passive imagers make measurements off-nadir, it is helpful to use the expression for Υ based on vertical height, z, rather than path length, s, and the vertical opacity, τ, such that

$$\Upsilon = e^{-\tau / \cos \theta}. \tag{8.2}$$

T_{SURF} is the surface brightness temperature, the parameter that we want to estimate. Whereas in the previous chapter the goal was to analyse T_{UP}, the term of interest in passive imaging is the one in brackets (which was also earlier described as the "background term", T_{BG}). Ideal conditions for viewing the surface are therefore at relatively low frequencies where the atmospheric transmissivity is very high, which is typically only the case well below about 20 GHz.

T_{SC} is the scattered brightness temperature from the surface. This contribution comes from the downward-emitted radiation from the atmosphere, T_{D}, that is scattered towards the radiometer and can be expressed as

$$
\begin{aligned}
T_{\text{SC}} &= \Gamma T_{\text{D}} \\
&= (1 - \epsilon) \, T_{\text{D}}.
\end{aligned}
\tag{8.3}
$$

We have used Kirchoff's law to express this in terms of the emissivity, ϵ, so as to have an expression that can be related to the emission from the surface — as the emissivity of the surface increases the surface scatters less of the radiation emitted from the sky.

Unlike the passive sounding case in the previous chapter, the downwelling radiation at frequencies chosen for passive imaging need not always be dominated by atmospheric emission. As measurement frequencies drop below about 40 GHz the atmosphere becomes increasingly transparent (i.e. neither absorbing nor emitting) so that reflection of the cosmic background radiation (3K) may begin to be more important, for instance, and reflected solar radiation can also be a problem. Observation of direct specular reflection of the solar radiation from the surface is referred to as "sun glint" and is usually avoided through careful choice of orbits and viewing angles. The low frequency case is shown in Figure 8.1(b).

A simple forward model for passive imaging of the Earth's surface is therefore given by

$$T_{\text{TOA}} = \Upsilon \left(\epsilon T_{\text{s}} + (1 - \epsilon) T_{3\text{K}} \right) + T_{\text{UP}}, \qquad (8.4)$$

or when measurements frequencies are lower than about 5GHz it can be written:

$$T_{\text{TOA}} \approx \epsilon T_{\text{s}} + (1 - \epsilon) T_{3\text{K}}. \qquad (8.5)$$

In the general case, therefore, the microwave brightness temperature at the top of the atmosphere, T_{TOA}, of any planet can be dependent on four independent factors: the surface temperature and surface emissivity, the sky temperature and emissivity, and the surface roughness. The importance of the surface contribution is that at angles of incidence greater than zero it is also a function of the polarisation. If we can choose frequencies that minimise the sky contribution, we should then be able to derive the other three parameters given three independent measurements (i.e., at different polarisations or frequencies).

8.2 Oceans

8.2.1 The Need for Measurements

The oceans are a vital component of the Earth's climate system because of water's ability to store thermal energy. Its high heat capacity means that water heated by solar insolation at the tropics will retain much of its thermal energy while it is slowly transported around the globe. This transport of heat is primarily driven by the thermohaline[73] circulation, which ultimately distributes heat energy from the tropical Pacific, through the Indian ocean and up to the North Atlantic. The surface waters of the Pacific are both less salty and warmer than elsewhere and so they are more buoyant. In contrast, when this water ultimately reaches the North Atlantic it loses fresh water through evaporation (even more than it has already) and cools, resulting in more saline, denser water that sinks. This sinking draws warmer water from lower latitudes and so helps to drive the global circulation of the oceans, as well as providing the regional effect of keeping Northern Europe significantly milder than it would otherwise be without this extra input of thermal energy to the climate.

---

[73] From the Greek words, *thermos* meaning hot, and *hals* meaning salt.

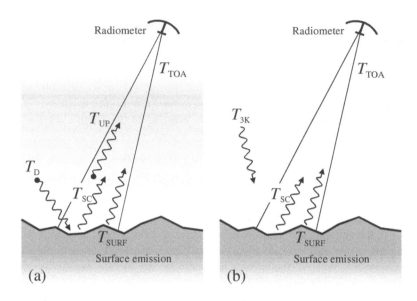

FIGURE 8.1 These diagrams summarise the two cases of the forward model for passive imaging of the Earth's surface. In (a), for frequencies $\gtrsim 40$ HGz, the atmosphere still contributes a strong direct (T_{UP}) and scattered (T_{SCAT}) component, whereas in (b), for $\lesssim 20$GHz, the atmosphere contributes very little and the scattered term originates from the cosmic background radiation.

The "shutting down" of the Atlantic conveyor, the name given to the process in the Northern Atlantic, is identified as one of the potential climate surprises that could occur as a result of global warming escalating Greenland ice melt rates and so increasing the freshwater input to the North Atlantic. Salinity patterns on a global scale are therefore a key input to debates on ocean dynamics and climate change.

Sea surface temperature (SST) is also important in the dynamic heat exchange between ocean and atmosphere, as well as the associated transport of water. Near surface winds drive water, inducing surface waves and producing near surface transport of water in the form of wind-driven currents. The surface temperature is also important in the processes that exchange water and heat — evaporation removes fresh water from the surface, but also makes available latent heat of evaporation, which is often a key part of the energy balance, especially in the tropics. Conversely, rainfall transports water back to the oceans.

The temperature of the ocean surface also influences the amount of radiative energy lost by the oceans, and combined with the wind speed it influences the rate of uptake of chemicals such as atmospheric CO_2. A significant influence on the global sea level is the thermal expansion of the oceans, which in the short term is likely to be the biggest impact on sea level as a result of global warming.

Frequent (daily or better) sea surface temperature measurements and surface winds are therefore a fundamental contribution to our understanding of the dynamic processes that depend upon the energy exchange between the oceans and the atmosphere. Since the oceans cover 70% of the Earth's surface, in situ measurements of SST (from moored buoys and ships that measure incoming water for the engine) are sparsely distributed around the globe, and measurements tend to be clustered around common trade routes or easily accessible locations. Remote sensing from satellite therefore offers an unparalleled view of the world's oceans that is vital for global weather predictions and climate change assessment.

8.2.2   Principles of Measurement: SST

The basic principle of ocean measurements is that the surface term, T_{SURF}, in (8.1) is a function of two of the three important properties that govern the physical attributes of the oceans: namely the temperature and the salinity, the density being the third property (but this does not influence the brightness temperature). In Chapter 5 we saw that temperature is positively correlated with emissivity across the entire microwave region of the spectrum, whereas salinity influences the emissivity significantly only at long microwave wavelengths (<6 GHz). Brightness temperature is directly related to the physical temperature times the emissivity. Additionally, since the atmospheric attenuation is low for microwaves, and hydrometeors have a decreasing influence for lower frequencies, then measurements can be acquired even when there is cloud cover.

For an idealised ocean, it should therefore be possible to infer an SST and/or salinity from microwave brightness measurements alone. However, there are a number of practical considerations that limits the effectiveness of passive microwaves measurements for ocean measurements. For instance, an idealised ocean would be free of surface debris, including sea ice and foam, and be free of the roughening effects of the wind. For most of the world's oceans we can safely assume the lack of debris, but not the lack of foam or wind. Of course, that roughening effect of the surface winds also means that microwave measurements can be used to infer wind speed, assuming we can account for the other factor. (The polar oceans and the effects of sea ice are dealt with later in this chapter).

The main limitation is therefore the non-linear effects of variations in surface conditions, so that direct estimation of SST from passive microwaves is not straightforward and not practical for the kinds of spatial and radiometric resolutions that are now required. Long wavelengths require larger apertures, so antennas in the region of about a metre are not uncommon (the diameter of the SSM/I antenna is about 0.6m) but still only achieve footprints in the region of tens of kilometres. There are two important consequences of the large footprint — firstly, it means that you lose detail in the patterns of SST. Secondly, it limits how close your measurements can be to the coastline. This is especially the case for microwave systems because, as we saw in Chapter 6, there are potentially large side-lobes in the antenna sensitivity pattern. Not only might these sidelobes observe the "bright" land surface at the coast, but they are also suscepti-ble to picking up stray noise from other microwave sources (your mobile phone, for instance, works in the L-band region). Although there are strict controls on which frequency bands are used for what purpose, even small stray signals can disrupt the sensitive SST estimations.

On balance, atmospherically corrected thermal IR measurements are the method of choice for regular SST measurements, despite the reduced coverage due to cloud cover. Thermal IR will get accuracies below 1K by using various techniques to take account of the intervening atmosphere. In the 10–12μm wavelength range that is usually used, clouds and water vapour are the greatest unwelcome impact on the measurements. But this region is well suited for SST as it is near the peak of the Planck function for the ocean at typical temperatures and since the emissivity for these wavelengths is almost unity, there is very little sky-scattered contribution.

Despite the limitations, a number of passive microwave instruments have been used to estimate SST, including the short-lived Seasat satel-lite and the SMMR (Scanning Multichannel Microwave Radiometer). The Microwave Imager (TMI) on the Tropical Rainfall Measuring Mission (TRMM) and the Advanced Microwave Scanning Radiometer (AMSR) in-strument on the NASA EOS Aqua satellite are two other recent examples.

8.2.3   Principles of Measurement: Ocean Salinity

SMOS

Whereas thermal IR is probably better suited to SST measurements, mi-crowaves are the only means of estimating ocean salinity. A number of airborne campaigns have demonstrated this application, but the first ded-icated spaceborne sensor will be the Soil Moisture and Ocean Salinity (SMOS) mission. This instrument has been specially designed to mea-sure these changes utilising the difference in dielectric constant that re-sults from changes in salinity. It uses a frequency of 1.4 GHz (L-band) and so is not substantially influenced by atmospheric attenuation. Such systems present a particular technical challenge because the long wave-lengths require a large antenna to provide useful spatial resolutions — it would require a circular antenna more than 10m across to provide a spa-tial resolution of 10km from a 800km orbit, which would not be practical or at least cost-effective. SMOS, however, will use a novel interferomet-ric technique to achieve spatial resolutions better than 50km by using three smaller (<5m long) rectangular antennas together (see Chapter 11 for fur-ther discussion on this topic).

The challenge of this instrument is that both SST and salinity both change the emissivity, so that it is necessary to know one to estimate the other. For SST measurements, frequencies can be chosen in the range be-tween 5–10 GHz — the lower boundary avoids the influence of variations in salinity and the upper boundary minimises the impact of clouds. For salinity, lower frequencies must be used but will be used in conjunction with high frequency measurements to constrain the solution based on SST estimates.

8.2.4 Principles of Measurement: Ocean Winds

One of the primary influences on the brightness temperature of the oceans is the surface roughness caused by the near surface winds. While this is an important limiting factor when trying to infer SST, it can clearly also be used as a means of estimating the surface wind speed. Since the wind roughening is on a scale of millimetres (in terms of the amplitude of the capillary waves), it is the higher frequency microwaves that are useful for this application. Frequencies below about 5 GHz are not very sensitive to wind-roughening.

In addition to roughening the surface, high wind speeds will also generate surface foam which can complicate the retrieval process.

The SSMI surface wind speed is generated from a combination of the   *Surface Wind Speed* following channels: 19 GHz vertical polarisation, 22 GHz vertical polarisation, and the 37 GHz vertical and horizontal polarisations. These channels detect ocean surface roughness, which has been related to surface wind speed. The output is usually in meters per second, but be aware that American agencies may still quote wind speeds in miles per hour! The wind speed is calibrated to relate to the wind speed at 10 meters above the ocean surface. The accuracy of the wind speed is about ± 5 mph and the spatial resolution is in the order of 30 km. The data gives only wind speed and not direction of motion but such data is a valuable complement to the ship and buoy reports that are available, which supply wind direction.

8.3 Sea Ice

8.3.1 The Need for Measurements

The mapping of sea ice coverage on a regular (weekly) basis is probably the major success story of passive microwave sensing from satellites. Sea ice forms during polar winters so that the lack of solar illumination and frequent cloud cover dramatically limits the use of optical sensors in comparison to microwave imagers. The development of spaceborne microwave imaging of sea ice was initially driven by military requirements during the Cold War. The strategic importance of the Arctic and the need for up-to-date information on ice conditions for naval vessels (particularly submarines carrying ballistic missiles) resulted in the Special Sensor Microwave Imager (SSM/I) instrument being flown as part of the Defense Meteorological Space Program (DMSP) from 1987 until present. Understanding the formation and patterns of sea ice is also important for merchant shipping and off-shore engineering, whereas in recent times it has become increasingly important from the perspective of studying the dynamics of the Earth's climate. Sea ice formation and melt are important factors in the exchange of heat between the oceans and the atmosphere so that regular information on density of coverage and ice type is extremely important. Even though first year ice will normally be less than 1m thick it is still a powerful insulator that reduces the exchange of energy across the air-sea interface. Frozen water in all forms has a very high albedo in contrast to open water, meaning that changes in ice coverage has a further influence on the Earth's radiation budget. Both these aspects have in turn important influences on the climate of the entire Earth as well as the polar regions in particular.

In a region where *in situ* measurements are difficult and expensive (as well as hazardous), using satellites to provide coverage over the vast Arctic and Antarctic oceans has obvious advantages that more than make up for the poor spatial resolution afforded by the long wavelength microwaves. Even at 50km, the SSM/I offer unsurpassed ice coverage maps on a frequent (daily) basis, in near real-time, every winter. The frequency of measurements is important because sea ice cover is very dynamic and has a very large inter-annual variability (on a par with the variability of seasonal snow cover on the land masses of Northern hemisphere). The limit of the sea ice is therefore a sensitive indicator of climate change.

8.3.2   Sea Ice Concentration

Sea ice has a different emissivity than open ocean, which means they have distinctly different brightness temperatures, even when they have similar physical temperatures. In the very simplest case we consider only the horizontal heterogeneity of the ice distribution, and assume a degree of consistency in the vertical profile over the ice type. We can then model the brightness temperature of a scene that is composed of some fractional cover of sea ice, c_i, at emissivity ϵ_i, surrounded by water with fractional cover, $(1 - c_i)$, at emissivity ϵ_w, as

$$
\begin{aligned}
T_B &= c_i \epsilon_i T_i + (1 - c_i) \epsilon_w T_w \\
&= \epsilon_w T_w + c_i (\epsilon_i T_i - \epsilon_w T_w),
\end{aligned}
\tag{8.6}
$$

where T_i and T_w are the physical temperature of the ice and water respectively. The fractional ice cover is then estimated from

$$
c_i = \frac{T_B - \epsilon_w T_w}{\epsilon_i T_i - \epsilon_w T_w}.
\tag{8.7}
$$

If we can estimate the physical temperature (from meteorological models, for instance) and have a good laboratory-based estimates of the emissivity of ice and water, then we can use brightness temperature to estimate fractional ice cover.

Unfortunately, the variety in dielectric properties of ice as it ages means that we also have to consider the relative proportions of first year and multiyear ice. Since these properties also depend on the frequency and polarisation a single frequency is therefore not sufficient to separate ice types, although two or more frequencies would allow trends to be observed.

A modified forward model for sea ice would therefore consider first year ice *and* multi-year ice separately, as well as the open water, each with cover fractions within the measured scene of c_f, c_m and c_w, respectively (so that $c_f + c_m + c_w = 1$). Each of these in turn may have a physical surface temperature T, and an emissivity ϵ, so that the total brightness temperature, T_B, is given as

$$T_B = c_f \epsilon_f T_f + c_m \epsilon_m T_m + c_w \epsilon_w T_w, \qquad (8.8)$$

where the calculation is made for a given frequency and polarisation.

Note that these expressions assumes that the contribution of the atmosphere has been accounted for or is sufficiently small for the given frequency. For different frequencies and polarisations, the value of T_B may be different even when the coverage conditions are the same (i.e. the same values of c_f, c_m and c_w). It is just such differences that form the basis of measurements aimed at determining percentage ice coverage. For example, the difference between H and V emissions are much smaller for ice than water (since the water is a much smoother boundary). And if we consider two frequencies such as 19 and 37 GHz, the difference between these signals is very low for first year ice, whereas for multiyear ice 19 GHz signal is larger and for open water the 37GHz signal is stronger still.

From Equation (8.8) we can see that differences in the physical temperature of the different surfaces will also influence the absolute measured signal. It is not unlikely that surface temperatures can vary — for instance, open water in a lead within the ice pack may be near freezing at 270K, but the surrounding ice can be much colder. Any assumption of uniform physical temperature will therefore result in a systematic error.

The value of assuming a constant temperature across a resolution cell is so that temperature factors out of (8.8). To accommodate variations in the temperature between resolution elements, the analysis of the data is based on a *normalised* difference (which gives a proportional variation, rather than an absolute difference). As long as the physical temperature of the three surfaces can be considered approximately equal, then the temperature is factored out of equation. The differences are based on either variation between polarisations *or* frequencies, so that the signals are characterised using what is referred to as the *polarisation ratio* (PR) and a *frequency gradient ratio* (GR). For the two frequencies commonly used from the SSM/I instrument, they are defined as:

GR & PR

$$PR = \frac{T_B^{19V} - T_B^{19H}}{T_B^{19V} + T_B^{19H}} \qquad (8.9)$$

$$GR = \frac{T_B^{37V} - T_B^{19V}}{T_B^{37V} + T_B^{19V}}$$

where T_B^{19V} and T_B^{19H} are the brightness temperatures for the verti-

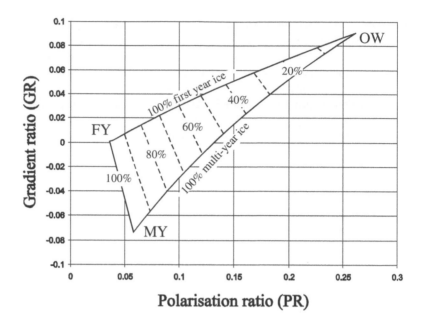

FIGURE 8.2 The trends in PR and GR used to classify sea ice type and concentration. OW is open water. The percentages mark increasing ice coverage. Within the ice coverage, the proportion of first year (FY) to multi-year (MY) is governed by where the data lies within the two 100% proportion lines.

cally and horizontally polarised 19 GHz signal respectively, and T_B^{37V} the brightness temperature for the vertically polarised 37 GHz signal. If we assume these three brightness temperatures are independent measurements this should allow the three ice parameter values in (8.8) to be inferred. If individual measurements are plotted on a PR/GR plane, we find that MY, FY and open water tend to cluster around three different areas, as illustrated in Figure 8.2. Measurements that lie in-between are assumed to be indicative of proportional coverage, so the combined use of the PR and GR can be used to estimate the proportional coverage of the three main cover types.

This approach is relatively effective at a regional scale, but is reliant on the underlying assumptions. The method is therefore prone to error when the surface is not composed of only the discrete classes of open water, first year ice and multiyear ice, or when there are increased variations is surface roughness (for both ice and water) or there are compression ridges or ponding. All of these characteristics can influence the value of the PR and GR.

The influence of the atmosphere is also generally ignored. The presence of heavy cloud cover, high winds or high water vapour can mask areas of sea ice and make them appear to be open water. In such cases, other methods are used to apply a weather filter so that such data can simply be ignored. This filter is normally generated by using other available frequency bands to characterise the response in such a way as to identify when it is atmospheric effects rather than open water than is causing the high GR value.

8.4   Land

8.4.1   The Need for Measurements

While the use of passive microwave sensors to look at the Earth's surface is mostly driven by ocean and ice applications, it is also the case that microwave emission from land surfaces contains information that is indicative of soil and snow properties, and in particular the amount of water they contain. Understanding the exchange of water at the boundary between the land surface and the atmosphere is of crucial importance for weather prediction and global climate models, and for regional hydrology and vegetation studies. Water transport is directly linked to the transport of energy. Solar insolation heats the Earth's surface, and that heat energy is then transferred to the atmosphere either through conduction to the air above (which influences both local and global atmosphere dynamics through convection) or in the form of latent heat of evaporation when the surface contains moisture. Moisture in the soil that is not lost through evaporation will mostly contribute to short timescale hydrology through drainage to another part of the system. Some of it may eventually seep through to deeper aquifers where it will remain for a significantly longer time. In vegetated areas the amount of soil moisture will also influence the rate of uptake by the plants, and the subsequent transpiration from the leaves back into the atmosphere. Soil moisture is therefore a key variable in modelling the climate and the hydrological cycle, and since it varies on a daily basis, it requires constant monitoring.

Snow is also a key factor in the hydrological cycle, the hydrological properties being characterised by a "snow water equivalent" (SWE). This avoids trying to represent the complex structural differences in the snow by simply quantifying how much water the snow contains. The value of SWE is most apparent when considering the potential run-off from a snowpack after it has melted — forecasting snowmelt runoff within a catchment is a fundamental input to flood prediction models.

The percentage of snow cover is important from a regional climate context and for weather prediction. Snow forms an insulating layer between the soils and the atmosphere. This layer also has an extremely high albedo that reflects a far greater proportion of the incident solar radiation than the underlying soil would do were it exposed. The dynamics of global snow cover is therefore a key variable in the Earth's radiation balance.

8.4.2   The Forward Problem Over Land

The key difference here compared to the forward model for land surfaces is that we need to take into account that the surface may be covered by a layer — a layer of snow, or a vegetation canopy layer above a soil surface. Unlike the case of sea ice, the approach for understanding the signal from land surfaces is to concentrate on the vertical complexity, rather than the horizontal. To do this in a manner that does not get too complex, we assume that the layer is a volume emitter/scatterer that lies above a well-defined surface and use the results of Section 3 to characterise the response from the volume.

In the current case, however, we generalise to a surface layer with physical temperature, T_s, and emissivity, ϵ_s, lying beneath a sparse volume scattering layer with a brightness temperature of T_V and vertical opacity of τ_v. When the volume is sparse with low transmissivity, Υ_V, the effects of the volume are threefold. Firstly, it modifies the surface emission term by attenuating the signal by the factor Υ_V. Secondly, it contributes a direct term of its own, given by T_V , as defined in (5.36). And lastly, it contributes a scattered term that corresponds to that part of the emission from the volume that goes downwards and scatters off the surface, and so is given by $T_V\Gamma$. This is summarised in Figure 8.3. For the final expression for the top of the atmosphere, this last expression is also attenuated by the volume so that we have the following model for the brightness temperature of a random volume over a surface:

$$
\begin{aligned}
T_B &= \text{(attenuated surface emission)+(direct volume emission)} \\
&\quad \text{+(scattered volume emission)} && (8.10) \\
&= T_{\mathrm{SURF}}\Upsilon_v + T_V + T_V\Gamma\Upsilon_v && (8.11) \\
&= \epsilon_s T_s \Upsilon_v + T_v(1-a)(1-\Upsilon_v) && (8.12) \\
&\quad + T_v(1-a)(1-\Upsilon_v)\Gamma\Upsilon_v && (8.13) \\
&= \epsilon_s T_s \Upsilon_v + T_v(1-a)(1-\Upsilon_v)(1+\Gamma\Upsilon_v). && (8.14)
\end{aligned}
$$

Although not shown explicitly in this expression, remember that both the surface emissivity, ϵ_s, and the volume transmissivity are dependent upon the zenith angle, θ.

For a zenith angle of θ, the observed brightness temperature is then given by

$$
T_B = \epsilon_v T_s \left[\cos\theta \exp(-\tau_v/\cos\theta)\right] + T_V. \qquad (8.15)
$$

Remember that it is the $\cos\theta$ term that accounts for the difference in path length when the measurements are not taken vertically, and be careful to spot the difference in the above expression between the brightness temperature T_V, and the physical temperature T_v.

For soil moisture, low frequencies (3 GHz and lower) are the most useful. The Soil Moisture and Ocean Salinity (SMOS) mission, for example, will passively sense L-band with the intention of being much less constrained by the vegetation cover than at previously measured frequencies.

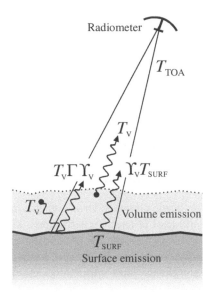

FIGURE 8.3 This diagram summarises the main components of the forward model for passive imaging of the land surface, including a volume layer. The surface layer lies beneath a sparse volume scattering layer with a brightness temperature of T_V and low transmissivity, Υ_V.

8.4.3 Empirical Approaches to Snow Depth

When operating at frequencies above 10 GHz the forward model is increasingly complex since it becomes appropriate to also include a scattered sky term as well as an atmospheric transmission term. While we might generally avoid the high frequency measurements, there are some cases where high frequencies are necessary to make the measurement of value. For mapping snow cover, for instance, longer wavelengths will have little trouble penetrating through deep layers of dry snow so that there is very little difference between the brightness temperature of ground with or without snow cover. 37GHz, for example, is one frequency where the brightness temperature shows a sensitivity to snow thickness and SWE over a uniform snowfield. At these frequencies an increasing snow depth tends to decrease the emission from the underlying surface by scattering the emitted radiation out of the field of view — brightness temperature therefore decreases as snow depth increase.

An alternative to trying to analytically retrieve snow parameters using a forward model is to simply utilise an empirical correlation. One such approach used a difference index to estimate snow depth, h_{snow}, from the (horizontally polarised) 18GHz and 37GHz channels on the SMMR instrument (Gurney et al, 1993). It is given by

$$h_{\text{snow}} = 1.59(T_B^{18H} - T_B^{37H}) \qquad (8.16)$$

where the factor of 1.59 was derived from a linear fit of empirical data. Note that while it is the 37GHz channel that is sensitive to snow depth, the 18GHz effectively provides a reference value (since this term is "snow-insensitive" so it encapsulates the variability that one would expect from the range of other influencing parameters, such as the atmosphere, the temperature of the snow and the ground, and the presence of clouds).

8.4.4   A Final Comment on Passive Polarimetry

Currently the satellite-borne passive microwave systems are designed only to measure the magnitude of H and V polarisations. Microwave polarimeters, on the other hand, can measure the phase difference of the signal so that the complete Stokes vector can be derived. From airborne studies this appears to provide significant extra information, especially when considering the asymmetric responses from ocean surfaces with oriented wave patterns. This is currently a topic of interesting research but it is not considered in detail here.

8.5   Further Reading

The "Atlas of satellite observations related to global change" by Gurney, Foster and Parkinson provides an extensive (if now a little dated) overview of many different types of satellite measurements are their role in monitoring the Earth system. Sections on oceans and the cryosphere are particularly relevant for passive microwave systems and includes examples of global data sets from such instruments. Ulaby, Moore and Fung's Vol. III (1986) provides a plethora of additional information with far more detail than is covered here. And although Janssen (1993) is mainly about atmospheric sensing, it does contain useful discussions on the emission from land surfaces and oceans when considering the unwanted "background signal" for sounding.

9
ACTIVE MICROWAVES

Light travels at 186,000 miles a second.
Any faster would be dangerous.
— B. Kliban, *The Biggest Tongue in Tunisia.*

In Chapter 8 it was the unique information content and independence of cloud cover and solar illumination that meant it was still worthwhile to make passive microwave measurements, despite the limited spatial resolution of satellite sensors due to the long wavelengths. Even at spatial resolutions in the order of kilometres, the data are invaluable for monitoring a number of global-scale dynamic processes.

In this chapter, we again consider the atmospheric window in the microwave region with the intention of measuring surface properties but will utilise quite different measurement techniques to effectively bypass the whole issue of angular resolving power. We do this by generating our own illumination using microwaves generated by the instrument itself and then measuring the properties of the echo after it has been scattered by the target. This is the principle of *active* microwave remote sensing.

Active sensors incorporate an additional factor that can be utilised — the timing of the return pulse to measure distance. Active radar systems can therefore provide two very different types of information: firstly the time delay in the echo which provides distance information, and secondly the properties of the echo such as intensity and polarisation. Instruments designed to focus on making accurate measurements of range (or, more exactly, *alti*tude) using the time delay from nadir-directed echoes are known as *altimeters*, whereas instruments that focus on accurately recording the echo characteristics (such as radar cross-sections) are known as *scatterometers*. Rain radars are included in the latter category. We will see in this chapter that while this distinction is often helpful, it is also the case that all these types of instrument actually utilise both kinds of information in some way or another.

221

Both of these types of instruments are distinguished from imaging radar (which is covered in the next chapter) merely as a arbitrary convenience, as the distinction is not always entirely appropriate. We might argue that neither altimeters nor scatterometers are explicitly imaging systems in that they tend to make sequential measurements on a grid through a relatively wide antenna beam, but such a definition is rather arbitrary, since these instruments can be used to build up a map or an "image" by making a collection of measurements over some area. However, the distinction can be helpful in reinforcing the difference between basic radar devices (and their applications) and the imaging radar systems described in the next chapter.

The concept of sub-beamwidth resolution is also introduced in this chapter. In the passive case, the poor spatial resolution was accepted as trade-off for the unique data, but in active systems it is possible to utilise some clever processing to improve the spatial resolution by a few orders of magnitude. The same basic technique will be considered in more detail in Chapter 10. An alternative approach to sub-beamwidth resolution is further discussed in Chapter 11.

9.1   Principles of Measurement

As with passive imagers, the low attenuation of microwaves makes them well-suited to frequent monitoring of the surface using satellites. Microwaves are usually preferred for active sensing over optical wavelengths because of the minimal impact of cloud cover, although laser-based LIDAR (LIght Detection and Ranging) can also operate from low Earth orbit, as demonstrated by the Lidar In-space Technology Experiment (LITE) that flew on the Space Shuttle in 1994[74]. The low level of ambient microwave radiation is also advantageous, since very low intensity returned echoes must be measurable within the background signal.

That longer microwaves can penetrate most cloud cover and do not rely on solar illumination is a clear advantage for a satellite instrument, but this is not necessarily the reason that microwave systems are flown. Indeed, in this chapter we will also consider rain radar, where frequencies are chosen so that weather *does* influence the signal. No, the reason great expense is invested in such systems is that they are sensitive to variations in parameters that are of *interest*, often measuring parameters that cannot be measured by any other way.

Active microwave systems suffer from the same limitations of spatial resolution as passive sensors, the long wavelengths meaning that large antennas or high frequencies must be used even for moderate resolutions. Altimeters and scatterometers can normally be expected to achieve ground footprint sizes in the order of tens of kilometres. As with passive sen-

---

[74]   In this case the influence of clouds means that cloud top-height becomes a key retrievable parameter from spaceborne LIDAR.

sors, the non-imaging radar systems are therefore also primarily used for global-scale measurements.

Operational properties of microwave altimeters make them suitable for very small average height variations over large horizontal distances, so are most appropriate for surface areas where local slopes are comparatively low and measuring small changes in average height (in space or time) are important. It is for these reasons that the value of spaceborne altimeters and scatterometers has mostly focused on applications in oceanography and the mapping of the two large ice-sheets: Greenland and Antarctica.

9.1.1   What is RADAR?

Radar is based on the principle of *echolocation* — transmit a signal and measure the echoes that return sometime later. By measuring the time it takes for an echo to return we can estimate the distance, as long as the speed of the signal is known.   *Echolocation*

Echolocation is used in a number of forms of remote sensing (sonar, lidar, seismology, ultrasound). It is a technique that is also quite common among many animals (bats, sea mammals and some birds use echolocation, for instance). The best example of the use of echolocation for scene analysis can be found in bats — natural active remote sensing systems. Bats use their extremely well-developed hearing to listen to the echoes of their own vocalisations that are reflected by surrounding objects or by their prey. In fact bats incorporate a number of features in their echolocation techniques that are similar to those found in radar systems, including pulse chirping (that will be described later in this chapter) and interferometry.

Although it may be hard to do, it can be useful when trying to understand radar to imagine the manner in which an echolocating bat perceives its environment. For a bat, the world is perceived not in terms of colour, but in terms of surface roughness, acoustic reflectivity, and relative motion (for the bat can infer motion from the frequency changes of the return signal as a result of the Doppler effect). Such a perception of the world is very close to the way in which radar systems sense the environment. In fact I would go so far as to say that many of the difficulties encountered by those new to radar remote sensing are due to the dominance of visual metaphors within the wider remote sensing literature. An auditory perception framework can allow students and newcomers to form a clearer appreciation of some of the differences between passive systems and active microwave systems.   *Bats*

As humans, we also use audio perception to tell us something about our environment, although the human dependence upon sight (when we are lucky enough not to have lost it) to construct an image of our local environment means that we are often quite unaware of the role hearing plays in forming our perception of the world around us. We often fail to consciously notice the acoustics of a room, for instance, even though we obtain from it various clues as to the shape, size and material of a space,   *Human perception* especially from the echoes of our own voices — in this sense, humans also act as active sensors[75].

Active remote sensing   The transmission of a signal to illuminate the target scene is, of course, what makes radar and audio echolocation an *active* form of remote sensing. The term "radar" is reserved for echolocation using radio waves (or microwaves) — RADAR is, after all, *RAdio* Detection and Ranging — but the basic principles and problems are identical to those in other forms of echolocating system. Sonar refers to echolocation using sound waves, and Lidar when using wavelengths in the visible or NIR regions.

This chapter describes the principles of ranging and the analysis of the echo parameters so as to extract information from within the radar beam. The approaches described will then be expanded in the following chapter to include the generation of data products that also include the spatial variability of such properties — i.e. an image.

9.1.2   Basic Radar Operation

The basic approach for the operation of a radar system is to effectively accept the poor resolution associated with the long wavelengths, but to apply a strategy of optimising the use of the accurate distance measurements to either provide novel information, or to improve the effective spatial resolution. This approach relies on two key properties of microwave systems as an alternative basis for making measurements. These are the ability to transmit relatively large-power microwave pulses to actively illuminate the surface (rather than relying on the much weaker emitted or reflected microwave radiation) and the ability to make extremely precise echo-delay timings.

The basic operation of a radar system is therefore based on the transmission of a pulse of duration, τ, which will travel at almost the speed of light, c. It will only travel at exactly c when there is no atmosphere between transmitter and target. In Section 4.1 this was discussed in the context of the refractive index of the material — in this case, the material is the atmosphere.

The two-way range distance between antenna and target and back again is therefore the velocity times the time delay:

$$R_{2\text{-way}} = c\tau \qquad [\text{m}],$$

so the range distance to the target is half of that,

$$R = \frac{c\tau}{2}. \qquad (9.1)$$

This is the fundamental measurement upon which all radar systems are based.

---

[75] Singing in the bath is the clearest example of how listening to how our voices echo off our surroundings provides some perceptual information on the size and material of the space around us.

9.2 The Generic Equations of Radar Performance

There are two basic equations that govern the performance of a radar system — the signal-to-noise ratio of the returned echo that tells us about the radiometric performance, and the range resolution that tells us about the ability to distinguish small changes in range. There are many different ways to use radar, either as an imaging or non-imaging device, but since they all use the principle of echolocation, their performance in measuring range is always governed by the same relationship — likewise, the measured power of the received echo is always governed by the radar equation.

9.2.1 The Radar Equation

A practical limit for all radars is the question of whether or not it is technically possible to measure the faint echo that returns from the target. More importantly, will the return echo be detectable within the measurement noise discussed in Section 8. The so-called "radar equation" does exactly this by determining what proportion of the transmitted energy is returned from a target.

Unlike in Chapter 5 we are here not interested in the details of the interaction of the signal with the target, so the properties of the object under observation can first be generalised with the use of some backscattering cross-section term, σ.

Deriving the general radar equation requires taking the following parameters into consideration: the range, R, of the target from the radar antenna (since the intensity of the radiated microwaves will drop-off with increasing distance), the directional sensitivity of the antenna (the gain, G), and radar cross-section, σ, for the observed target.

Derivation

Consider first the question of how much of the transmitted energy is intercepted by the target and redirected back towards the instrument (backscattered). The power (energy per unit time) leaving the antenna, P_t, is "shaped" by the antenna sensitivity pattern (characterised by the gain, G), and is reduced by a fraction $\sigma/4\pi R^2$ — the proportional area of that intercepted by the target to the surface area of a sphere at radius R. The power intercepted by the target and redirected is therefore given by

$$\text{(power scattered by target)} = P_t G \frac{\sigma}{4\pi R^2} \quad \text{[W]}. \quad (9.2)$$

This is the power that the target scatters. Since we cannot tell otherwise, we assume this power is scattered equally in all directions (which follows directly from the definition of the scattering cross-section, σ). Next, we have to determine the power density of the scattered signal at the receiving antenna, P_r. We again take account of the fact that the radiation drops off with increasing return distance, and so the power at the antenna is reduced by a further factor of $4\pi R^2$, where this time we include the proportion of the *effective area*, A_e, of the antenna to the surface area of a sphere with radius R. We can therefore consider A_e to be equivalent to the cross-section of the antenna.

The radar equation is then,

$$(\text{signal received at instrument}) = P_r = \left(P_t G \frac{\sigma}{4\pi R^2} \right) \frac{A_e}{4\pi R^2} \quad [\text{W}].$$

(9.3)

The transmitting gain and the receiving effective area of an antenna, are related by the following expression:

$$A_e = \frac{G\lambda^2}{4\pi},$$

(9.4)

which can then be used to rewrite (9.3) as

$$P_r = \frac{P_t G^2 \lambda^2 \sigma}{(4\pi)^3 R^4}.$$

(9.5)

There are several things about this equation that merit attention. The first, and perhaps most important thing to note is that the signal drops of *very* quickly with range — as $1/R^4$. If we double the range distance to the target the return power will drop by a factor of sixteen! If we treble the range, the power drops by a factor of eighty one! This puts a very severe limit on the operationality of an instrument, and clearly must be considered in detail when designing a new radar system. In particular, a design engineer must make sure that the signal from a typical target or interest is larger than the instrument noise, N_0. For this reason it is often

SNR

convenient to rewrite the radar equation as a *signal-to-noise ratio* (SNR):

$$\left(\frac{P_r}{N_0} \right) = \frac{P_t G^2 \lambda^2 \sigma}{(4\pi)^3 R^4 N_0} \quad [\text{-}].$$

(9.6)

From a similar perspective, it is also worth stressing, however obvious it will seem to say it, that the return power (and therefore the SNR) increases linearly with transmitted power. Doubling the transmitted power doubles the return power in the echo. This simple result has a significant impact on all aspects of a radar system — one such impact will be considered in the next section with regards to maximising the power within a single pulse.

Antenna size

If the effective size of the antenna is increased, the SNR also increases — but this time, since the same antenna is being used both to transmit and receive, the increase is with the area squared (from substituting (9.4) in (9.3)). A doubling of the antenna area quadruples the return power. From the discussion in Section 4 on antenna patterns, this should seem reasonable, since increasing the antenna size narrows the central beam of the antenna pattern and so directs (and receives) more of the available power to a smaller area.

Now, you might now be thinking that this idea is contradicted by Equation (9.5), since it seems to imply that an increase in wavelength gives a better SNR, even though an increase in λ should result in a broader antenna pattern. In fact, the way this equation is written is a little misleading since an increase in λ also has an impact on the antenna gain G, from Equation (9.4), so that the resultant influence of wavelength is that the SNR is inversely proportional to $1/\lambda^2$. Shorter wavelengths have better SNR.

The final important thing to note is that the only part of this equation *Cross-section* that is dependent upon the target properties is σ, the radar cross-section. Every other element within (9.5) should be known by the instrument engineers: the transmitted power and wavelength are defined by the transmitter, and the antenna gain and effective area are properties of the antenna that should be determined from calibration. This emphasises the importance of the instrument calibration discussed in Section 9 since it directly influences the ability to estimate the radar cross-section. The range is known because the instrument uses the time delay of the echo to estimate R directly — this is the basic measurement of a radar. The limits of the range are often governed by the particular instrument configuration and how it is used (e.g., flying height or viewing geometries).

The target radar cross-section is therefore estimated from:

$$\sigma = P_r \frac{(4\pi)^3 R^4}{P_t G^2 \lambda^2} \qquad [\text{m}^2]. \qquad (9.7)$$

In the context of Earth observation radar, the target is usually a distributed area rather than a discrete object, so it is often appropriate to use the normalised radar cross-section, σ^0, rather than a cross-section, σ, the two being related by:

$$\sigma^0 = \frac{\sigma}{A} \qquad \left[\frac{\text{m}^2}{\text{m}^2}\right]$$

where A is the area over which the measurement is made. This is the parameter most appropriately inferred from imaging radar measurements whereby the image products are aimed at providing gridded values corresponding to accurate estimates of the normalised cross-section value for a given pixel. Equation (9.7) then becomes,

$$\sigma^0 = \frac{\sigma}{A} = P_r \frac{(4\pi)^3 R^4}{A P_t G^2 \lambda^2} \qquad [\text{-}]. \qquad (9.8)$$

9.2.2   Range resolution

Since the basic operation of any radar is to estimate target distance from a measured echo time delay, the accuracy with which this distance can be measured in a fundamental quality indicator of a radar system. Equation (9.7) also tells us that measuring the distance, R, to a target is not simply a useful parameter in itself, but an accurate range measurement is also fundamental if we want to make accurate estimates of the target cross-section.

The limit of the range resolution of a radar system is defined as the *Definition* ability to distinguish (in time) the return pulses from two idealised point targets. If two identical objects are moved closer together (in the range direction) their separation will eventually be so close that their return pulses

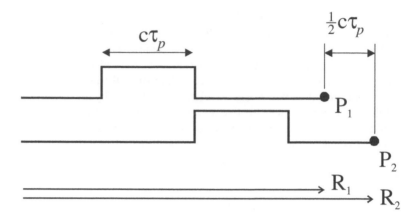

FIGURE 9.1 In this diagram a pulse of length $c\tau_p$ interacts with two point targets. When the returning echoes return such that the tail end of one coincides with the leading edge of the other (as shown) then the system is at the limit of the range resolution. As shown, the echo will appear as one single, long, echo, rather than two separate ones. A very small increase in the separation would separate the two echoes in time.

overlap and they are no longer "resolvable" — the two targets are not separable and since their echoes overlap they would be indistinguishable from a single broad target. The range resolution is therefore dependent upon the duration of the pulses transmitted by the radar system, τ_p.

We can consider two point targets, P_1 and P_2, with range distances R_1 and R_2 (with R_1 being the closer of the two). The range resolution limit is when the front edge of the echo from P_2 reaches P_1 just as the tail end of the echo from P_1 has left. This situation is illustrated in Figure 9.1.

The separation of the two targets is then equal to half the total distance the pulse has travelled so that the range resolution, ρ_r, can be written:

$$\rho_r = \frac{c\tau_p}{2} \qquad \text{[m]}. \qquad (9.9)$$

The factor of 2 appears because there is a "there and back again" situation with the pulse.

The range resolution of a radar system is therefore limited by the length of the pulse: the shorter the pulse the closer the two objects can be before their echoes overlap, and the better the range resolution. This introduces a new technical challenge since in practice it is rather difficult to make transmitters that can generate short rectangular pulses with a high peak power. Range resolution considerations tell us to keep the pulse as short as possible, yet the radar equation with its $1/R^4$ drop-off in power tells us that we need as much power as possible in the transmitted pulse. This is clearly a particular problem for spaceborne platforms with $R \approx 800$ km.

Chirp   For many radar systems, therefore, a linear *frequency modulated* (FM) or *chirped* pulse is used[76]. By sweeping the signal over a small range (*bandwidth, B_p*) of frequencies the transmitted signal is essentially encoded, so that even though a whole collection of overlapping return signals may be received, they can be distinguished in time to an accuracy much shorter than the length of the pulse. This is illustrated in Figure 9.2.

This technique makes use of the spectral filtering capabilities of the radar system and is analogous to similar frequency modulation techniques which are a common feature among echolocating bats. It is not hard to imagine being able to distinguish overlapping pulses if we consider the ability of the human ear to isolate sounds with specific temporal frequency patterns, such as being able to identify different musical instruments that are played in an orchestra.

The process is known as *range compression* and involves a correlation of the received signals with the original chirped pulse. The decoding of the collective return signal into individual returns is called *matched filtering*. Different parts of the pulse are then identified by their individual frequency, or more exactly, their frequency *pattern*, so that even overlapping pulses can be distinguished. For example, this technique allows the use of pulses in the order of $40\mu s$ but with a range resolution comparable with a 60ns rectangular pulse.

Range compression

Matched filtering

As an analogy, imagine trying to hear the sound of a single piano key being repeatedly played within a very loud background noise, such as random piano playing — a very difficult task. However, if instead of a single note, a pre-defined sequence of notes is played (i.e., a melody) then you stand a much greater chance of identifying the signal within the noise because you are listening for a pattern, rather than just a single note. This pattern, or melody, is unlikely to happen by chance within the random notes being played. You still might not hear all of the signal within the noise, but portions of it should be sufficient to identify the tune. And if you were trying to separate overlapping melodies, you would stand a good chance of both identifying the two tunes, as well as recognising which part of the melody is playing (assuming you knew the melody before hand).

The key property that determines the range resolution is therefore based on how well the matched filtering can locate the position of a particular echo. The "sharpness" of the pulse is given by the *effective pulse length*, τ_e, which is governed by the bandwidth of the pulse B_p, such that

$$\tau_e = \frac{1}{B_p} \qquad [\text{s}]. \tag{9.10}$$

The bandwidth is the frequency range over which the FM chirp is made, with $B_p = \nu_2 - \nu_1$.

The range compression ratio, C_R, relates the actual to effective pulse lengths:

$$C_R = \frac{\tau_p}{\tau_e} = \tau_p B_p \qquad [\text{-}]. \tag{9.11}$$

---

[76]   If the same thing was done to an audio signal it would sound like a "chirp".

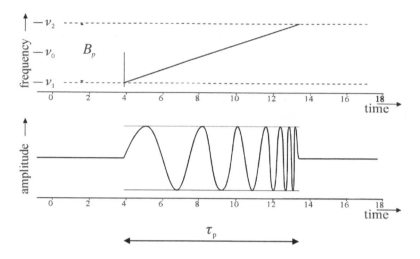

FIGURE 9.2 In order to maintain a long pulse (to keep the transmitted power high) but also have a signal that can be identified to a fraction of a pulse (to keep the range resolution as short as possible), the pulse is "chirped". This entails sweeping the frequency of the pulse over some bandwidth, B_p, which ranges from ν_1 to ν_2.

The range resolution for a chirped pulse radar is therefore

$$\rho_r = \frac{c}{2B_p} \qquad [\text{m}]. \qquad (9.12)$$

Note that the range resolution is now not dependent upon the pulse length at all, but rather the bandwidth of the signal. The range resolution improves as the range of frequencies increases as this allows individual echoes to be more precisely located within the collective return signal[77]. For this reason some technical descriptions of a radar performance will quote the bandwidth directly rather than a range resolution in metres.

It is also possible with this set-up to transmit pulses end-to-end with no gap between them. As long as the signal is continuously re-swept in frequency, the echoes can still be identified and the time delay determined. *CW* This is the principle behind what is called *continuous wave* operation.

Note also that this *range* resolution is a system limited parameter that is totally independent of the target under observation or the observation geometry. In particular, note that the range resolution is also independent of the distance over which the measurement is made. For readers more familiar with optical systems, this may seem counter-intuitive since it means that spatial resolution in range is not governed by flying height of the instrument. A radar system's range resolution will be the same whether it is flown at an altitude of 3km or 300km. The equivalent in optical systems in the angular resolution, which is also a system parameter — it is the ground

---

[77] In the context of the analogy above, the wider the range of keys used on the piano, the greater the range of complexity (and therefore uniqueness) of the melody.

resolution that is dependent upon the flying height and view angle, not the system's angular resolving power.

Echolocating bats use FM pulses, and some bats, by using brief broadband (FM) signals can measure object distance to a staggering degree of accuracy, discriminating between echoes that return with a difference in their arrival time of less than $100ns$, which corresponds to a difference in range distance of the two objects of less than 17 μm (Metzner, 1991).

9.3 Radar Altimeters

The primary function of a radar altimeter is to use the time delay relationship in Equation (9.1) to determine the range distance R with great accuracy. Spaceborne altimeters are sufficiently precise with their measurement of time delay, that measurement precision of less than a few centimetres can be achieved, even from an altitude of 800km!

The difficulty of translating such high precision to high accuracy estimates of the surface topography is a combination of knowing exactly where the instrument is at the time of the measurements and being able to characterise all the other variables that influence the delay time for the echo.

Radar altimeters are probably the most common type of radar system — they are used on many aircraft and on terrestrial satellites, and they have been used on manned space missions, on lunar missions and on interplanetary spacecraft. They range in complexity from systems that are designed only to give an indication of altitude as a guide to flight, to instruments that allow intricate analysis of the return echo and accurate estimates of surface topography for applications over land, sea and ice.

The value of a radar altimeter is the extremely high precision with which it can measure the average height of a surface, and given long enough wavelengths these measurements can be regardless of cloud cover. Given sufficient ancillary data and error compensation, it is possible to make average height measurements down to centimetre-scale accuracy. The microwave altimeter onboard the TOPEX/POSEIDON mission, for example, had a precision of 2cm and a single-pass accuracy of about 13cm.

The development of accurate spaceborne altimeters has been driven by a need for height measurements over two important regions of the Earth: the oceans and the ice sheets of Greenland and Antarctica. As with passive microwave imagers in the previous chapter, the large horizontal footprint is a severe limitation on microwave altimeters and causes some particular limitations on its use. The result is that they are largely constrained to observations over the open oceans and larger seas, and the vast ice sheets of Greenland and Antarctica. In both scenarios, the extent, inhospitability and inaccessibility are factors that favour remote sensing techniques over extensive in situ networks.

9.3.1   The Need for Altimeter Measurements

The use of altimetry in Earth observation has focused on oceans and ice because their topography is characterised by small vertical changes in the region of cm–m's over horizontal distances of many kilometres. We will see later in this section that such conditions are required for optimum performance of radar altimeters. These surfaces are also dynamic, with ocean surfaces changing due to tides, currents and winds, and ice sheets varying seasonally as well as interannually due to variations in the ice balance. Frequent and repeated altimetry measurements over a long time scale (more than a few years) are therefore a significant source of information about planetary scale processes in the oceans and cryosphere.

Height measurements made on a horizontal scale of many kilometres are not that useful over land surfaces, such information being more reliably and effectively made by other means[78]. One context where such measurements are useful is in the mapping of other terrestrial planets and moons. Venus, for instance, has a thick, visibly opaque atmosphere, so microwave altimeters have provided an invaluable tool to understanding its surface geology by providing otherwise unobtainable information on surface topography. While the atmosphere of Mars is visibly transparent, the upper layers of the surface are sufficiently dry that very long wavelength microwaves (in fact radio waves) are able to penetrate hundreds of metres beneath the surface. The penetrating power (of atmospheres or dry dust and sand) of microwaves and the relative technological simplicity of altimeters (in comparison to imaging radars) make them worthy candidates when considering some extraterrestrial applications.

On Earth there is one context where spaceborne altimeters have been used to map solid Earth topography and that is the topography of the ocean floor. This is achieved by analysis of long-term (yearly) averages of ocean surface topography — this should corresponds to the *geoid*, a virtual surface of equal potential energy. (Long averages are required to even out the effects of atmospheric forcing (wind and pressure), currents and tidal effects.) Assuming that the Earth's surface water rotates at the same angular speed as the solid Earth the ocean surface would match the equipotential surface that equates to mean sea level and we define this to be the geoid. Practically, the geoid is a very effective reference surface for defining vertical location because it relates to a measurable physical property, namely the mean level of the sea.

The geoid

The geoid is not always the best reference surface, however — what is "mean sea level" to someone living in Switzerland? The geoid is also difficult to use for mapping purposes since it is a real property that must be measured, rather than being a mathematically well-defined reference surface. The latter are referred to as *ellipsoids*, which are created to provide a common reference surface in order to define vertical location and to

---

[78] Later in this chapter we will see that it is also relatively difficult to make accurate measurements over land.

form the basis of map projections. In Chapter 10 we will discuss the need for some reference surface onto which we can project our radar image. WGS84 (the World Geodetic System defined in 1984) in one such global ellipsoid that is commonly used in planetary-scale studies and Earth observation.

Ellipsoids are usually defined so that they approximate the shape of the geoid, but at both local and regional levels there are a number of factors that will change the geoid such that the difference between geoid and ellipsoid will be as much as 100m in both positive and negative directions. These discrepancies are due to changes in the gravitational force caused by fluctuations in rock density and surface (or ocean floor) topography — changes in density or topography of the lithosphere (or even in the mantle) correspond to changes in mass, so that the local gravitational pull will vary. The density of rock beneath the oceans is greater than the overlying water, so that changes in ocean floor topography influence the local lines of equal gravitational potential. More rock means higher gravitational pull, so that the equipotential lines increase in altitude. Variations in the mean sea level therefore correspond to changes in seafloor topography (assuming the density of the lithosphere remains relatively constant). When there are observed discrepancies between the topography inferred from such a technique and topography measured by other means (such as ship-based acoustic bathymetry) then the information can be used to infer changes in the rock density. *Ellipsoids*

Mapping variability in the geoid is also important from the perspective of satellite systems since even small variations in the gravitational field can perturb satellite orbits, which is of particular concern for global positioning satellites.

The study of the geoid using ocean altimetry is a long affair. In practice the ocean is a fluid, providing a dynamic interface between atmosphere and ocean. On short timescales, the surface topography is constantly being modified by large-scale currents, tidal forces and atmospheric forcing through variations in air pressure and wind. Despite appearances, the sea is not flat and altimetry provides a tool to measure and map these short term fluctuations on a regular (and global) basis.

Throughout the last quarter of the 20th century, the role of spaceborne altimetry has focused on oceanographic applications, with ice sheet studies playing a supporting role, albeit an increasingly important one in recent years. *Oceans*

The reason for this is that measuring the surface topography over oceans is possible to high enough relative accuracies (a few centimetres) to allow analysis of global ocean circulation patterns. One of the effects of flowing currents in the oceans is large scale slopes with a gradient of about 1m in 100km associated with the movement of the water. Surface winds will also tend to blow water so that it piles up downwind, especially in shallow waters near coasts, but also on broader scales such as the normal Pacific conditions whereby the strong equatorial Easterlies blow surface water over to the Western Pacific causing about a metre's difference in average ocean height.

Highly accurate measurements of ocean topography at the centimetre level are therefore extremely important for studying ocean surface dynamics on a global scale. Being able to do so repetitively from space potentially gives a global view of ocean dynamics every 1 or 2 days.

One further oceanographic application is in determining wind speed across the oceans. As discussed in Chapter 5, surface roughness influences microwave backscatter in such a way that increasing surface roughness will result in less specular reflection from the nadir direction returning to the instrument, and more diffuse scattering across the entire footprint, including some scattering into the backscatter direction. The characteristics of the altimeter echo intensity as a function of echo delay time is therefore indicative of ocean roughness, which can be related to significant wave height and wind speed.

Ice sheets

The processes of ice sheets and glaciers are similar, but the mechanics can be different since the latter are constrained within a linear channel. From a global perspective the ice sheets of Greenland and Antarctica are the most significant since they approximately contain 9% and 91% respectively, of all the frozen fresh water on the planet (glaciers and ice caps containing <0.5%). Within the context of radar altimetry is it also more appropriate to focus on ice sheets because of the large horizontal footprint which is not appropriate for narrow glaciers.

The high vertical precision but low horizontal resolution of microwave altimeters is therefore well-suited to making estimates of average height change that can be used as input to ice balance models and aid in our understanding of the large ice sheets. Importantly, such instruments can inform us of the state of the ice on a regular (daily) basis over a long enough time period to be valuable for climate studies (>10 years).

Mass balance

The change of height of an ice sheet is an important indicator of mass balance, the net change in the mass of the ice due to variations in accumulation and ablation. Accumulation is the addition of ice, mostly from precipitated snow, and ablation refers to the loss of ice mass through melting and sublimation (evaporation) of snow and ice, as well as the calving of icebergs at coastal margins of the ice sheet. Ablation is always greatest at the ice margin. For smaller ice caps accumulation always increases with increasing altitude (away from the margin) but on the Greenland and Antarctic ice sheets the accumulation eventually drops again towards the centre (due to a decrease in precipitation) so that the maximum accumulation may be somewhere nearer the margins rather than in the centre.

To understand the scales at play here, it is worth pointing out that the accumulation rate in Greenland is about 500 Gt per year[79], whereas in Antarctica it is estimated as being greater than the staggering value of 2000 Gt per year. The total ablation rates are similar, but in Greenland it is approximately equally distributed between calving and sublimation/melting,

---

[79]   Giga tonnes is 10^9 tonnes.

whereas in Antarctica the loss is almost exclusively by calving (which might be more appropriately monitored by imaging systems rather than altimeters).

Since the Greenland ice sheet has a turnover time of around 5,000 years, three times less than Antarctica, it is thought to be more sensitive to short term climate change and is therefore an important area of study by remote sensing.

The mass balance of the ice is defined as the total difference between accumulation and ablation, such that it is positive if the ice is growing and negative if shrinking. The sign and magnitude of the ice balance has major long-term implications for the entire planet. If the entire ice sheets covering Greenland and Antarctica were to melt, there would be an estimated increase of sea level of 72m (7 and 65m respectively).

There are also potentially more subtle, but no less dramatic, effects of the loss of these ice sheets. Currently, Northern Europe is kept significantly milder than other parts of the globe at similar latitudes by the warm current that flows across the North Atlantic. This circulation is driven by evaporation of the ocean, which results in saltier, and therefore denser, water which sinks, drawing warmer water from lower latitudes as it does so. The increase in fresh water draining into the North Atlantic due to increased ablation in Greenland could upset this delicate "conveyor belt" and result in a severe cooling of Northern Europe.

The importance of sea ice was described in Section 3. Altimeters can *Sea ice* also measure relevant properties of sea ice as demonstrated by the European Space Agency Cryosat mission. Cryosat is a dedicated satellite altimeter system designed to conclusively determine whether polar ice cover is increasing or decreasing. As well as measuring changes in the height of the Greenland and Antarctic ice sheets, it will also be sensitive enough to measure the "freeboard" — the height by which sea ice rises above the water. Regular measurements of sea ice extent and thickness will allow the first regional-scale assessment of whether polar sea ice is thinning or not.

9.3.2 Altimeter Geometry

The geometry of a basic altimeter is shown in Figure 9.3. An antenna system directs a pulse of microwave energy towards the nadir and it measures the time delay of the return echo. A spaceborne microwave altimeter will have a footprint that is very large — in the order of tens of kilometres. However, the altimeter has an extra advantage in this regard, since the time delay is the key measured parameter, not a radiant power, and so it can be assumed that in this case the leading edge of any measured echo always comes from directly below the instrument. The first part of the echo will always be from the ground point nearest the instrument, which for a flat Earth is always at the nadir. The width of the beam is therefore not necessarily the determining factor on the size of the measurement footprint, but it may instead be determined by the length of the pulse and the equivalent area of ground the pulse illuminates at any specific time. Figure 9.4 illustrates the two cases — the difference is that in the first case the pulse does not illuminate the entire beamwidth footprint at the same time, whereas the second case does. The size of the corresponding footprints are therefore defined by the pulse length (pulse-limited) and the beamwidth (beam-limited), respectively. The beam surface footprint, F_b, is given approximately by multiplying Equation (6.3) by the flying height, H, so that

$$F_b \simeq \frac{\lambda H}{D} \quad \text{[m]}, \tag{9.13}$$

where D is the antenna diameter[80].

The pulse surface footprint, F_p, is found using the equivalent distance of the pulse length, $c\tau_p$, and using Pythagoras, such that,

$$(H + c\tau_p)^2 = H^2 + x^2 \tag{9.14}$$

$$c\tau_p = (H^2 + x^2)^{1/2} - H,$$

where we have used $x = F_p/2$ as the radius of the footprint (as shown in the figure). From this we can use the fact that for $x \ll H$, the following approximation is valid (from the Binomial theorem),

$$(H^2 + x^2)^{1/2} = H \left(1 + \frac{x^2}{H^2}\right)^{1/2}$$

$$= H \left(1 + \frac{x^2}{2H^2} + ...\right)$$

$$\simeq H + \frac{x^2}{2H},$$

so that

$$c\tau_p \simeq \frac{x^2}{2H}. \tag{9.15}$$

giving

$$F_p = 2x \tag{9.16}$$

$$= 2\sqrt{2c\tau_p H} \quad \text{[m]}. \tag{9.17}$$

An altimeter that operates with $F_b > F_p$ is *pulse-limited*, since the pulse determines the footprint size, whereas an altimeter with a large enough antenna will give $F_b < F_p$ and is *beam-limited*. Since most satellite altimeter antennas are restricted in physical size, they tend to be pulse-limited.

9.3.3   Instrumentation

Before considering the limitations of a radar altimeter in the context of natural targets, it is useful to first consider some engineering issues that arise

---

[80] We have made the assumption here that the beamwidth is in radians, and is very small, so that $\sin \theta_{3dB} \approx \theta_{3dB}$.

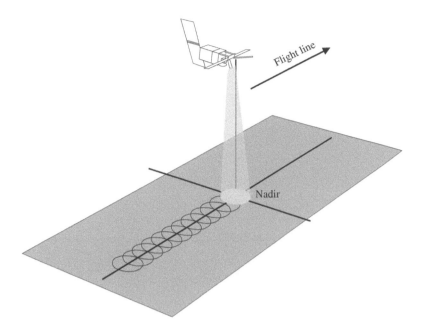

FIGURE 9.3 The basic geometry of a simple altimeter. The beam is pointed downwards (at nadir) and a single line of points is acquired.

in altimetry, even for an idealised flat-Earth target. One practical consideration of all radar systems is the requirement for a measurable power in the return echo using the radar equation given in Equation (9.5).

For a pulse limited altimeter the radar equation is given by

$$P_r = \frac{P_t G^2 \lambda^2 A \sigma^0}{(4\pi)^3 H^4}. \tag{9.18}$$

In this case the area A is is found using F_p, the pulse surface footprint, so that

$$A = \pi \left(\frac{F_p}{2} \right)^2 = 2\pi c \tau H. \tag{9.19}$$

Note that we have used H for height rather than R for range, on the assumption that the measured range corresponds to the actual height.

For a beam limited altimeter, the derivation of this equation is slightly altered since all the power that is transmitted is scattered (in some direction) by the footprint.

9.3.4 Echo Shape Analysis

The echo that an altimeter measures is not an exact copy of the transmitted pulse, but is "smeared out" in time due to a number of factors.

Firstly, the underlying target surface is not likely to be perfectly smooth

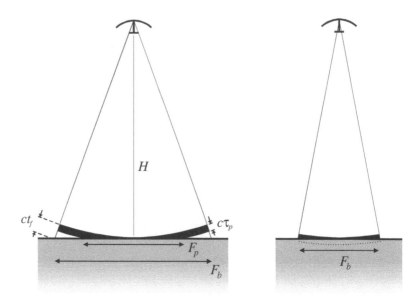

FIGURE 9.4 An illustration of the difference between pulse-limited (left) and beam-limited (right) altimetry. τ_p is the length of the pulse and F_p and F_b are the pulse footprint and beam footprint, respectively.

(even for an ocean in calm conditions) so that there is always going to be some energy that is scattered back to the instrument from all parts of the footprint. A flat, perfectly smooth, specular reflector would return an exact replica of the transmitted pulse — a mirror image, you might say. The signal only returns from directly beneath the sensor in the centre of the footprint, and the only difference from the transmitted pulse is that it will have lost some of its original intensity but its shape will have remained the same. In such a scenario signals that are incident on the surface outward from the centre of the footprint are reflected away from the sensor and so are not detected.

In the case of a slightly rough surface, however, these same signals are scattered, with some of the energy being scattered back to the sensor. These echoes take marginally longer to return because of the extra distance they have to travel, the short delay meaning that all the individual echoes merge together in the received signal, stretching out the tail of the echo. This is illustrated in Figure 9.5, which shows the pattern of the echo as a function of time delay for an short transmitted pulse.

The length of the echo, t_f, for an idealised infinitely narrow pulse is given by the time the echo is in contact with the ground within the footprint. This is found from the same geometry as shown in Figure 9.4 by considering the time lag, t_f, between the first echo from nadir and the last echo from the footprint edge. This distance can be found by recognising that the angle that ct_f (the distance travelled in t_f) subtends at nadir is equal to half the antenna beamwidth, which is $\lambda/2D$ for small beamwidths expressed in radians. The sine of this angle is then ct_f divided by $F_b/2$, which for small angles gives

$$\frac{\lambda}{2D} \approx \frac{2ct_f}{F_b}. \tag{9.20}$$

We can then substitute the expression for F_b to give

$$t_f \approx \frac{H}{c}\left(\frac{\lambda}{2D}\right)^2. \tag{9.21}$$

An even rougher surface not only scatters more energy into diffuse directions, but even at nadir the signal has a short delay between the highest peaks of the surface and the lowest troughs. The local height differences will therefore tend to smear out every localised echo by a factor proportional to the average height variation (usually expressed as a root mean square height of the surface).

The final factor that contributes to the smoothing of the echo is the effective pulse length. This links to the discussion on range resolution in Section 2.2 since the detail discernible in this echo shape cannot be better than the range resolution.

The actual echo shape is therefore the combination of these three factors: the shape of the original pulse, the scattering properties across the footprint, and the localised height variations. It is the latter two that make the link between echo shape and surface properties. With increasing roughness, the echo length remains approximately constant, but the peak intensity decreases so that the observed rate of decay gives information on the backscatter properties with incidence angle across the footprint. This can subsequently be related to near surface wind speed over the ocean, or it can be used to characterise the statistical distribution of surface slope.

The peak of the echo power decreases as the surface roughness increases, so that the peak power also gives some indication of the surface roughness.

Figure 9.5 shows what happens when the surface topography becomes too varied and irregular — in this case the echo is no longer a clear response and it is now unclear from the echo which delay time corresponds to the altimeter height since the leading edge of the echo no longer corresponds to nadir.

The influence of the height variations (as opposed to wavelength-scale roughness) is apparent in the slope of the leading edge of the echo. As the variety of heights increases there is a more gradual rise of the leading edge of the echo. The rise-time, t_r, therefore changes in proportion to the RMS height of the surface. Over oceans, where the height variation can be modelled as a regular perturbation of an otherwise flat surface, this relationship can be used to infer the significant wave height.

The waveform profile over the ocean surface is sufficiently well understood to permit real-time estimates of ocean parameters to be carried out on-board the satellite. For other surfaces (such as ice) the waveform

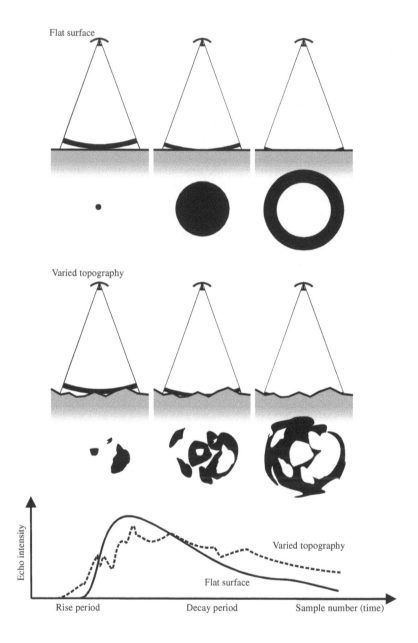

FIGURE 9.5 Over a flat surface the altimeter pulse illuminates progressively larger rings on the surface, centred on the nadir point. If the surface is rough on the scale of the wavelength then some energy is scattered back to the instrument from all areas within the footprint and the echo return looks like the solid line in the lower figure. When there is significant topography, the pattern is not so well defined and the echo shape is less clear.

shape does not always conform to a simple model and further data analysis is necessary. The return echo from sea ice appears more specular than from the ocean and has a peaked trace, since the ice surface remains flat in comparison to the open water.

9.3.5   Range Ambiguity

In order to make the most of the faint echoes, most radar systems encountered in remote sensing use a sequence of pulses, to maintain a constantly updated estimate or to improve the accuracy of the measurement by averaging over many echoes. The rate at which these pulses are transmitted is termed the *pulse repetition frequency* (PRF). In many practical situations   *PRF* there is a relatively long delay between the initial pulse and the return echo. From 800km, an echo would take about $5ms$ to travel down to the Earth's surface and back[81]. But in the microwave region it is possible to generate sufficient power to satisfy the SNR requirements with any individual pulse lasting only a few micro-seconds. On the scale of the time measurements relevant to an altimeter, this is a very long wait between pulses if each subsequent pulse is only transmitted after the previous echo has been detected. The short period over which measurements are received is known as the range window — it is only for this well-defined short period of time that the instrument will record the return echo in a finite number of range "bins". It would be quite inefficient to record data even when there is no echo signal to measure.

A more efficient method of operation is therefore to transmit one or more pulses even before the first echo has returned. This is illustrated schematically in Figure 9.6 where a second pulse is transmitted before the return of the first echo. This technique allows for a more efficient use of the available time, but also introduces a potential problem known as *range ambiguity* whereby the echo associated with one pulse is measured as if it was an echo from a later pulse. With reference to the figure, we can see that, for example, the first echo includes a signal from a very distant target so it extends into the receiving window of the second echo. At time $t = 10$ the final measured signal includes a contribution from both the first and the second echo. The distant target is then mapped at a range much nearer than it actually is — the target echo appears to have arrived very quickly from the second pulse, instead of arriving very late from the first pulse. The range of this target is thus "ambiguous".

The real practical limitation that this arrangement produces is that while the PRF and the range bin recording may be optimised for a given range distance, any quick change in range will shift the pattern. The recording bins would certainly have to shift, and the PRF might also have to change to accommodate the change in echo delay time.

By way of example, the radar altimeter (RA) on the ERS-1&2 satellites was an altimeter that exhibits features common to most spaceborne

---

[81]   To travel 2×800km at $3 \times 10^8\mathrm{ms}^{-1}$ would take approximately 5×10^{-3}s.

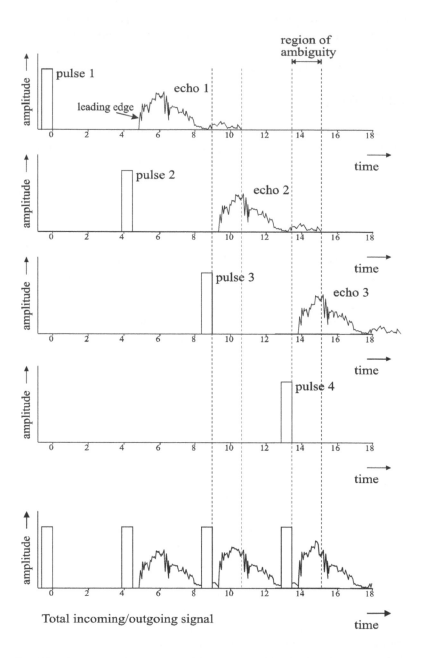

FIGURE 9.6 The top four graphs show independently the first four pulses from a radar system and their corresponding echoes. Note that in the total incoming/outgoing graph at the bottom, some of the echoes from the tail-end of one set, are actually picked up in the next receiving window — simultaneously with other echoes. This is known as range ambiguity.

altimeter instruments. The RA was a nadir-pointing active microwave sensor designed to make precise measurements of the echoes from ocean and ice surfaces. It operated at 13.9 GHz with a 1.2m diameter antenna giving a beamwidth of 1.3° (20km footprint) and provided more than a decades worth of information on significant wave height, surface wind speed, sea surface elevation and various parameters over sea ice and ice sheets. The RA operated in three modes. The first is the *acquisition mode*, during which time the radar finds the approximate distance to the surface in order to determine the appropriate PRF for that range. It then switches to one of two *tracking modes*: the ocean tracking mode that is optimised for only slowly varying changes in topography, and the ice tracking mode, which has a lower PRF that can accommodate larger changes in topography.

It is this restriction on rapid changes in topography that has tended to constrain the use of radar altimetry to oceans and ice sheets.

9.3.6   Accuracy of Height Retrievals

The range ambiguity provides a good illustration of one of the practical challenges of altimetry, namely that the Earth's surface is never perfectly flat. There are two issues here. The first is that a varied topography increases the length of the return echo, increasing the possibility of a signal being measured in the wrong reception window. The second is that if the average topography changes significantly over successive pulses, as the instrument moves in its flight path, then the echo will change its timing within the reception window and may start to impinge on another reception window. If the topographic change is predictable or slow enough to pick out as a trend, then the PRF can be adjusted to compensate. These are the technical challenges about detecting the right echo. A more pertinent issue from the perspective of application of the measured data, is the accuracy with which the true ground height can be estimated.

There are four major sources of uncertainty in the height measurements: the quality of the instrumentation, the location of the platform, the estimate of the speed of travel of the pulse, and variations in the surface topography.

The instrument uncertainty should be well-characterised. The clock accuracy can be estimated from calibration studies and the range resolution, as given in Section 2.2, is defined by the effective pulse length of the system. This will likely give a range *precision* in the region of a few centimetres, but this high precision (the repeatability of the measurements) does not equate directly to high accuracy (closeness to the actual value).

A key requirement for accurate altimetry is knowing the location of the *Position* instrument! Without accurate knowledge of the instantaneous location, or flight path, the measurements from an altimeter are merely arbitrary range values that include information on both surface topography and instrument height variations. Inertial navigation and real-time differential GPS make it possible to locate aircraft with high accuracy. Satellites have the advantage of keeping fairly steady and predictable flight paths, but platforms with altimeters usually carry a laser retro-reflector (effectively an optical trihedral corner reflector) so that they can be tracked by ground-based laser range-finding instruments. The satellite position can then be known relative to some terrestrial reference system (such as WGS84) so that the range measurements can be translated into a height in a geographical coordinate system.

Environmental factors
After the instrumentation and platform location are well-defined, there are still a number of environmental factors that have to be accounted for when translating the measurements into a accurate height. These can be separated into atmospheric transmission effects and geometrical effects.

Atmospheric effects
All altimeters require their pulses to travel through the atmosphere, and additionally for satellite systems, the ionosphere. Both of these media impose a delay on the velocity of the EM wave in proportion to its refractive index, n, as given by Snell's law, Equation (??).

The atmosphere has a refractive index that varies in proportion to the temperature, pressure and water vapour content. The wave velocity assumed in (9.1) is therefore incorrect, resulting in a small overestimation of the range. The difference between the speed of light in a vacuum and in the atmosphere is very slight and so this error will only be significant when high accuracy is required. A dry atmosphere, for instance, will contribute a range error in the order of a few metres if not accounted for using some model of the atmosphere. A "wet" atmosphere contains water vapour, which is much more spatially and temporally variable than temperature or pressure and may contribute an additional error of up to 50cm. This may not seem like a large error for an instrument at 800km, but they are significant when trying to measure changes of only a few centimetres in ocean topography or ice sheet balance. Such errors can be corrected if accurate knowledge of the state of the atmosphere is known at the time of the measurement. The most effective method is therefore for the same platform to carry a microwave radiometer that complements the altimeter by providing simultaneous estimates of the state of the atmosphere. The main function of the passive microwave sounder on the ERS satellites, for instance, was to provide atmospheric water vapour information for correcting the altimeter measurements.

Ionospheric effects
The error on the range due to the ionosphere is given by

$$\Delta R_i = \frac{40E}{\nu^2} \qquad \text{[m]}, \qquad (9.22)$$

where E is the total columnar electron content and ν is the frequency. E varies from about 10^{16} to 10^{18} e/m^2, meaning that ΔR_i is in the region of tens of centimetres for most microwave altimeters. The high sensitivity of the ionospheric influence as a function of frequency means that one way to minimise this effect is to use a dual frequency system, whereby the small discrepancy in the measured range between the two frequencies allows one to estimate the influence of the ionosphere. Alternatively, using as high a frequency as possible will minimise this effect.

The geometric factors include surface slope and the impact of the sur- *Topographic errors*
face roughness of the leading edge of the echo.

Variation in topography influences the accuracy of the height measure- *Slope*
ment since the nearest point (and therefore the location of the first echo)
will no longer necessarily be located at nadir. We will consider here the
simplest case of a regular slope of angle α, as illustrated in Figure 9.7. The
nearest location is at the perpendicular to the surface, shown as R_n in the
diagram. It is this distance that will be the measured "height" according
to the altimeter. The relationship between the measured height, R_n, and
actual height, H, is given by the cosine of the angle α,

$$\cos \alpha = \frac{R_n}{R},$$

which can be approximated by an expansion of the secant ($\sec x = 1/\cos x$) term if we assume that α is small (and given in radians), so that

$$R = \frac{R_n}{\cos \alpha} = R_n \sec \alpha s \simeq R_n \left(1 + \frac{\alpha^2}{2} + ... \right). \quad (9.23)$$

This can be rearranged to give the expected error, ΔR, such that

$$\Delta R = R - R_n \simeq \frac{R_n \alpha^2}{2} \quad \text{[m]}. \quad (9.24)$$

The impact of surface slope can be minimised by using a beam limited
altimeter so that the footprint is as small as possible. For pulse-limited
altimeters there is some scope to account for the slope in the flight-path
direction by combining sequences of measurements to build up a transect
of the surface height.

Figure 9.5 illustrated that the variation in surface topography can also *Topographic Variation*
introduce height errors. As the topography becomes increasingly more var-
ied and complex, the echo shape loses its well-defined form and the correspondence between echo and instrument height becomes lost. The first
return of the echo may no longer correspond to the actual height of the instrument above nadir but rather the nearest point on the surface, which may
be located anywhere in the footprint. In such cases the altimeter height
measurement is therefore both incorrect and misplaced (since the altimeter will normally assume the location of the first return corresponds to the
nadir). This can be overcome by narrowing the beamwidth to constrain the
location of the echoes.

9.3.7 Scanning Altimeters

Scanning altimeters are designed with narrow beamwidths that are scanned
in the across-track direction, as shown in Figure 9.8. This allows an increase in coverage without introducing errors due to slope or topography,
but the narrow beam requirement would normally mean the need for a
large antenna and high frequencies. The variation of incidence angle for
the backscatter across the swath is not an issue here because the primary

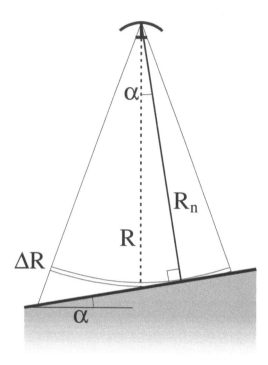

FIGURE 9.7 This figure illustrates the error associated with a surface of slope, α. The closest part of the surface is no longer at nadir with range R, but is at a range R_n.

measurement is the time delay. For this to be effective, there must be an accurate mechanism for determining the pointing direction, since the assumption that the first return is always at nadir no longer applies.

9.3.8   Calibration and Validation

As you might imagine, there must be a great deal of ground-based effort to validate altimeter measurements that claim to be of centimetre accuracy. Much of the calibration and validation of a spaceborne instrument will be based on measurements made by fixed and mobile laser tracking stations complemented by in-situ measurements of various parameters from off-shore instrument towers.

A further important consideration is the inter-calibration between successive satellites — in the interests of continuity of data sets, it is vital that data from one satellite can be compared to another. In the ideal case, this is done by comparison of data during an overlapping period of operation at the end of one satellite's lifetime and the beginning of another.

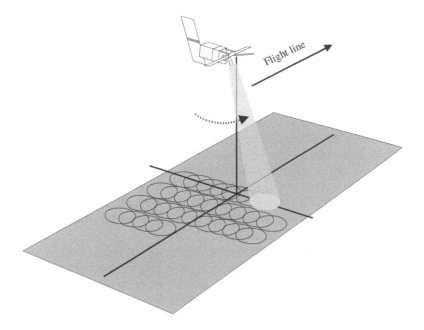

FIGURE 9.8   The geometry of a scanning altimeter. Here a narrow beam is used to scan across a wide swath. Accurate pointing information is required to locate the origin of the backscattered echo.

9.4   Improving Directionality

Basic altimeters are very useful instruments, but due to their large beamwidth they usually have relatively large footprints. For many global and regional applications this is not a significant problem, but when considering abrupt changes at ice margins or the topographic effects over ice sheets, there is some advantage to be gained from improving our knowledge of the direction of the returning echo.

The simplest way to determine direction is to use a narrow beam to limit the direction of the transmitted and return signal. Acoustically this is also a good strategy for determining the direction of a sound source — cats have very directional ears, for example, and can point them (independently) towards sounds of interest.

For radar, the direction of the target can be determined by the pointing of the antenna, but the antenna beamwidth will be of finite dimensions (re- *Antenna Beamwidth* fer back to Section4) — a good estimate of its width is the 3dB beamwidth approximated (in radians) by the expression:

$$\theta_{3dB} \simeq \frac{\lambda}{D}. \tag{9.25}$$

This is a good "ball-park figure" for getting a first estimate of antenna beamwidth. As we have repeatedly seen, this is a strict limit on the application of microwave systems to tasks requiring high resolution. In the following sections we find that by some clever processing it is possible to improve on this theoretical limit.

9.4.1   Sub-Beamwidth Resolution

In certain circumstances, directional information can also be extracted from *within* the antenna beam. The general principle relies on the fact that for some targets, the echo has different properties depending upon which part of the beam the scattering took place. We will see in the next chapter how such techniques can be used to develop high spatial resolution imaging radar, but it is instructive to describe here a more simple case of sub-beamwidth information extraction using planetary radar imaging from Earth.

Planetary Radar   In radar based imaging of planetary surfaces the antennas used are very large — for instance, Arecibo[82] is a staggering 300m in diameter, operating at 12.5 and 70cm wavelengths. The earliest radar observations from Earth were of Mars (1963) and Venus (1965), but have also been obtained from other bodies in the solar system including the moon, mercury and many asteroids (and by the time this is published, Titan, one of Saturn's moons). Even though the Arecibo antenna is extremely large and so provides a very narrow beam, it will still illuminate the entire visible disk of Mars or Venus. Some other method must therefore be employed if it is to distinguish features on the planetary surface.

The radar echo returning from a planet such as Venus is recorded and analysed in a number of different ways. First the echo can be broken up into components as a function of time — i.e. into range bins, as was the case with the echo shape analysis above. A specific echo of a time period will consist of all the reflections from planetary surface elements that are all equidistant from the antenna — the equidistant area forms a ring around the planet known as a *time delay ring*, or *range circle*. This is similar to the circular pattern seen in Figure 9.5 except it is for an entire visible hemisphere of the planet.

E-W ambiguity   After this first analysis, there still remains ambiguity — ambiguity of which part of the ring the echo comes from. There is no way to tell from the time delay alone whether the signal comes from the East or Western part of the ring. The East-West ambiguity is resolved by the second clever piece of data analysis, which makes use of the fact that a planet rotates around its axis of rotation. This is very slow for planet Venus, at about 240 Earth days per rotation, but this rotation alters the frequency of the reflected

---

[82]   The Arecibo receiving antenna in Puerto Rico is still the largest curved reflecting antenna in the world. Built in 1963 this bowl shaped reflector sits in a natural depression in the limestone landscape. The rotation of the Earth and a moveable detector gives some control over the pointing direction. The spectacular location of the antenna has made it a favourite with film makers – it appears in the movie Contact, an adaptation of Carl Sagan's novel, and the Bond movie Goldeneye, as well as an episode of the X-files.

electromagnetic energy enough to be measured as a Doppler shift[83]. Those parts rotating toward the Earth add to the frequency whereas those moving away reduce the frequency. The line from pole-to-pole facing the earth has no rotational velocity component in the direction of the antenna, so the reflected wave has the same frequency as the original transmitted wave.

By analysing the return signal as a function of the frequency of the returned wave, the echo can be broken into components of equal frequency as well as equal range. They represent Doppler rings in a plane parallel to the Earth-planet line, and to the axis of rotation of the planet (illustrated in Figure 9.9) . The result of this range and Doppler processing is a set of time-delay/Doppler-shift boxes. If these boxes are small enough, we can build up a picture of the scattering properties of the planet surface.

Note, however, that there is an unresolved ambiguity; the Northern and Southern intersections of the range and the Doppler circles coincide (*A* and *B* in the figure). Early planetary radar images did not resolve this North-South ambiguity, except in the case of the moon, where that radar beam could be kept small enough to avoid the ambiguity between the northern and southern echoes.

One method of resolving the N-S ambiguity is to use a second receiv-
ing antenna. The distance between each surface feature on the planet, and the antenna is slightly different for each antenna location. This small path difference can be measured by determining the difference in phase of the two detected signals. At the point on the surface where the range to the two antennas is identical, the phases of the two echoes are the same. Moving away from this point results in a change of path length, and hence phase, of the echoes. This phase shift can be measured and used to resolve the North-South ambiguity in the return signal. This technique, known as interferometry, is described in more detail in Chapter 11.

Phase Difference

9.4.2 Synthetic Aperture Altimeters

The discussion above gives us an important insight into how the Doppler shift can be used to provide a surface resolutin greater than that given by the aperture limit of the antenna. For Venus we utilise the rotation of the planet, but there is no reason why this same principle cannot be used in other circumstances where there is a relative velocity between instrument and target area. In Earth observation, this includes platforms that are moving along a flight-path. In this case the relative motion is a result of the instrument motion, rather than the target, but the effect is the same.

For instance, an altimeter travelling at a constant velocity and illuminating a very large footprint at nadir, will find that signals from in front of the instrument will be Doppler shifted to slightly higher frequencies, while signals behind will appear to have slightly lower frequencies. Directly below the flight line there will be a direction we can call the *zero*

---

[83] We will see in the next chapter that in fact there is a more appropriate explanation than the Doppler shift, but for simplicity, that discussion is left for later.

Doppler line. If the return echo is analysed as a function of the signal frequency, the echoes from difference parts of the footprint can be distinguished. In altimetry the technique is exploited by analysing the Doppler shift of the echoes such that the wide beam is broken down into bands of equal Doppler shift, from the front to the back of the footprint. The return echoes are therefore decomposed into their sub-beamwidth location and an improved ground resolution (in the along-track direction only) can be achieved. This technique is called *aperture synthesis* (since it synthetically produces the same result as a larger "virtual" antenna) but is also known as *Doppler beam sharpening* (depending on the manner in which the technique is implemented).

In the following chapter this concept will be considered in more detail, but it is worth introducing it here to illustrate how this technique applies to more than just the side-looking synthetic aperture radar imagers described in the next chapter.

The interferometric technique used to resolve the North-South ambiguity in Venus imaging can also be used to improve the across-track resolution of altimeters, and this will also be explained in the Chapter 11 under the heading of interferometric altimetry.

9.5   Scatterometers

Scatterometers differ from altimeters in that they are designed to sacrifice range accuracy and spatial resolution in order to provide very accurate measurements of the radar cross-section of the target. The time delay is still used to estimate range, but usually as an aid to locating the cross-section measurements, rather than as the final product itself. A good example of when both attributes are equally useful is when a scatterometer is pointed downwards over some target region that allows some penetration of the microwaves — forest canopies and snow or ice being good examples. In this case the data provides accurate measurements of the backscatter response as a function of the depth into the target.

Scatterometers are vital research tools as they allow detailed analysis of how microwaves interact with surface features. They are flown by aircraft and satellites, and scatterometer systems can be flown by helicopter (to allow longer integration time for the measurements) and so that they can be directed in different look directions to get estimates of σ for a number of different observation angles.

In order to make accurate estimates of σ, or σ^0, all the other parameters in Equation (9.5) must be well defined. Of these, the antenna gain is perhaps the most difficult parameter to determine with a high degree of accuracy and its value is often regularly updated once in operation. P_t and λ are predetermined systems parameters, while R is found from the echo delay time.

Spaceborne scatterometers now tend to fall into one of two categories: *wind*-scatterometers or rain radars. In both cases they are scatterometers

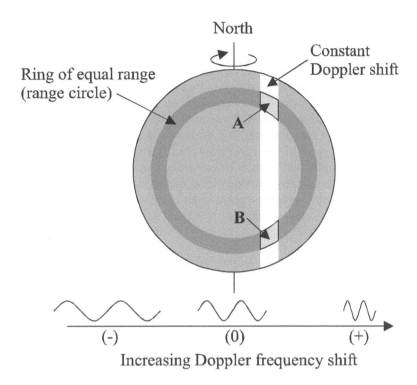

FIGURE 9.9 This figure shows a face-on disk of a planet such as Venus, showing a particular range and Doppler segment. The dark circle represents the area of the surface that lies equidistant from the antenna. The vertical strip shows the line of equal-Doppler shift caused by the rotation of the planet. (After Leberl, 1990).

designed for a specific application.

9.5.1   The Need for Scatterometer Measurements

Scatterometers are important instruments because they can provide well-calibrated cross-section measurements over a variety of targets that can be used to help provide insight into the way in which microwaves interact with natural objects, or as operational data for monitoring aspects of the environment. Chapter 5 described in some detail the backscattering properties of natural targets. Good quality scatterometer measurements should therefore help to provide information on properties such as sea surface roughness, soil moisture or vegetation cover. A great deal of important work has been done with ground-based scatterometers, both inside the lab and in the field to verify scattering models and generate databases of scattering properties. Scatterometers are also flown on aircraft or helicopters, and a scatterometer was flown on the NASA Skylab mission in the 1970s and made measurements across the globe.

*Wind*scatterometers are specialised instruments for measuring near-surface wind speed and direction over oceans. These instruments are designed to take backscatter measurements from different (two or more) look (azimuthal) directions across open water, from which it is possible to determine the direction of the wind-induced waves on the surface and hence the wind direction. With sufficient measurements and a suitably tested model it is also possible to additionally determine the wind speed. As well as being important for parameterising weather forecast models, near surface wind over the oceans is also important for modelling the exchange of energy, water vapour and carbon between the atmosphere and the oceans. Windscatterometers now provide global measurements of wind fields on a regular basis and provide regular input to climate models and weather prediction on an almost daily basis. This is a vast improvement on the traditional data from a sparse collection of tethered buoys and individual ships.

Rain radars are also a type of scatterometer, where the goal is to determine properties of rain clouds through accurate measurements of their radar cross section. The importance of global precipitation has been covered already in Section 5.1. The value of scatterometer measurements over the passive techniques discussed before is that radar can measure the scattering properties as a function of range, and therefore height, allowing the precipitation to be discriminated from background.

9.5.2   General Operation

Scatterometers operate in a similar manner to altimeters: a series of pulses (at a fixed PRF) are transmitted and the return echoes are measured, although more effort is now put into quantifying the properties of the echo, rather than timing the echo delay. Nadir scatterometers are usually used to explore the radar cross-section as a function of height through either vegetation, ice or precipitating clouds and have the same configuration as Figure 9.3. Scatterometers need not always be directed towards nadir, however, since the altitude is no longer of primary concern. For spaceborne scatterometers the beam is more usually directed obliquely from the instrument — pointed towards some location far from the nadir. The beam direction may be perpendicular to either side of the flight path, or along the flight path in either the forward or reverse directions. Depending on the intended application the beam may point at any arbitrary look angle, or a combination of different look angles and distances from nadir. Multiple beams can be used simultaneously by either taking advantage of differences in Doppler frequency shifts corresponding to different relative velocities of the beams and the surface, or by choosing slightly different frequencies for each beam.

As was the case for passive microwave imaging, there is some advantage to maintaining a constant angle of incidence for the backscatter measurements, so that for scatterometers that employ a narrow scanned beam,

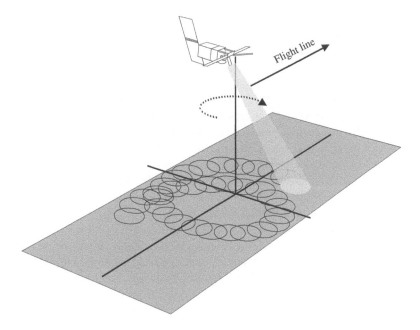

FIGURE 9.10 A conical scanning geometry. The advantage of the conical scan is that the incidence angle remains constant for all the measurements. Note also that each area of the surface is measured twice (except the very edges) from two different look directions.

the scanning is done so in a conical fashion, as illustrated in Figure 9.10. In this way, all the measurements will be for the same incidence angle and from two different directions (except at the edge of the swath).

If no scanning is employed, then area coverage is a consequence of the movement of the platform only. In this case the antennas are often designed to give footprints that are wide (in the across-track direction). Differentiation of the scattered signals across-track are then approximated using the echo delay time and by assuming the surface is relatively smooth. When viewing obliquely, the delay in the echo is directly related to the distance along the ground.

9.5.3   Rain Radar

When operating a radar system using frequencies above about 10GHz scattering from large hydrometeors (water droplets) in the atmosphere starts to become significant since the wavelength is now small enough that it is similar in size to the circumference of the droplets. This is a major drawback from instruments trying to measure the ground surface, such as altimeters, but such scattering can be utilised to estimate the rainfall rate. When hydrometeors become so large that they can no longer be supported by the atmosphere, they fall to the ground as rain.

Radars specifically designed to measure scattering from such hydrometeors are called *rain radars* or *precipitation radars* (PR). Although they have been used since the very first development of radar (and are now used to provide impressive graphics for television weather reports) it was not until 1997 that a PR was flown on a satellite, onboard the Tropical Rainfall Measuring Mission (TRMM). This instrument was simply called the PR (precipitation radar) and was used in conjunction with a passive imaging system also designed to measure rainfall. The system focused on measuring rain in the tropics, since two thirds of global rainfall falls within this region. The tropics are sparse in measurements from ground stations, although some of the major gaps are filled with rain radar on the occasional ship or island. Until TRMM there was a serious lack of information on intensity and amount of rainfall in this region.

PR on TRMM

The PR itself operates like a scanning altimeter in that it measures the intensity of the backscatter as a function of time delay, but it focuses on measuring the backscatter accurately across a wide swath (about 215km) with a range resolution of only 250m. The system is based around a 2.1m square active phased-array antenna and operates at two frequencies just slightly above and below 13.8 GHz. With the high frequencies being used it can achieve a relatively narrow beam giving a 4.3km footprint at nadir. The phased-array means it can electronically scan across the swath over an angle of $\pm 17°$. The 250m range resolution does not sound impressive compared to an altimeter, but combined with the scanning capability, the PR was able to generate data describing the three-dimensional distribution and intensity of rain within the swath.

9.5.4   Windscatterometers

Windscatterometers are scatterometers specifically designed to measure the wind speed and direction over water surfaces. Recall from Section 7.1 that oceans have an asymmetric backscatter pattern as shown in Figure 5.10. Measurements made viewing towards the up- or downwind direction have higher off-nadir backscatter returns that measurements made across the direction of the wind due to the presence of small wind-induced capillary waves. Windscatterometers exploit this relationship by making a number of measurements over the ocean from different viewing (azimuthal) directions. The minimum number of look angles is three. Three measurements allow some characterisation of the asymmetry pattern, but with a 180° ambiguity in the wind direction — the orientation of the wind ripples gives you a wind vector, but a wind vector without an arrow. The solution is to have at least one (but preferably more) locations where the wind direction is known from other (*in situ*) measurements, even if only approximately. Even a pretty rough estimate is sufficient to overcome the directional ambiguity at one location in the scatterometer data, from which the adjacent areas can then be determined (since the wind direction will not change significantly between adjacent measurement cells).

The extent of this asymmetry and the intensity of the echo are also used to determine the strength of the wind. This is achieved by semi-empirical modelling of the response as a function of *in situ* measured wind fields.

Such models are also necessary because the three (or more) observations are not necessarily made at the same incidence angle. This is apparent if we consider the possible operational geometries of windscatterometers.

The key constraint on the viewing geometry is the need to measure the *same* region of ground from different azimuth look directions. There are two basic ways in which this is achieved. The first is illustrated in Figure 9.11 for the example of the ERS Wind Scatterometers (WSC), which were mounted on the ESA ERS-1 and 2 platforms. These instruments consisted of 3 antennas producing 3 wide fan-beams looking 45° forward, sideways and 45° backwards with respect to the satellite's orbit direction. By range processing the echoes returning from the fan beam the signal is divided into 19 segments corresponding to 25km spacings on the surface[84]. Frequency processing of the echoes also allows for discrimination between the three beams, since the Doppler shifts are different for each beam. This is similar to the technique used for resolving the East-West ambiguity when imaging Venus, except this time the Doppler shift arises from the motion of the satellite. Echoes from the front beam are therefore shifted to a higher frequency, while those in the aft beam to a lower frequency. Those echoes from the centre beam have no Doppler component. *Geometry*

ERS-WSC

Note that across the swath the local incidence angles vary, ranging from 18–47° for the mid-beam and 25–59° for the forward and aft beams. The three backscatter measurements at each grid point are therefore obtained at two different incidence angles and so some model-based analysis of these triplets is required.

Over successive orbits these beams were used to continuously illuminate a $500km$ wide swath using microwaves at 5.3 GHz (C-Band) with vertical transmit and receive (VV) polarisation (although it should be noted that the SAR instrument onboard the ERS and the Wind Scatterometer were not able to operate simultaneously so that complete global coverage was not always possible). Any particular region on the ocean surface would be viewed sequentially (but in such quick succession that it was assumed instantaneous) by the fore, mid, then aft beams, providing 3 distinct azimuth views. These triplets were then used routinely to extract wind speed and direction over sea surfaces for over a decade.

The alternative approach to the fan-beam geometry is to use conical scanning, such as that used by the QuickSCAT scatterometer on the Sea-Winds platform. This geometry employs the conical scanning set-up as shown in Figure 9.10, but this time with two narrow pencil-beams scanning at different scan radii. The reason for this is that a single conical scan only affords two look directions for any ground area, so that a second beam is needed to provide additional viewing angles. A high frequency is used (Ku-band, 14 GHz) to allow a narrow beam, and both HH and VV *QuickSCAT*

---

[84] Note however, that the final measurement cell size is actually 50km.

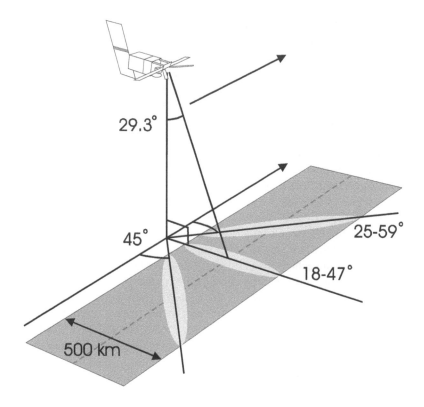

FIGURE 9.11 The geometry of the ERS windscatterometer. Note that the instrument has three beams in order to infer the wind speed and direction from the asymmetric ocean surface scattering. The angles refer to the local incidence angle ranges across the three footprints.

polarisations are recorded, so that for each grid point four measurements are recorded: a VV from the outer cone (at high incidence angle of 54°) and an HH from the inner cone (at lower incidence angle of 46°). For all the swath except along the nadir track and at the edge of the swath, these four measurements are also made at four different azimuth directions. For QuickSCAT this gave a 1600km swath with 12.5 km grid spacing between retrieved points.

Land Measurements   Measurements from both these scatterometers have also been studied over land and ice surfaces. Despite resolutions of tens of kilometres, the instruments provide novel datasets. They are novel because they have high radiometric accuracy and stability, they are made over a range of incidence angles, they provide global coverage in a matter of days, and they provide azimuthal information not usually available from other radar instruments. As a consequence, these scatterometers have been used to study soil moisture dynamics, ground thaw in Siberia, ice sheet properties, and even wind ripple orientation over sand seas in the Sahara.

9.5.5   Polarimetric Scatterometers

Scatterometers that are also polarimetric can additionally provide information on wind direction since the relative orientation of the surface ripples will impact on the polarimetric response. Fully polarimetric data would therefore minimise the requirement of three look directions. From the perspective of satellite design, this looks appealing, but it brings with it the added complication of making an well-calibrated instrument that is fully polarimetric and one that can achieve a sufficiently good SNR for the cross-polarised channels.

9.6   Further Reading

The most comprehensive source of information on radar operation that is still accessible to a wide audience, is Elachi's "Spaceborne Radar Remote Sensing: Applications and Techniques" (1988). Even though some of the missions he discusses are long passed, the text is thorough enough to cover techniques that are right up to date. As usual, Ulaby Moore and Fung also provide full examination of the topics within this chapter. More information on imaging the solar system can be found in Frank Leberl's "Radargrammetric Image Processing" (1990).

10
IMAGING RADAR

'What is the use of a book,' thought Alice, 'without pictures or conversations?'
— Lewis Carroll, *Alice in Wonderland*, (1832–1898).

The last chapter demonstrated that active microwave radar systems in the form of altimeters and scatterometers provide quite different types of information than is available from other sensors. They can make very precise measurements of distance, are sensitive to fine scale roughness patterns and have the potential to provide useful information on geophysical properties such as sea ice extent and soil moisture content, amongst other things. All this without the requirement of solar illumination and, for the longer wavelengths (frequencies lower than 10GHz) complete, weather-independent, coverage. There is therefore good reason why we might want to consider methods that would considerably improve the spatial resolution of such instruments beyond the kilometre-scale footprints of altimeters and scatterometers, down to scales of metres.

Given the relationships between natural materials and microwaves described in Chapter 5, being able to produce a very high spatial resolution map of radar backscatter potentially allows detailed ocean wind and sea ice mapping, soil moisture estimation and vegetation discrimination, and a range of other possible applications. The question that this chapter has to address, therefore, is how to go about generating the necessary high-resolution images using radar principles.

10.1 The Need for Imaging Radar

A key question to address before describing how to create radar images, is whether there is a real need for high resolution maps of microwave backscatter. As well as being large and power hungry, imaging radars also have the capacity to produce vast amounts of data, which must then be elaborately processed to generate an image (this is particularly true for the synthetic aperture radars). All of these qualities equate to considerable expense, especially for satellite systems, so there should be compelling reason to invest in such systems.

There were probably two main driving factors for the introduction of imaging radar into the civilian context. The first is coverage. The fact that cloud cover is not an issue and solar illumination is not required, gives a clear advantage over optical-based systems. This was the justification of the mapping project in the Darien Province mentioned in Chapter 2 — this area had never been mapped before due to persistent cloud cover. And as we saw in the case of passive imagers, the ability to measure through the frequent polar cloud is a necessity when mapping sea ice — the lack of dependence upon solar illumination being a further key advantage in polar regions.

The second driving factor is the unique information about the ocean surface that can be inferred from radar backscatter. Both the fine-scale capillary waves observed by scatterometers, but also longer wavelength gravity waves and swell (going from tens to hundreds of metres) will be observable at high spatial resolution because of the influence of the changes in local slope and surface roughness.

It should be no surprise, therefore, that it was Seasat, an oceanographic satellite launched in 1978, that carried the first civilian imaging radar system into orbit, and the most successful commercial imaging radar satellite has been Radarsat, a Canadian system with the primary goal of mapping Arctic sea ice.

More recently, the impetus for future radar satellites is the high precision topographic data available from interferometric imaging radar and the statistical relationship between forest biomass and L- and P-band radar backscatter, as described in Chapter 5, which offers the potential to help map and monitor carbon stocks and terrestrial carbon dynamics.

10.1.1   Oceans

Although not the first use of active microwave instruments, studying the seas and oceans with microwaves certainly matured faster than other applications. This is due largely to the fact that scattering from open water surfaces is, in principle at least, relatively straightforward since it is governed almost entirely by the surface roughness alone, and the roughness properties can be considered relatively constant across a large area (i.e. many pixels over part of an image). The dielectric constant is high and so there is little or no penetration into the water, even at long wavelengths and so the overall total scattered power is also relatively high. The salinity of the water will also change the dielectric constant of the water, but in a radar imaging context this will not usually vary significantly over small areas and is also very difficult to detect since the effect is small compared to the impact of changes in surface roughness.

Perhaps the most obvious influence on the surface state is the local   *Surface conditions*
weather conditions. As a result of friction between the water and the moving air, the wind sets up an oscillation on the surface, which results in short
wavelength ($\sim$cm's) *capillary waves* on the surface in addition to the long
wavelength gravity waves. These waves are aligned perpendicular to the
wind vector, and their wavelength is dependent on the wind speed. As was
mentioned in the discussion on windscatterometer instruments, the orientation of these capillary waves with respect to the incoming radar pulse is
very important, and has a big influence on the scattered energy. Rainfall
may also roughen the water surface, so that patterns of rainfall may also
be observed in radar images.

Although it suffers from more directional ambiguity than the windscatterometers described in the previous chapter, imaging radar can none-the-
less provide higher spatial resolution information on wind speed and significant wave height for input to forecasting models. Indeed, most satellite
imaging radar will operate a "wave mode" designed specifically to measure the wave spectrum of the ocean surface.

10.1.2   Sea Ice

We saw in Chapter 8 that passive microwaves can discriminate sea ice
from open water, and different ice types, based on changes in dielectric
properties and surface roughness. Active microwaves are also sensitive to
the same changes, so high resolution imaging can provide detailed sea ice
maps.   From a global perspective, high resolution maps of the ice extent
are not necessary — the coarse resolution is sufficient. There are, however, two situations where high resolution imagery of the ice is extremely
important.

The first is for navigation for merchant shipping. There are three natural trade routes from the Atlantic to the Pacific by sea: South, round the
southern tips of Africa or South America, and North through the Canadian
Arctic (the "Northwest Passage")[85]. Canada's Radarsat was a dedicated
imaging radar satellite with the aim of providing high resolution, fast turn-
around, ice maps as a commercial product to aid shipping.

The second role for imaging radar is mapping the edge of the ice shelves
in Greenland and Antarctica. The extremities of these ice sheets eventually make their way to the sea, often ending up as a sheet of floating ice
on the water. The ice can eventually break off from the edge (calving)
to form icebergs. Detailed maps of these regions can provide information
on the ice balance of the ice sheets and so help identify continental scale
responses to global climate change.

---

[85]   The two "un-natural" routes are through the canals of Panama and Suez.

10.1.3   Terrestrial Surfaces

The two areas where imaging radar warrants greater use in the terrestrial context is soil moisture estimation and forest biomass mapping[86].

Soil moisture

The relationship between dielectric constant and soil moisture has been discussed elsewhere, and was the basis of the regional soil moisture estimation from passive sensors (Chapter 8). The key advantage is this chapter is the potential for high resolution soil moisture maps, particularly at longer wavelengths which can afford some penetration through surface vegetation. High resolution soil moisture mapping is probably most relevant at the level of precision agriculture, but medium resolution mapping (circa 500m–1km, which would not be feasible with passive sensors) would offer information that would be vital for weather forecasting, climate modelling and hydrology.

The one limitation is that radar backscatter is very sensitive to surface roughness, so that it is very difficult to develop robust methods of estimating soil moisture content over agricultural fields that might continually change in roughness. As yet, while it is a potential future driver of satellite technology, it remains a secondary application of instruments designed with other purposes in mind.

Forestry

Forestry applications, on the other hand, are now increasingly becoming a primary consideration for instrument design. As well as being important ecosystems, forests exert a strong influence on the environment from local to global scales, both in their impact on boundary layer climate and their importance as a carbon sink. Timely and accurate mapping of forest parameters such as canopy height, structure, species and biomass density are therefore invaluable for conservation studies, mapping vegetation dynamics and modelling climate processes at a range of scales. Since seasonal and short-term changes (such as forest fires) are important this can only be done effectively over large areas using some form of remote sensing.

Biomass estimation

The need for high resolution estimates of forest biomass and biomass change is now extremely important, both for understanding the global carbon budget, as well as for potential carbon storage assessment for supporting international treaties on carbon emissions now that the Kyoto Protocol has taken effect. The economic value of data on forests is therefore becoming increasing sufficient to warrant the expense of a dedicated sensor and there is a wealth of information to support the use of imaging radar in this application. By the late 1980s it was quite clear that there was a correlation between the above-ground biomass of a forested area and the measured backscatter. As an indication of biomass distribution in remote areas, this can be a very useful relationship to exploit, but it can be limited since different forest types, ground parameters and weather conditions (in terms of surface water) can all influence the backscatter. The correlation estimated

---

[86]   There are also imporant terrestrial applications of interferometric SAR but these are discussed in the following chapter.

at one region, over a set period of time, is therefore not always applicable to a different region, or a different time of year, although long-term monitoring should help identify biomass *change*. As discussed in Chapter 5 the longer the wavelength the better the penetration into a forest canopy and therefore the greater sensitivity to standing biomass. L-band, for instance, is generally sensitive to biomass up to about 70–80tha^{-1} whereas the longer P-band microwaves will give biomass up to about 100 tha^{-1}, with VHF being sensitive to even higher values. To date, this approach has been difficult to implement on a satellite because of the practical problems of P-band, in particular the Faraday rotation from the ionosphere.

10.1.4   The Water Cloud Model for Vegetation

The long-standing simplified model used for exploring the basic microwave response of vegetation canopies is the "water cloud model" (WCM) whereby the canopy is modelled as a cloud of identical, randomly oriented scatterers, as described in Section 3 (Attema and Ulaby, 1978). The WCM was developed from earlier models used to characterise backscatter from clouds of hydrometeors, the link being that a vegetation canopy could, as a first order approximation, be described as a collection of identical, randomly oriented scatterers (i.e. leaves and branches).

If we take a similar approach to Section 4.2 we can generalise to a surface layer with a NRCS of σ_s^0 lying beneath a sparse volume scattering layer with a NRCS of σ_v^0 and vertical opacity of τ_v. When the volume is sparse with low transmissivity, Υ_V, the effects of the volume are threefold. Firstly, it modifies the surface scattering term by attenuating the signal by the factor Υ_V^2 (the exponent of 2 a consequence of the signal originating from above the canopy then transmitting through the canopy to the ground and back out again). Secondly, it contributes a direct term of its own, given by σ_v^0 , as defined in (5.36). And lastly, it contributes a scattered term that corresponds to that part of the wave scattered by the volume that goes downwards and scatters off the surface. This "forward scattering" term is often ignored in the formulation of the WCM and we do so again here, but it is mentioned for completeness since in some circumstances it can make a significant contribution to the final measured signal.

For the final expression for the NRCS for the top of the canopy viewed at an incidence angle of θ, we therefore have

$$\sigma_{\text{canopy}}^0(\theta) \;=\; \text{(attenuated surface emission)+(direct volume emission)}$$

$$=\; \sigma_s^0(\theta)\Upsilon_v^2(\theta) + \sigma_v^0(\theta) \tag{10.1}$$

$$=\; \sigma_s^0(\theta)\Upsilon_v^2(\theta) + \cos\theta \frac{N_V\sigma}{2\kappa_e}(1 - \Upsilon_v^2(\theta)). \tag{10.2}$$

The dependence on incidence angle, θ, has been left explicit to emphasise that this is an important consideration in radar observations.

The water cloud model does not explicitly include polarising effects, but a separate model can be generated for each polarisation. If the volume is composed of randomly oriented scatterers then the only variability in the polarisation arises from the ground scattering term. Changes in polarisation of the measured signal can therefore change both as a consequence of changes in the ground characteristics, but also from the relative contribution from the canopy versus the ground — if the random canopy dominates the signal, then the polarisation dependence of the ground will be less noticeable.

In non-random volumes the polarisation will change as a consequence of the preferential orientation in the canopy, but also as a result of preferential transmission through the canopy.

10.1.5   Other Uses of Radar Imagery

The use of radar imagery in the above examples are the important issues that have justified investment in such expensive systems (about 3 billion Euros for Envisat, for instance). That is not to say, however, that radar images do not have a wide range of other uses, simply that these other applications are often not sufficiently well-established or economical to warrant launching a dedicated satellite.

Oil slick monitoring

One such application that follows on from the oceanographic context is oil spill mapping. Wind and rain are not the only influences on the surface conditions of the ocean — an oil spill will decrease the surface tension properties of the water and so "dampen" the wind-induced waves. The result is a smoother surface which acts more like a specular reflector than the surrounding water and therefore the patch of oil will appear dark in the final radar image — that is, as long as there is sufficient wind to create the contrast between the open water and the oiled surface. It is not just the headline-grabbing large oil slick disasters that are a problem globally. Oil on the ocean surface may occur naturally due to seepage from the ocean floor and from algal blooms, but will also occur from ships illegally cleaning tanks out on the open ocean. Imaging radar provides one method by which such oil pollution can be monitored on a regular basis.

Additionally, imaging radar can provide a means to locate and identify the individual ships that are acting illegally.

Ship detection

As ships move through water they generate a wake, the turbulence churning up the surface behind them. This results in an increased surface roughness that can be detected in radar imagery even by spaceborne radar systems. Ships are also very bright radar targets as they are usually made of metal (a conductor) and have many right-angled corners. Additionally, if a moving ship is imaged by a radar system utilising the Doppler effect of the backscattered echoes then the motion of the ship changes the Doppler shift of the echo and so relocates the image of the ship to a slightly different location, the distance of the displacement being in proportion to the relative velocity of the ship and the radar. The relative distance between the end of the imaged wake (the "true" location of the ship) and the image of the ship therefore gives a measure of the ship velocity, and an analysis of the ship backscatter can often narrow down the type of ship to a particular class. For obvious reasons the navies of the world have been interested in radar for this purpose for quite some time.

In coastal regions imaging radar can have a further application. Although microwaves do not penetrate a water body, under shallow-water ($< 10m$) conditions, radar images of sufficient spatial resolution ($< 50m$) can provide information that can be used to estimate water depth. The measurements are based on the fact that when water is relatively shallow, the subsurface topography will influence the water surface if the water is flowing (as a current, usually in an estuary) such that the flow will be faster where the water depth is shallower. The resultant change in water speed will also result in a corresponding lengthening of the wave features on the surface. The final result in the image is that for shallower regions, the radar image appears darker.

Bathymetry

10.2   What is an Image?

The term "image" is frequently used in a remote sensing context, but rarely do we stop to think what actually constitutes an image and what might distinguish an imaging from a non-imaging sensor. A microwave sounder can provide a map of upper atmospheric chemistry. Microwave altimeter data may be gridded to form a graphic representation of the spatial variability of the measurements. But in neither case do we refer to these methods as "imaging".

An imaging system could be defined as one that *directly* measures properties related to the spatial variability, rather than *gridding* independent measurements, but optical scanning imagers would then not be consider as an imaging system. And an image might not be any kind of picture at all — a hologram appears to the viewer as an image when properly illuminated, but if you look closely at the photographic plate you will see only interference fringes of alternating bright and dark lines, not an picture. A hologram is not a pictorial representation of the light reflected from the object, but a recording of the actual scattered light field and when this field is reconstructed by proper illumination we observe the object in the hologram *as if it were actually there*[87]. Imaging radar has much more in common with holograms than they do with optical cameras.

The point to note here is that there is no unambiguous distinction between "imaging" versus "non-imaging". In the case of radar imaging the key distinction is that, unlike, say an altimeter, each of the individual

---

[87]   If you need convinced about the distinct difference between a photograph and a hologram, you need only consider the fact that if you were to break a glass-plate hologram, each individual piece of the glass would still contain the image contained within the original hologram (albeit a much poorer version with a restricted field of view compared to the original). Even more remarkable is a hologram on display at the University of Edinburgh's School of Physics. In this hologram there is a microscope that appears to protrude from the glass and if you place an eye as if to look down the eyepiece, you can view the details of the microchip that is on the microscope slide!

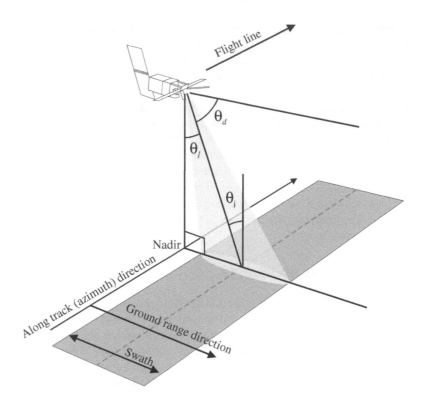

FIGURE 10.1 The geometry of a side-looking airborne radar imager. Note that the radar beam is relatively wide in the range direction (providing a large swath) and narrow in the azimuth direction (giving a thin "slice" of data). The movement of the instrument allows the construction of a two dimensional image.

echoes are used to generate a number of image data points (rather than a single measurement) and these data points are in some way related to the spatial dimensions of the target scene. Collecting a sequence of echoes therefore allows the construction of a two dimensional radar image of the scene, normally with each echo contributing to a "slice" of measurements across a swath, while the movement of the instrument allows a collection of echoes to be measured at different points along the swath.

Unlike optical imagery, an imaging radar gives no information on direction to the target — we have to rely on simplifying assumptions such as the imaged surface being smooth. Of course, this assumption need not always be valid, and in this chapter some of the unique image features that arise from the Earth not being smooth will be discussed.

A flash camera is often used as an equivalent example of an active optical system, yet this comparison distracts from the key features that make optical and radar images different. Some of these are discussed in the following sections.

10.3   Radar Image Construction

So how do you construct an image from a radar system? At various points throughout this text it has been emphasised that the problem with microwave systems in that they cannot practically achieve the narrow beamwidths you would require to produce a detailed image of the surface. If we took the same approach as an optical scanner we would need a very narrow beam to provide a good instantaneous field of view (IFOV). Chapter 9, on the other hand, discussed the impressive ability of radar systems to measure distance. So the approach for imaging radar is to compensate for the poor angular resolving power by using a technique that takes full advantage of its excellent range detection abilities. This is achieved by taking a basic radar instrument and pointing the antenna sideways, i.e. with an oblique viewing angle as shown in Figure 10.1. This basic idea was introduced in Section 4.1 to make spatial measurements of the backscatter from the Earth's surface using a scatterometer. Distance on the ground is then approximately related to the time delay of the echo. The subsequent construction of a 2-D radar image results from utilising the motion of the instrument to scan the ground along the flight path.

In this way the problems associated with the poor angular resolution of microwave sensors are avoided and we can utilise the high quality range measurements — one dimension of the final image is derived from the range distance to the target. If in the first instance we assume a smooth surface for the Earth we have a direct relationship between range from the instrument and distance along the ground, which is ultimately what is required in any geoscience application.

The important point to stress is that a radar imager, like a bat, principally obtains information as a function of distance from the instrument, rather than look direction, as is the case for optical systems.

The key geometric variables are identified on the figure. The look angle, θ_l, is the angle from the nadir direction to the instrument line of site, whereas the depression angle, θ_d, is measured from the instrument's local horizontal. The incidence angle, θ_i, is measured between the line of sight and the normal to a predefined local reference surface. This surface may be a local datum[88] or a site-specific surface such as an average surface elevation. Note that the *local* incidence angle refers to the angle between the line of site and the normal to the actual ground surface, rather than a reference surface. For airborne systems it is usually possible to assume a flat Earth, in which case the look angle and incidence angle are the same for any given line of sight. This is not the case for satellite systems where the curvature of the Earth has to be taken into account.

Radar geometry

The term *slant range* refers to the actual range measured by the radar derived from the time delay of an echo. This range can be translated into

Slant range

---

[88] A datum is the now out-dated term for a terrestrial reference surface (TRS), which is the projection surface for a particular mapping system. The WGS84 is a common one for global applications. Individual countries, regions and continents will use their own TRS.

a ground range if we assume the ground follows the chosen reference surface. The fact that the Earth's surface actually deviates from this surface is the source of some quite dramatic topographic distortion in the final image — this is discussed in more detail later in this chapter.

The ground range is sometimes referred to as the "true ground range", but note that for curved reference surfaces the projected ground range may not be linearly related to the slant range.

Monostatic vs bistatic

For the rest of this book it will be assumed that the signal is transmitted and received by the same antenna. This is the *monostatic* case. When the transmit and receive antennas are located in different positions it is called *bi*static. Although most remote sensing is done by monostatic radar, the potential for bistatic radar is becoming increasingly interesting as the bistatic scattering cross-section provides very different target information to the backscattering cross-section.

10.4   Side-Looking Airborne Radar

Before diving straight in at the deep end with a description of a Synthetic Aperture Radar (SAR) system, it is customary, and useful, to first give a brief description of the original radar imaging system, *Side-Looking Airborne Radar* (SLAR) or *Real Aperture Radar* (RAR). The geometry of the situation is that shown in Figure 10.1. The radar equipment is mounted on an aircraft moving in a straight line at altitude H above some reference surface. As described above, the antenna points sideways and down at the ground so that the radar beam illuminates a small part of a continuous *swath* across the imaged area. We refer to the area of illumination as the *antenna footprint*.

Range

The radar system transmits a sequence of short microwave pulses (of length τ_p) which allows discrimination of slant-range R since a radar pulse travelling at the speed of light, c, will take t_1 seconds to travel the distance to an object at range R_1 and back again. The received power as a function of time delay therefore gives one dimension of the final image. This dimension is referred to as the *range* direction.

In the range direction the system operates much like the altimeter in Section 3 except that for a flat surface the shortest range distance is now to the near edge of the antenna footprint, and not the centre of it, as was the case for the altimeter. The returned echoes return progressively later from ground that is further away from the instrument.

This system is limited in its ability to measure range by the same constraints as the altimeter, namely the effective length of the pulse. The radar equation introduced in Section 2.1 applies equally to RAR systems, but is far more restrictive than for an altimeter: the ranges are longer since the radar pulse has an oblique path length; and the radar cross-section, σ, is smaller because the larger incidence angle results in much of the incident energy being scattered towards the specular direction, away from the radar system. There is therefore an even greater need to optimise the power in a pulse while keeping a short pulse length for high range resolution. Chirped pulses and range compression techniques are therefore standard on imaging radar systems and the slant range resolution is therefore given by Equation (9.12), repeated below:

$$\rho_r = \frac{c}{2B_p}, \qquad [\text{m}], \qquad (10.3)$$

where ρ_r is the slant range resolution in metres, c the speed of light and B_p the bandwidth of the chirped pulse. As with the altimeter the factor of 2 is because it is a there-and-back distance. The range resolution is ultimately defined by B_p, not the pulse length, so very often the range resolution of radar imagers is quoted as a pulse bandwidth (in units of frequency) rather than a distance. Larger bandwidths give better range resolution.

10.4.1 Ground Range resolution

One good reason for quoting a pulse bandwidth rather than a distance is that you will not be fooled into thinking that this slant-range resolution is the same as the spatial resolution of the final image. Equation (10.3) is the *range* (or *slant range*) resolution and is a system-limited parameter that is totally independent of the target under observation. In the Earth observation context, we are more interested in the *ground range resolution,* ρ_g — the ability to discriminate features that actually lie on the Earth's surface. The range resolution on the reference surface is then determined by the local angle of incidence, θ_i, as illustrated in Figure 10.2. Some straightforward trigonometry gives the range resolution on the reference surface as,

$$\rho_g = \frac{\rho_r}{\sin\theta_i} \qquad [\text{m}]. \qquad (10.4)$$

There are two things to note here. The first is that the ground resolution is progressively better as you go further from nadir (increasing incidence angle) — the near edge of an image swath has poorer ground resolution than the far edge of the swath. This is the exact opposite to what you would get with wide-swath optical imagers. In a single image the variation will be greatest for low altitude sensors (e.g. aircraft rather than satellite) since the incidence angle range increases as the altitude decreases, if we assume the swath width remains the same. The best achievable ground range resolution is achieved at grazing angles ($\theta_i = 90°$) when $\rho_g = \rho_r$.

The second point to note is that ρ_g is the range resolution projected onto a reference surface. For terrain that deviates from this reference surface the local slope will influence the local incidence angle and therefore the local range resolution over the terrain. A mountain, for instance, will have a better range resolution on the far side, than it does on the near side because of the differences in local incidence angle (look forward to Figure 10.7 if you need to be convinced of this).

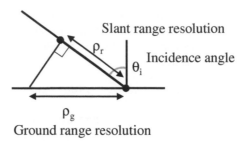

FIGURE 10.2 The relationship between the instrument limited range resolution, ρ_r, and the effective ground resolution, ρ_g. Note that the resolution on the ground is dependant upon the local incidence angle, θ_i.

10.4.2   Azimuth Resolution

The *along track* direction of the final image is called *azimuth* direction. In all types of imaging radar the movement of the platform scans the terrain in the azimuth direction to build up a 2D image. In RAR the azimuth resolution (which is the ability to differentiate two point targets at the same range but different azimuth angles) is determined by the width of the beam in the azimuth direction, which in turn is limited by the length of the antenna. This is the diffraction limit and is identical to the limitations of an optical system. Using Equation (9.25) a ball-park estimate of the spatial resolution in azimuth is therefore given by

$$\rho_a = \theta_{3dB} R \approx \frac{\lambda}{d} R, \qquad \text{[m]}. \qquad (10.5)$$

This does not pose particular problems for airborne radar which might reasonably be able to achieve a spatial resolution of about 50m from an altitude of 3km with an 3m long antenna, operating in X-band (3cm) (and remembering that R would be about 1.5 times the altitude for an incidence angle of 45°).

For higher altitude aircraft or for satellites, RAR has a very limited potential due to the constraints of building the large antenna necessary to counteract the considerable increase in R (about one hundred times higher). Consider trying to achieve a similar azimuth spatial resolution of 50m with a RAR at 800km, again at X-band: you would require an antenna over 700m long! To achieve a resolution similar to contemporary optical imaging satellite sensors (circa 30m), would require an antenna longer than a kilometre (and proportionally longer for longer wavelengths). This is clearly not a very practical proposition unless you are satisfied with azimuth spatial resolutions in the order of hundreds of metres. Spaceborne imaging radar must therefore employ a different strategy to achieve the required azimuth resolution.

10.5   Synthetic Aperture Radar (SAR)

The alternative to the real aperture is the *synthetic aperture*. The "synthesis" term refers to a method of processing the returned echoes to improve the azimuthal resolution allowing practical spatial resolutions of the imaged scene, even from spaceborne instruments ($H \approx 800$km). The approach can be explained with reference to two techniques already mentioned within this book. The first is the Doppler beam sharpening approach explained in the context of imaging Venus from Earth and synthetic aperture altimetry, discussed in Sections 4.1 and 4.2. The second is the principle of coherently combining a collection of low-resolution antennas in order to provide an greatly improved resolution, as discussed in Section 4.5 and illustrated in Figure 3.12.

The key factor that is utilised in SAR is to synthesise a much larger effective antenna using a small antenna which is *moving* along a flight line. The following two sections offer two models of explaining the aperture synthesis, and although the Doppler explanation is sufficient for many readers (and you may even wish to skip the geometric interpretation) I do recommend you give both explanations at least a brief read-through so that you are aware of some of the more complicated elements of aperture synthesis.

The geometry of a SAR system is shown in Figure 10.3.

10.5.1   Aperture Synthesis: A Doppler Interpretation

The SAR platform is continuously moving so that we can utilise this effect to locate which part of the beam the echo returns from — the echoes returning from objects in the front part of the beam are Doppler shifted to higher frequencies, while echoes from the aft part of the beam are shifted to lower frequencies. The greater the frequency shift, the further fore/aft the echo is located, so that the antenna footprint can be divided in to bins of equal Doppler shift (as well as bins of equal range). For any given target the return signals will change in frequency as it passes through the radar beam.

Unlike RAR, SAR does not need a narrow beam, but instead it makes very precise measurements of the frequency properties of the returned echoes. One way to consider the process is to compare it to the pulse compression technique described for chirped pulses in Section 2.2 since (in the Doppler model) it utilises the frequency analysing capability of the radar system.

An expression for the frequency shift of a moving target was given in Equation (3.32) such that the change in frequency is given by:

$$f_D = \frac{V_{\text{rel}}}{\lambda} \qquad [\text{s}^{-1}], \qquad (10.6)$$

with V_{rel} the relative motion between source and detector and λ the wavelength of the signal. We can utilise this relationship by considering

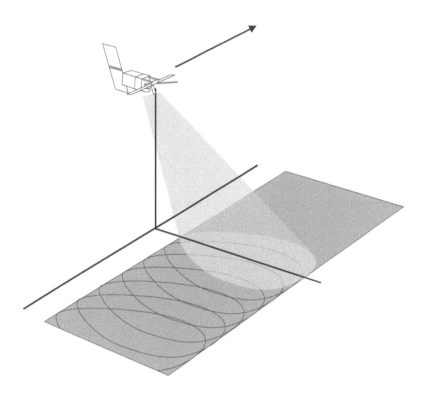

FIGURE 10.3 The geometry of a synthetic aperture radar (SAR) system. Note that the beam is now designed to be very wide, making adjacent footprints overlap. By using the Doppler shift of the return signals the echoes from different parts of the footprint can be discriminated. By combining the signals over many footprints, it is possible to synthesise a larger antenna.

Figure 10.4, which shows, in plan view, the geometry of the illuminating beam shown in perspective view in Figure 10.3. Although it is the instrument that is moving, it is convenient in this discussion to consider the instrument as stationary while the ground moves through the beam. *Azimuth compression* The frequency "history" of a signal from a target that has moved through the beam has an approximately linear shift in frequency. In fact, it has the same form as the linear FM chirp used in range compression shown in Figure 9.2 (although on a different scale) — *i.e.,* the frequency of the signal appears to sweep through a range of frequencies (bandwidth) from high to low. Indeed, the processing step of *azimuth* compression is just another correlation but this time it is over a *set* of returned signals rather than over a single returned echo.

The system must therefore be able to transmit consistent (coherent) pulses and accumulate the echo information over successive pulses to allow the synthesis of a virtual antenna that is much longer than the physical antenna.

There is, however, a potential difficulty with azimuth compression. Un-

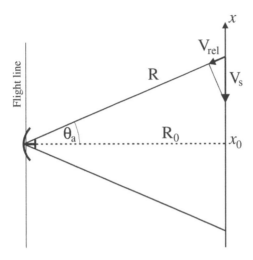

FIGURE 10.4 A plan view of the geometry used for the Doppler explanation of aperture synthesis. The figure is drawn from the perspective of the instrument, so it is the target that seems to move through the beam with velocity, V_s, along the x-axis (the azimuth direction).

like range compression, azimuth compression has no definite reference function to do the correlation since it requires some assumption about the imaging geometry and some correction of the effects of irregular instrument motion. The reference function is usually determined by a combination of known platform parameters and correction functions that repeatedly update the reference function as the data is processed.

We can now determine the azimuth resolution using the same approach *Azimuth resolution* given in Section 2.2. The equivalent pulse width after range compression was determined by the inverse of the bandwidth of the chirp. In azimuth compression we can write a similar expression that tells us how well the instrument can differentiate signals in time, ρ_t, such that

$$\rho_t = \frac{1}{B_D} \qquad [\text{s}], \qquad (10.7)$$

where B_D is now the Doppler bandwidth. But we need to determine what this time resolution will equate to in terms of azimuth distance. To do this, and hence determine the azimuthal spatial resolution, we must include the azimuthal speed of the sensor, V_s, (rather than the speed of light as was used in the range resolution) measured in the coordinate system of the target, such that

$$\rho_a = \frac{V_s}{B_D} \qquad [\text{m}]. \qquad (10.8)$$

The azimuthal speed is just the speed of the platform, so to determine an expression for the azimuth resolution we now have to determine the Doppler bandwidth. To do this we have to consider in a little more detail what happens to a target as it moves through the radar beam (as illustrated in Figure 10.4). The maximum relative velocity between the sensor and the target is when it enters or leave the beam. This is given by the velocity, V_s, scaled by the angle between the velocity vector and the direction to the target, such that

$$V_{\text{rel}} = V_s \sin \theta_a \qquad [\text{ms}^{-1}].$$

When the beam is centred on the zero-Doppler line (i.e., directly perpendicular to the flight path at x_0) the minimum relative velocity is simply $-V_{\text{rel}}$, as it leaves the back of the beam. The range of echo frequencies is then $(f_0 - f_D)$ to $(f_0 + f_D)$, where f_0 is the original transmitted or "centre" frequency. Using the expression given in Equation (3.32) the Doppler frequency is then

$$f_D = 2\frac{V_{\text{rel}}}{\lambda}, \tag{10.9}$$

where λ is the wavelength of the transmitted wave. The extra factor of two is necessary because the signal is Doppler shifted twice: the transmitted wave is Doppler shifted relative to the (apparently) moving target, and the return echo is Doppler shifted again relative to the moving instrument. The full bandwidth is then given by

$$
\begin{aligned}
B_D &= (f_0 + f_D) - (f_0 - f_D) && (10.10)\\
&= 2f_D && (10.11)\\
&= \frac{4V_{\text{rel}}}{\lambda} && (10.12)\\
&= \frac{4V_s \sin \theta_a}{\lambda} \qquad [\text{Hz}]. && (10.13)
\end{aligned}
$$

We know the value of the angle θ_a because it is half of the full beamwidth of our real antenna of length D, given by

$$\sin \theta_a \approx \theta_a = \frac{1}{2} \cdot \frac{\lambda}{D} \tag{10.14}$$

where we have utilised the fact that if we deal with units of radians, then when θ is small ($\ll 1$ radian, which is the case for most microwave radars) the sine of the angle is approximately equal to the angle itself. If we now use (10.14) and (10.13) in (10.8), we get the remarkable result that:

$$
\begin{aligned}
\rho_a &= \frac{V_s}{B_D} && (10.15)\\
&= \frac{V_s \lambda 2D}{4V_s \lambda} && (10.16)\\
&= \frac{D}{2} \qquad [\text{m}]. && (10.17)
\end{aligned}
$$

This result implies that a properly optimised SAR system will have an azimuth spatial resolution that is equal to half the length of the antenna! It is not dependent upon the distance from instrument to sensor, so is the same whether it is a spaceborne or an airborne platform. Nor is it dependent upon the wavelength of the radar, so the same size of antenna provides the same azimuth resolution regardless of the wavelength. Most surprising is the counter-intuitive condition that the azimuth resolution actually improves as the physical antenna gets smaller.

At first glance such an idea would appear to contradict the expressions already given for resolution from an aperture of length D, but remember that in a SAR we are no longer relying on the angular beamwidth, but rather the range of Doppler frequencies such that a large spread of Doppler frequencies will give us a better azimuth resolution. As the antenna gets smaller, the beamwidth gets larger, meaning that θ_a in Equation (10.13) increases and so the bandwidth also increases, and in a linear fashion (i.e., halving the antenna doubles the Doppler bandwidth). In a wider beam, a target goes through a much larger range of relative velocities. This also explains why there is no range dependence in ρ_a since as the antenna moves further away from the target the azimuth beamwidth (and therefore the bandwidth) on the surface increases. Again, it does this in a linear fashion (i.e., doubling the range doubles the bandwidth).

Now, before getting too excited it should be realised that to optimise *Limitations* a SAR in order to achieve this kind of performance is not easy. There are a number of trade-offs that must be considered in terms of, for instance, power requirements (a small antenna cannot generate as much power as efficiently as a large one) and azimuth ambiguities, which are described later in Section 8. Suffice it to say that a practical SAR system will rarely achieve this optimum limit, even though many sensors do come close — the ERS-SAR, for example, achieves an azimuth spatial resolution of about 25m with a 10m antenna.

10.5.2   Aperture Synthesis: A Geometric Explanation

The Doppler explanation of SAR is straightforward and provides a convenient way to understand how a wide beam can provide a fine spatial resolution. However, it does contain two features that would appear to contradict the situation of operational systems. Firstly, radar systems are not actually sensitive enough to measure the shift in frequency associated with the Doppler shift of individual echoes, not even for a fast moving satellite system. Secondly, SAR processing makes use of the "start-stop" approximation whereby the aperture is synthesised using the coherent addition of lots of echoes at numerous *fixed* positions along the flight path — i.e., it assumes that the instrument is stationary every time it makes a measurement. The following discussion will hopefully make it clear how these points are not inconsistent with the Doppler explanation.

The basis of the geometric approach is not that there is a Doppler shift *Stop-start approxima-* to be measured but that a large antenna can be synthesised using an array of *tion* small antennas working together. If each of their individual measurements are added together coherently then we have the same situation as in Section 4.5 and Figure 3.12, whereby we generated a large directional antenna from a collection of many small less-directional transmitters and receivers.

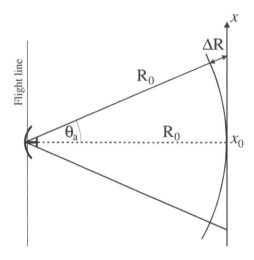

FIGURE 10.5 For the geometric explanation of aperture synthesis we use a different version of the plan view shown in Figure 10.4. Instead of the relative velocity of the target, we now consider the phase change due to the extra range distance ΔR, as the target moves through the beam.

This is also the principle of a phased-array antenna discussed in Section 4.

The trick with a SAR is that the collection of antennas is a sequence of measurements made by a single small antenna that is progressively moved along a flight path, rather than many small antennas working simultaneously. As with the Doppler explanation, it is a technical necessity for SAR to be able to transmit coherent pulses, meaning in this case that they all have the same initial amplitude and phase. When this is the case, a sequence of echoes can be processed coherently (by taking both the amplitude and phase into account) just *as if* they had all originated from a collection of many stationary antennas working in unison to form a single large antenna. Using this technique is then possible to synthesise a very large antenna — up to many kilometres!

Geometry   Qualitatively we can simply relate the idea of a synthetic aperture to the phased-array, or to the many point sources of Figure 3.12, but in order to explain how such a technique acquires the extra directional information to differentiate targets in azimuth it is necessary to consider the imaging geometry in a little more detail.

Figure 10.5 demonstrates the geometric properties of a target as it moves through the wide SAR beam from some arbitrary point on the x-axis to the centre of the beam at x_0, where the target is closest to the antenna. If we consider point x_1 just before this point (or just after) the distance from the target to the antenna is slightly longer by a factor ΔR

such that $R = R_0 + \Delta R$. From Pythagoras we get

$$
\begin{aligned}
R^2 &= (R_0 + \Delta R)^2 = R_0^2 + x^2 &\quad (10.18)\\
\Delta R &= \left(R_0^2 + x^2\right)^{1/2} - R_0.
\end{aligned}
$$

From this we can use the fact that for $x \ll R_0$, the following approximation is valid (from the Binomial theorem):

$$
\begin{aligned}
\left(R_0^2 + x^2\right)^{1/2} &= R_0\left(1 + \frac{x^2}{R_0^2}\right)^{1/2}\\
&= R_0\left(1 + \frac{x^2}{2R_0^2} + \ldots\right)\\
&\simeq R_0 + \frac{x^2}{2R_0}
\end{aligned}
$$

so that

$$
\Delta R \simeq \frac{x^2}{2R_0}. \quad (10.19)
$$

If the transmitted pulse has to travel this extra distance then it will result in a change in phase of the measured echo, so we can rewrite ΔR as a corresponding phase shift ϕ for each x location, such that,

$$
\phi(x) = -2\Delta R \frac{2\pi}{\lambda} = -\frac{2\pi x^2}{\lambda R_0} \quad [\text{-}]. \quad (10.20)
$$

As an individual target moves through the beam this phase progressively changes with azimuth position x, so the echoes from the target exhibit a "phase history" (compare this with the frequency history of Figure 9.2). This is the formula for a parabola. It is now the phase that tells us what part of the azimuth beam the target lies in, rather than the Doppler frequency. More exactly, it is the phase *history* that tells us about the target direction since the phase of any single echo is ambiguous — it is the *pattern* of phase change over a series of echoes that is identifiable. It is the pattern of the phase history that forms the reference function required to carry out the azimuth compression (in the same way that the chirped pulse is matched to a reference chirp). This emphasises why a SAR system needs to be able to generate pulses with a well-defined phase — without it, this regular pattern of phase changes over a collection of pulses would be lost.

There is one last thing that needs done here — the estimation of the *Azimuth resolution* azimuth resolution from the geometry. To do this we need to consider the maximum size of the synthesised antenna, D_s, which is based on the distance that the antenna travels while illuminating a single point target on the ground. From Figure 10.6 it should be apparent that this is given by the width of the illumination footprint at range R_0 such that

$$
D_s \simeq R_0 \frac{\lambda}{D}. \quad (10.21)
$$

Between the time the target enters the beam until it leaves, the antenna travels a distance D_s, so it is only these echoes that can be used to synthesise the antenna.

Now, in fact we already considered the theory of how to derive the angular resolution of an antenna composed of many smaller antennas in Section 4.6. It would be prudent at this stage to re-read that section to remind yourself of how the farfield situation was represented as a combination of n vectors each of amplitude A, each with a relative phase difference of ϕ. The minima of the beampattern occurred when the vectors added in such a way as to bring the point of the last vector back to the origin. By definition, this was when the cumulative angle was 2π, so that we were looking for when $n\phi = 2\pi$. In that section it was said that the first minimum is when the two ends of d (first to last n) are exactly in phase again. However, the important difference in the synthetic aperture case is that each antenna is both transmitting *and* receiving so that there is a double distance that contributes to the phase difference. The result is that the beamwidth for the synthesised antenna D_s is half the width of the beamwidth of a real antenna of the same size. Equation (3.26) is then rewritten to give the angular resolution, θ_s, of the synthetic aperture as

$$\theta_s \simeq \frac{\lambda}{2D_s}. \tag{10.22}$$

Hence the best possible resolution which can be achieved by a synthetic aperture is:

$$\rho_a = \theta_s R_0. \tag{10.23}$$

Substituting Equation (10.21) and (10.22) into (10.23) gives the result:

$$\rho_a = \frac{D}{2}, \tag{10.24}$$

which is the same as Equation (10.16) derived using the Doppler explanation.

10.5.3   Geometry vs Doppler

It is important to realise that we use the term "history" in both discussions above only because the sensor is moving and there is a progression in time. But the geometry of Figure 10.5 applies equally well whether the antenna is continuously moving, or if it is discretely moving in small stop-start jumps, or whether it was composed of a collection of many *stationary* antennas. It is also the case that the geometry is not dependent upon the direction of movement of the platform — it applies equally in both flight directions and there is not even a requirement to process the collection of echoes in the forward direction. It also has direct corollary to the geometry of Figure 3.11. This appears to contradict the Doppler explanation. Combine this with the technical issue that radars are not sensitive enough to measure the Doppler frequency shift, it is easy to start to wonder whether the Doppler explanation is valid at all!

The reconciliation of these two explanations derives from Equations (3.30) and (3.32) — a Doppler shift is also a phase shift in time since by

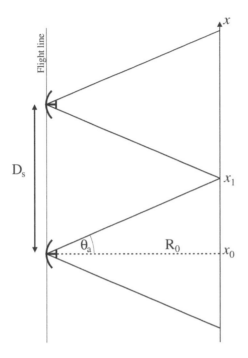

FIGURE 10.6 A plan view showing the extent of the synthesised aperture, D_s. This is found by considering how long the target remains within the broad SAR beam footprint.

definition it is a rate of change of phase. Qualitatively you might imagine that if you make a small change to a signal frequency the measured phase will also change as a consequence. The SAR system cannot measure such small changes in frequency, but it can measure the associated small change in the phase between successive pulses.

But what about the symmetry and lack of motion contained in the geometry of Figure 10.5. In fact the velocity is contained within this diagram because, along with the PRF, it defines the distance between each acquisition point. When we explicitly include the motion of the platform then the x-axis of Figure 10.5 is directly related to relative velocity and the Doppler frequency shift is given by the rate of change of the phase shift (i.e., the slope) along the trajectory, given by the derivative with respect to x of Equation (10.20):

$$\frac{d\phi}{dx} = -\frac{4\pi x}{\lambda R_0}. \tag{10.25}$$

This rate of change of phase is what would be detected as a frequency shift (and called a Doppler shift) since the shift in location of the antenna is due to the platform movement. In the geometrical context, this rate of change is positive in one half of the beam and negative in the other (the phase history being a parabola) so the rate of change of phase is not symmetric — the rate of change is the slope, so is positive on one side of a parabola and negative on the other, the sign only depending on which direction the data are processed. This positive-negative change corresponds to the positive-negative frequency changes of the Doppler explanation.

In a SAR system these changes of geometry are only evident if we measure the phase changes throughout the whole acquisition — the Doppler shift is not detectable on individual echoes, but now manifests itself as a phase change across a sequence of many successive echoes.

10.5.4   SAR Focussing

In fact there is a further problem with the Doppler approach, which is that the echo is not simply shifted in frequency, which it would be were the target a source, but the leading phase of the echo is shifted as a consequence of the changing range of the target as it passes through the beam — i.e., the phase shift associated with the change in ΔR in Figure 10.5. The Doppler explanation above does not address this directly and so is known as *unfocused* SAR. As a consequence it would not actually achieve the azimuth resolution of Equation (10.16) but would be significantly worse. The echoes would therefore require a phase adjustment before a Doppler frequency analysis could be optimally achieved. It is just this very phase adjustment that the azimuth compression of the geometric explanation achieves since it looks for the phase changes across the footprint. The geometric description above is therefore a description of a *focussed* SAR.

10.6   Radar equation for SAR

The form of the radar equation for real-aperture radars was given in Equation (9.6) in the context of altimetry and scatterometry. For SAR this equation is relevant only for a single received echo and not the whole system since we now know that SAR processing requires the coherent addition of all n received echoes and not just a single echo. The SNR is therefore now given by,

$$SNR_{\text{processed}} = n \left(\frac{S}{N} \right)_{single}. \tag{10.26}$$

For any individual target, the number of echoes received is simply the length of time it remains in the radar beam multiplied by the PRF, *i.e.*,

$$n = \left(\frac{R\lambda}{2V_s \rho_a} \right) PRF.$$

As was the case for deriving (9.7) we introduce the area over which the RCS will be measured in order to represent it as a σ^0. By applying the substitution $\sigma = \sigma_0 \rho_a \rho_r$ the radar equation for SAR becomes:

$$SNR_{\text{SAR}} = \frac{P_t G^2 \lambda^3 \sigma^0 \rho_r PRF}{(4\pi)^3 R^3 N_0 2V}. \tag{10.27}$$

This is an important result because we now have a few more instrument parameters determining the signal-to-noise ratio compared to Equation (9.6). In addition to the other constraints, if we want to keep a high SNR we will also want a high PRF and a small platform velocity, both of which increase the number of echoes received for any target within the scene.

10.7   Geometric Distortions in Radar Images

As a consequence of their side-looking geometry, real and synthetic aperture radar systems generate images with a number of unusual image features, which are often readily apparent when observing a radar image. Consideration of such features is fundamental to designing a practical RAR or SAR system, for effective data acquisition planning and for the proper interpretation of the image data. These image features are dependent upon a number of instrument characteristics, such as the choice of PRF and the shape of the antenna pattern, as well as the viewing geometry of the particular sensor.

10.7.1   Lay-over and Foreshortening

The most noticeable feature of radar imagery arises when the ground surface diverges significantly from the reference surface onto which the slant-range data are projected, as described in Section 3. In the side-looking geometry of a radar system (described from the nadir) upward slopes are projected as narrower than identically shaped downward slopes. This is seen in the ground range image in Figure 10.7 where the nearside of the middle pyramid appears shorter than the far side in the projected data. In the extreme case, when the slope is steep enough, the top of an object, say a mountain, may even be closer to the instrument than the bottom of the object nearest to nadir. From the perspective of an audio analogy, this would mean that one would hear the echo from the mountain peak *before* the echo from the foot of the mountain. In the final image (which maps the return echoes as a function of time) the top of a mountain would then be mapped nearer to nadir than the base of the mountain — the mountain would appear as if it is "leaning over" (as is the case for the first pyramid in Figure 10.7). This effect is known as *layover* and is an extreme case of   *Layover* the foreshortening that results from the side-looking geometry of imaging radar.

There are two important points to consider here. First, note that this distortion is progressively worse as you get nearer to nadir— i.e., the foreshortening is greatest at small incidence angles (which is opposite to what you would expect in an optical image). For many satellite systems the range of incidence angles across the image swath will be relatively low (since the width of the image swath will be < 100 km with the satellite at 800km) so the distortion will be fairly consistent across an image scene.

Wide-swath satellite imaging radars as well as airborne systems, on the other hand, will have a large range of incidence angles between near and far edges of the image swath since in both these cases the swath is larger in proportion to the altitude. The degree of foreshortening would therefore vary across the imaged swath.

Secondly, it should be noted that it is the *slope* that is the important factor here, not the absolute height, or total variation in height, across the imaged scene.

In Figure 10.7 two different radar mappings (slant range and ground range) are compared to an optical (look angle) mapping that would result if an optical system were used in the same geometry (i.e. mapping the look angle to the same reference surface). Note that due to the oblique viewing angle, the projected image of the pyramids also appear to lean over in the optical image, but they do so *away* from nadir, rather than towards it. We are so familiar with visually interpreting such a "leaning over" as being a consequence of our viewing position that there seems nothing strange about it. That the radar projection results in the same pyramids appearing to lean towards the system rather than away from it, is because the projection is based on measurements of time (i.e distance) rather than look angle.

The underlying reason for both distortions is that a 2-D imaging system is being used to measure a 3-D surface — the underlying (false) assumption in both projections is that the surface is flat and has no topography.

10.7.2   Radar Shadow

Another feature of radar images is also illustrated in Figure 10.7: *radar shadow*. Unlike shadows in optical images, which are imaged regions that are weakly illuminated, radar shadows are areas of a radar image that are, in principle, completely black and sharply defined since they correspond to areas where there is a complete lack of received information[89]. They correspond to the region that lies behind objects in the imaged scene and from which there is no return echo. This hidden region is mapped as a discrete area since the final image is a function of time; *i.e.,* there exists a time delay between the echo arriving from the top of the obstructing object, a mountain for instance, and the far edge of the shadow region. Rather than considering it as radar "shadow", it is perhaps more useful to think of it as radar "silence" — a region of no measured signal.

Note that shadowing gets worst the further you go from nadir, opposite to the trend in foreshortening. Shadowing is also a factor of local slope in relation to the imaging geometry, not the absolute height — a surface will be in shadow if the local incidence angle is greater than 90° (in the direction away from the instrument).

---

[89]   In practice even shadows will not be completely black since they will contain the background instrument noise.

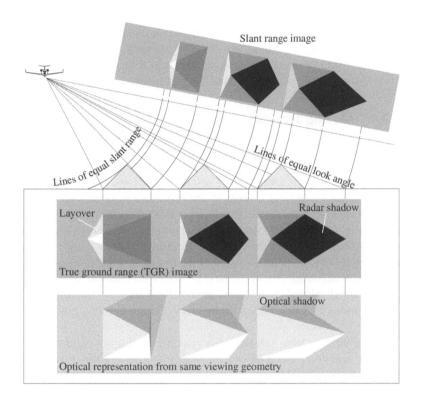

FIGURE 10.7 The mapping of a radar image is contrary to the intuitive mapping of an optical image. In this figure, two alternative mappings are shown — the slant range (as measured by the instrument) and the true ground range, which is determined using the assumption of a smooth Earth. The pyramid structures deviate from this assumption, and so are mapped rather strangely, and near nadir they exhibit layover — the pyramid looks like it is leaning over itself. In the optical mapping (as a function of look angle rather than range) the opposite is true — the pyramids appear to lean *away* from the instrument, and the effect is worse further from nadir.

10.7.3   Motion Errors

The azimuth compression based on the phase history requires an idealised phase history pattern as a reference, but this pattern is only known if the motion of the platform is well-defined. Satellites tend to be less of a problem here because the orbital path is relatively smooth and varies only slowly. Aircraft, however, are influenced by changes in wind speed and direction and are buffeted by turbulence. The aircraft will change in pitch, yaw and roll, and these movements will alter the direction of the centre of the antenna beam. Many of the parameters in Figure 10.5 are then likely to vary constantly along a flight path. The range pulse compression is not a problem because the frequency pattern of the signal is controlled by the transmitter and so the matching pattern is constant, but there is no such luxury with the phase pattern. The only way around this is firstly to try to both control *and* measure the motion of the platform, and then at the processing stage to essentially apply a process of trial and error — a range of reference patterns are applied until the best resulting image is obtained.

10.7.4   Moving Targets

There is also an implicit assumption in all the above discussions about aperture synthesis and azimuth compression, that the targets remain stationary throughout the entire image acquisition. However, this is not always the case — moving objects in the scene, such as cars or ships or surface waves on water, will introduce an extra Doppler component (or phase component) in proportion to the relative velocity between the instrument and the target. The result is that a specific target that is moving will be displaced in the azimuth direction of the image by a distance in proportion to the relative velocity of the target. This effect explains why moving ships appear displaced from their wakes in SAR images and in high resolution images cars may appear to be driving along beside a road, rather than on it.

Ocean surfaces have instantaneous velocities in proportion to the wave height — any given patch of water moves in a circle in the direction of the wave motion[90]. This circular motion means that the surface area being imaged has a distribution of velocities that together tend to defocus the resulting image. This defocusing, though, does something different for scattering surfaces moving in opposite directions. Echoes from water moving towards the instrument are Doppler shifted to higher frequencies, so that SAR places these regions further forward in the azimuth location of the final image, while water moving away from the instrument gets shifted backwards in the image. For waves moving in the range direction, this has no significant impact on the appearance of the waves since the wave patterns generally remain similar over many image pixels in the azimuth direction. However, waves moving in the azimuth direction will get "bunched up" as a consequence of the shifting in both directions (positive and negative parts of the wave get moved together in the final image). The degree to which such *velocity bunching* occurs is dependent upon the wave velocity relative to the platform velocity, and the incidence angle.

10.8   Operational Limits

Within the above sections a number of requirements were mentioned that are mutually incompatible — these are trade-offs that must be considered for any operational SAR system. For instance, the radar equation asks

---

[90]   A ball floating on the ocean surface does not just bob up and down, but moves forward and backwards as well, because it is actually moving in a circular manner.

for the transmitted pulse to have as much energy as possible, requiring a large antenna. But payload restrictions limits the physical antenna size and azimuth resolution improves if we have a shorter antenna.

Considering the geometry: low incidence angles increases the power returned from the surface (and so gives a better SNR), but low incidence angles give poorer ground range resolution than high incidence angles. And if the surface has a varied topography and relatively steep slopes there will be less radar shadow, but there will be more foreshortening and layover. Choosing an appropriate incidence angle is therefore dependent upon a number of factors that are target-area dependent as well as instrument dependent. In general it is often easier to deal with shadow than layover because at least in the areas of no shadow you can be relatively confident of where it relates to on the ground whereas layover is much more ambiguous. High incidence angles also give better ground resolution. The range of incidence angles, however, is defined by the choice of swath which is governed by flying height and PRF, the influence of which is discussed below.

In addition to these trade-offs are the real practical limits of power availability, data transmission (or storage) and physical size. That SAR systems are already the largest, heaviest, most power hungry and most data processing-intensive remote sensing instruments to be put into orbit means that careful thought has to go into every design decision to make the optimum use of what will ultimately be a very expensive item of equipment.

10.8.1   Ambiguities

The term "ambiguity" is used in a radar context to describe image features that do not belong at their imaged position, usually resulting when an echo from one location on the ground arrives at the same time as an echo from a quite different location. Ambiguities place some restrictions on the design of a SAR system, particularly on the choice of the pulse repetition frequency and the shape of the antenna pattern.

The idea of range ambiguities were introduced in Section 3.5 when discussing altimetry, whereby late echoes from one pulse were measured during a reception period for a later echo. In general ambiguities arise due to the assumption that all the echoes in the reception window come from the same footprint across the target scene whereas spurious echoes from outside the main lobe of the antenna pattern may sometimes be detected and processed as if it were a signal originating from the main beam. For an imaging radar this is mainly a consequence of the antenna pattern of a radar, which, as shown in Figure 9.4, is sensitive to signals from other directions not within the central beam. Signals from beyond the far edge of the imaged swath may also contribute, and if the signal arrives within an unexpected echo-reception window, the signal may be wrongly mapped within the image swath. Minimising range ambiguities is one of the main reasons for designing antennas with low sidelobes (as described in Section

4.4).

In the azimuth direction it is not simply the antenna pattern that is important, but also the Doppler bandwidth — in this regard it is important to understand the geometric interpretation of SAR given above. Since a SAR measures the rate of change of phase, rather than measuring frequency directly, ambiguous echoes can occur if the beamwidth is so large that the phase shift goes a full cycle — i.e., past 2π radians. Such a phase difference is equated with zero (since the absolute phase cannot be determined, only the *relative* phase) resulting in a target echo being misplaced in the azimuth beam. This places a theoretical limit to how short you can make the antenna.

10.8.2   Coverage vs PRF

The most significant potential source of range ambiguity comes from the point directly below the instrument — the nadir return. Although the antenna pattern will usually have very low sensitivity to the nadir direction it still has *some* sensitivity, albeit tens of dB below the main lobe. The reason the nadir response is significant is that it is at zero incidence angle and so has a considerably higher backscatter response than any other part of the ground — tens of dB higher, which can more than compensate for the low antenna sensitivity and potentially gives a signal as high as the ground area being imaged. In a worst case scenario (a flat water surface at nadir) the nadir return may even be powerful enough to damage the sensitive receiver designed to measure the much fainter signals from the image swath.

The solution to this is to use the time interval during which the nadir echo arrives as a time for *transmitting* pulses, rather than for receiving them. The duplexer in the radar system avoids any of the nadir echo energy getting into the receiver as it will be switched to transmit-mode. Indeed, the system cannot be receiving any echoes when it is transmitting pulses, so that for a given PRF and flying altitude there will be areas of the ground that cannot be imaged because they correspond to echo delays from a transmit period.

These two effects combine to limit the possible coverage to certain parts of the ground surface for a given combination of flying height and PRF, whereby the arrival of echoes from these areas do not coincide with transmission of a pulse.

Consideration of all the possible ambiguities puts strict constraints on the configuration of a SAR system. What this means is that there is not a completely free choice of swath width, incidence angle, ground range resolution, azimuth resolution and radiometric resolution. In a SAR system all these factors are linked in some way to the flying height, flight speed, antenna size, PRF and chirp bandwidth.

10.9   Other SAR Modes

10.9.1   ScanSAR Operation

In a standard SAR system flying at a given altitude, the PRF constrains the choice of image swath because of the risk of ambiguities and especially to avoid receiving a nadir return signal. In order to get very wide-swath SAR imagery of the kind that would be useful for regional- to global-scale monitoring (about 500km wide) it is therefore necessary to image a sequence of overlapping narrower swaths which are then subsequently combined together. To do this requires each narrow sub-swath to be imaged at a different PRF even though the images must be acquired virtually simultaneously. This is achieved by transmitting short "bursts" of pulses at a given PRF to image a small azimuth section of the swath, followed by a burst of pulses at a different PRF to image a small section of an adjacent swath.

ScanSAR operation is now very common on satellite sensors (such as ASAR and Radarsat) for providing wideswath coverage.

10.9.2   Spotlight Mode

The azimuth resolution of a SAR system is governed by the length of time the target remains in the beam, returning echoes back to the sensor. A SAR operating in spotlight mode does this by utilising the steering capability of phased array antennas to point a narrow beam onto a given ground area and letting the beam dwell on the same area while the sensor flies past. The geometry is illustrated in Figure 10.8. The trade-off for spotlight mode is that you lose along-track coverage — the high resolution is for one patch of ground, and while the SAR is focused on this one patch, it is not acquiring echoes from the ground before or after it.

10.10   Working With SAR Images

The particular nature by which a SAR system creates an image results in an image with equally distinctive properties. Most importantly, it must be remembered that the radar image is fundamentally a collection of data samples — quantified samples of power, or of amplitude and phase, at different polarisations. It is *not* a photograph made with microwaves.

The fundamental dimensions of a SAR image are time delay (range) and flight path distance (azimuth). Each pixel represents the microwave response for a particular dimension in range and azimuth, which do not necessarily equate directly to dimensions on the ground.

When the data describes amplitude and phase properties then each pixel is represented by a complex number, described by either an amplitude and phase angle, or as a real and imaginary component. Calibrated images may

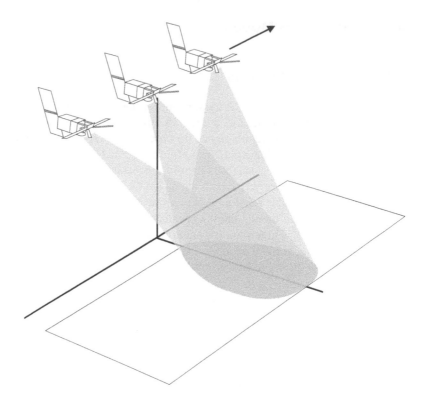

FIGURE 10.8 In spotlight SAR, the beam is steered towards a particular footprint for a much longer dwell time than a normal SAR overflight. This increases the size of the synthetic aperture length and so increases the azimuth resolution. The trade-off is that adjacent ground areas (before and after the footprint) are not imaged.

give pixels as NRCS, σ^0, or more often as a radar brightness, β^0 (when the topography is not known you cannot determine σ^0). It always pays to pay attention to the data description of a SAR image data set. Be aware of the particular values recorded in the data (σ, σ^0, or γ) and in what units (amplitude or power? unitless and linear? unitless and dB's?). Note any geometric correction that may have been applied to the data as this can influence the final properties of the image (discussed below).

Calibration   A key feature of SAR data is that it should be calibrated. Most SAR systems are extremely well calibrated — they provide more than just relative "brightnesses" in an image — they can provide physical, quantitative data. Good calibration is therefore fundamental to extracting the most effective use from a SAR system. The digital number (DN) in each pixel will be related to a calibrated value through some arithmetical expression provided by the data supplier and this will be different for each data set. As an example, the following expression converts a DN to a calibrated value of radar brightness in values of dB:

$$\beta^0 = 10. \times \log_{10}\left[(\text{DN} + 32768.)^2\right] - 60.0000 \qquad [\text{dB}]. \qquad (10.28)$$

In this case, the 32768 converts the DN from a signed integer to an unsigned integer. This value is squared to convert it from an amplitude scale to a power scale. The logarithm and the factor of 10 convert it to dB and the −60.00 is a scale factor[91].

SAR image data can be provided in a variety of different formats, depending on the application to which it will be used. The "raw" data from a SAR — before range or azimuth compression — is not easy to work with and requires specialised software to do anything useful. Even after this process a user of SAR imagery should also expect the data to be calibrated and to have had some motion compensation applied to it at an appropriate stage. Some typical data formats are described below.

Warning

The one common feature in all the radar imagery that most users are likely to come across is that they will all exhibit *speckle*.

10.10.1   Speckle

Radar images are distinctive in that they have a characteristic speckled effect which looks like "salt and pepper" noise. The resulting grainy appearance can be seen in the example shown in Figure 10.9 and is referred to as *speckle*. It is found in all coherent imaging system (including side-looking sonar, ultrasound imagery and holography) and is a result of interference among the coherent echoes of the individual scatterers within a resolution cell.

In the Bragg-type scattering utilised by wind scatterometers the scattered field is relatively ordered due to the periodic nature of the scattering surface (similar to the pattern in Figure 3.8). In the general case, however, where the surface is not periodic but is composed of randomly distributed scattering elements, the scattered field is not well-ordered. It is much more like the field shown in Figure 3.13.

Although these figures relate to sources rather than scatterers, a similar looking pattern would emerge for scatterers, their scattered waves acting as "secondary sources"[92].

The pattern (which is an interference pattern) that exists in the scattered field is effectively chaotic and unpredictable — if you only measure a small part of it, and have no knowledge of the positions of the scatterers, then it is *effectively* entirely random. This randomness manifests itself in the image as speckle. It is for this reason that speckle is often described as a "noise", but it is important to realise that, strictly speaking, speckle is not noise

---

[91]   It looks like a constant displacement because it is in dB. A subtraction in logarithmic units equates to a division in real values.

[92]   The one difference is that each scatterer would scatter a wave with a relative phase difference in proportion to the extra distances the incoming wave would have travelled to each scatterer.

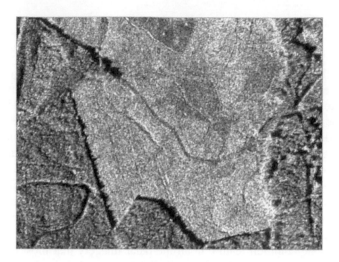

FIGURE 10.9 The grainy appearance of this high resolution X-band SAR image of a forest clearing is a result of speckle. (Image courtesy of Intermap Technologies Ltd).

since it is a deterministic and repeatable phenomenon. If a measurement is repeated with exactly the same imaging system, with exactly the same geometry, for a region of ground that did not change between image takes, then the speckle patterns in each image would be identical. True noise, such as the thermal noise within the sensor, would show a different pattern for each image.

One important aspect of the speckle effect is that it varies the backscatter over homogeneous targets. Let us consider the case of a distributed target, such as a bare agricultural field, that appears homogeneous on the ground, and has characteristics such as surface roughness that are statistically homogeneous across a large area (much larger than the image pixel size). In the radar image, adjacent pixels will exhibit a very different backscattered echo because the actual distribution of scatterers at the level of a wavelength is unlikely to be identical in the two patches of ground. Even small variations of the scatterer locations will be enough to change the scattered interference pattern so that, viewed from the same location, the combined effect results in a quite different amplitude and phase measured for each adjacent patch of ground.

The relative position of the instrument will also be slightly different for each ground cell (in range) as a consequence of the imaging geometry and topography. For a single image these variations would appear to be a non-deterministic fluctuation across an otherwise homogeneous target — this is speckle.

Vector explanation   Speckle can also be explained using the complex vector representation of the scattered waves. If we consider one resolution cell of a radar image to contain a random collection of individual scatterers, each with a slightly different location (and therefore phase) from the instrument, and each with a slightly different radar cross-section (signal amplitude), then we can represent the total return signal from this area as a coherent addition of all the individually scattered waves. This is illustrated graphically in Figure 10.10. It is a coherent addition because the incoming pulse is short with a relatively narrow bandwidth of frequencies. A very small random change in all of the individually scattered waves can result in a large change in the total resultant wave. It is this variation that causes speckle. The random change for each scattered wave can be caused by a number of different reasons:

- Randomly changing the position of the scatterers (note that a uniform change will merely rotate the resultant vector in Figure 10.10 as each individual contribution would be rotated by the same phase angle).

- changing the relative position of the instrument so that the individual path lengths change. For a perfectly flat surface this change will be predictable.

- randomly changing the amplitude of each scattered wave.

The first point is important because in a real image we might not make multiple measurements of the same particular patch of ground, but we do make measurements across extended areas that have similar surface properties. For a bare agricultural field for instance, it is because the distribution of scatterers is slightly different for each patch of ground that when we image the field there is a speckle-induced variation between pixels over what should otherwise be a homogeneous target area.

Speckle can be illustrated visually using a laser projected onto a wall — the laser light is coherent and the surface texture of a painted wall, for instance, is comparable with the size of the wavelength (100s of nano metres) and so we have the right conditions for speckle. For readers who are short-sighted, I also suggest that you take your spectacles off if you try this — even though you may see a very blurred wall, you will find that the speckle pattern is crisp and clearly in focus! The reason for this is that the speckle pattern is not like an image on the wall, but is a field of high and low intensities filling the whole room, just as in Figure 3.13. The pattern of light and dark that makes up the speckle pattern exists within your eye, not on the wall!

This might be more easily visualised by considering an experiment using sound. Given a combination of two or more loud speakers, both emitting the same (correlated) audio signal (*e.g.,* from a signal generator) we can set up an interference pattern. The emitted sound waves radiate from the speakers and interfere coherently, setting up an interference pattern in the space in front of the speakers. As the speakers are moved, or the listener moves their head, the pattern of constructive and destructive interference is apparent as variation in the signal volume. Just as in the case of the laser speckle pattern, the interference pattern exists throughout the room.

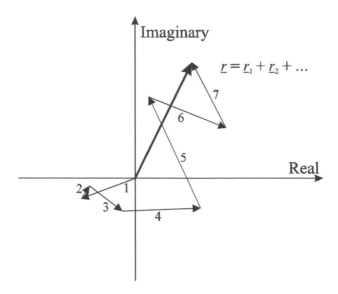

FIGURE 10.10 The random phase and amplitude from a collection of scatterers in a resolution cell will combine coherently to produce the measured phase and amplitude. This can be illustrated graphically using the complex notation, whereby the measured signal is a vector combination of all the contributing waves.

This is a standard high-school or undergraduate physics demonstration described in many introductory physics texts.

10.10.2   Speckle Statistics

It is often easiest to describe something that is either completely ordered or complete random. To theoretically describe the properties of speckle we want to consider a completely random collection of many scatterers which is more indicative of the types of surface we have when considering scattering from soil, ice and vegetation (but less so for water surfaces).

A single resolution cell usually contains a large number of randomly located identical scatterers, so we can assume that both the I and Q (real and imagery) components will be Gaussian distributed with a mean of zero and a variance of $m/2$, the value of which is determined by the scattering amplitudes of the individual scatterers. The joint probability distribution function (PDF) then describes the probability of any particular real (z_1) and imaginary (z_2) values occurring as the resultant combined wave, given the random location of the scatterers. Mathematically, this is given by

(Oliver and Quegan, 1998):

$$P(z_1, z_2) = \frac{1}{\pi m} \exp\left(-\frac{z_1^2 + z_2^2}{m}\right) \qquad (10.29)$$

where $P(z_1, z_2)$ is the probability that the resultant wave has real and imaginary components z_1 and z_2 and the πm term is used to normalise the PDF so that the total probability is equal to one.

In this scenario the phase angle is uniformly distributed so that they are all equally likely, which is what we should expect since there is no reason for any one phase angle to be favoured over another. The amplitude A of the resultant wave may, in principle, lie anywhere between zero and infinity. The PDF of the amplitude is given by the Rayleigh distribution such *Rayleigh distribution* that the probability that the resultant wave is of an amplitude A (regardless of the phase) is given by

$$P(A) = \frac{2A}{m} e^{-A^2/m}, \qquad A \geq 0. \qquad (10.30)$$

The mean value of the amplitude is then $\frac{1}{2}\sqrt{\pi m}$.

However, it is not the amplitude we are actually interested in, but the power, or intensity, since that is the quantity directly related to our definition of the radar cross-section through the radar equation. The observed power is the square of the amplitude, so that the PDF for power (represented by the intensity I) has the following form:

$$P(I) = \frac{1}{m} e^{-I/m}, \qquad I \geq 0. \qquad (10.31)$$

This is a negative exponential distribution and has some interesting *Negative exponential* properties. Note, for instance, that both the mean and the standard deviation are both equal to m. And although the power can have any value between zero to infinity the *most likely* value is zero!

The value we are ultimately trying to determine is m, since it is this average intensity that will give us our best estimate of the underlying average radar cross-section. While we are interested in making an estimate of m, it is also important to realise that since this is also the standard deviation, for a single sample we effectively have a measurement that has an uncertainty of $\pm 100\%$, whatever the underlying value of m (although note that this is a simplistic statement since the distribution is not normal). It is this large variation of intensity values that gives the image the extremely speckled appearance, since such an image is almost entirely dominated by this noise-like distribution.

Clearly we have to do better than this if we are to make useful estimates of the underlying radar cross-section. This can be done effectively only by making a number of sample measurements (the more the better) and averaging them to make an estimate of m. A first guess may be to suggest spatially averaging the pixels — simply applying a smoothing filter that averages the intensity values using a collection of L pixels within a small window of the image. This is not a bad idea, since it does generate an

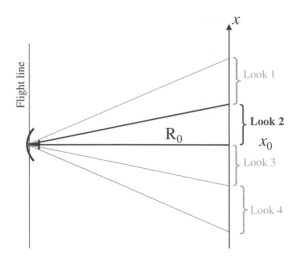

FIGURE 10.11 This illustrates the case of a 4-look process. The azimuth beam is processed as if it were 4 much narrower (and therefore poorer resolution) beams. The degradation in azimuth resolution is compensated by the improvement in speckle reduction.

average value that will provide an improved estimate of m.

However, to be theoretically precise we have to assume that the spatial average includes only pixels that are of the same target — otherwise we would effectively be taking many measurements of *different* targets (with different mean values, m, or the RCS). Note also that the application of such a filter degrades the spatial resolution of the data by the process of averaging across the filter window.

A different approach to the problem, which also helps to demonstrate the unique nature of SAR imaging, is to make the independent measurements during the measurement process itself — this is done by splitting up the azimuth beam into L sub-beams. Rather than using the full Doppler bandwidth of the azimuth beam to synthesise the full aperture capability of the SAR, a number of smaller "sub-apertures" are synthesised using smaller portions of the full bandwidth (as illustrated in Figure 10.11). Each sub-aperture is then used to create an image, albeit with poorer azimuth resolution than the original (by a factor L, which can be seen from Equation (10.8)). These sub-apertures are known as "looks" and the resulting image would be said to be of "L-looks". The difference with spatial averaging is that while spatial averaging takes samples of different (but adjacent) resolution cells on the ground, the looks correspond to multiple measurements of the *same* resolution cell. The measurements are therefore independent samples of the *same* underlying average RCS directly, without having to make an assumption about the similarity of neighbouring pixels.

Multi-looking

Averaging

It is important to note here that there is a difference between the ways in which you can average the complex numbers that make up the individual looks. An arithmetic average of L values is simply the sum of all the values divided by L. For vectors, therefore, you might be tempted to add the vectors, nose-to-tail, and then divide the length of the final vector by L. This would be a *coherent* average. For multi-looking, this is not what we are interested in since it does not tell us anything about the average power (i.e., the average vector length) being returned and the statistics of the intensity would be identical to Equation (10.31). Using the same type of diagram as Figure 10.10, one could imagine L vectors representing the looks, each of amplitude A but at different phase angles, adding together so that they all cancel each other out and provide an average result of zero!

What we actually need is the average of the power, given by A^2, of each of the different measurements. An *incoherent* average is given by the sum of the *amplitudes only*, ignoring the phase angles, and dividing by L. This distinction is an important one since a coherent average would simply give us a new speckled image — it is no different from adding together all the individual scattering components within an area, which is what causes the speckle in the first place.

We can now elaborate on (10.30) to take into account L independent *PDF for L-looks*
measurements of a uniform distributed scatterer so that the probability distribution function of the intensity is now given by

$$P(I) = \frac{L^L I^{L-1}}{\Gamma(L)} \exp(-LI) \qquad (10.32)$$

where $\Gamma(\ldots)$ is a special function known as the "gamma function". For *Gamma function*
integer values of L,

$$\Gamma(L) = (L-1)!.$$

The number of looks need not always be an integer since the sub-apertures may be processed in such a way that they overlap and are therefore not entirely independent measurements.

As the number of looks increases, the spread of the intensity values gets progressively narrower, but the average stays the same and so our estimate of the average becomes increasingly better. The measure of the spread can be given by the variance of the intensity (or the standard deviation, which is the square root of the variance) given by

$$\mathrm{Var}\{I\} = \frac{\bar{I}^2}{L}. \qquad (10.33)$$

In cases where we do not know exactly the number of looks of an image, we can use this relationship to give us an equivalent number of looks, L_e, such that $L_e = \bar{I}^2/\mathrm{Var}\{I\}$, over a collection of pixels.

The true value of L may not be known for a number of reasons: the data may have been resampled during geocorrection, for example, or a filter may have been applied that averages over a window with different sizes for different image features.

Increasing the number of looks improves the RCS estimate and reduces *Confidence* the speckle, but there always remains some uncertainty due to speckle. This is very important to consider if an application requires very precise estimates of RCS. For example, if you want to be 95% confident that your estimate of RCS lies within ± 3dB you require 9 looks.

Note that this has a significant impact on the practical limit to the spatial resolution. The spatial resolution quoted for an instrument is, by definition, based on the separability of idealised bright point targets within a dark background — more precisely it is related to the shape of the point spread function of the system. However, because they constitute a well-defined single scatterer, speckle effects are not included when quoting resolution (since point targets do not suffer from speckle[93]). Direct comparison with the resolution of optical systems is therefore very difficult since users are tempted to compare directly the quoted spatial resolutions of each instrument.

Comparison between SAR systems is also complicated by the fact that image products may have been resampled and multilooked to different pixel spacings. Pixel spacing (or size) and spatial resolution need not necessarily be equivalent, although clearly the resolution of an image cannot be better than the size of the pixel.

10.10.3   Speckle Filtering

In order to decrease the variability due to speckle some other kind of incoherent averaging may be applied at the post-processing stage, usually through the application of some kind of despeckling filter to an image that may have already been multilooked at the processing stage. The key to a good speckle filter is to try to average across areas that are relatively homogeneous, and not to average across boundaries — this helps to make sure that the averaging is over targets of the same average RCS. A simple box-car averaging filter, while providing some speckle reduction is not optimum because it simply averages the pixels irrespective of their properties within the imaged scene. In such cases you might be tempted to use a median filter, which is well known for its applicability to salt-and-pepper-type noise and is better at preserving edges. Note, however, that while it might improve the appearance of the image visually, a median filter does not provide an estimate of the mean RCS and so will alter the final image statistics in a manner which is not physically meaningful. Equation (10.31) tells us that a mean is the only appropriate operation to reduce speckle and obtain a better estimate for m.

Many speckle filtering approaches rely on two underlying assumption. The first is that spatial averaging of smaller elements over a large homogeneous region is the same as averaging many measurements of the whole

---

[93]   The explanation for speckle in the earlier section relies on a description of a pixel being composed of *many* point scatterers, each contributing to the signal and represented by a different vector. A single point scatterer is represented by a single vector. It therefore has an ordered, that is, predictable, response.

surface. This is known as *ergodicity*: that averaging over space, or time, is equivalent to averaging over many realisations of the same phenomenon.

The second assumption is that the scene we are measuring, whatever that may be, is composed of many large areas with homogeneous RCS. This is known as the cartoon model, referring to the way in which cartoons are composed of large blocks of colour. For some types of target, the cartoon model is very appropriate — agricultural fields, for instance, are a classic example, or large blocks of sea ice. The cartoon model is not necessarily as appropriate for heterogeneous landscapes or regions where point-target responses dominate the signals.

In Chapter 4 the nature of polarimetric data was described. In this *Polarimetric data* case the data are in different channels, each of which will exhibit speckle. The span image ($HH + 2HV + VV$) gives a result which has reduced speckle because it is effectively averaging over four measurements, but the full polarimetric information is lost. Speckle filters that filter each channel separately are an improvement on this, but they do not retain the complete polarimetric statistics. The details of full polarimetric speckle reduction filters is beyond the scope of this book, but suffice it to say, they do exist, and should be used when you want to keep the integrity of the full polarimetric information.

One further thing to note here is that in interferometric processing (discussed in more detail in Chapter 11) a coherent averaging aimed at reducing noise but at the same time retaining the phase information is often also called "multilooking". This is unfortunate as multilooking is really an incoherent processing operation, not just averaging, as you have hopefully seen from the above discussion. This can lead to confusion if you do not pay attention to the terminology used by different researchers.

A final word of warning is to be careful you are not applying an averaging filter to data that is in deciBels since in this case you are averaging logarithms. This is not the same as the arithmetic mean of the intensity since $\log A + \log B = \log AB$, and $\frac{1}{2}\log(AB) = \log(AB)^{\frac{1}{2}}$. Averaging logarithms therefore gives you the *geometric* mean, rather than the *arithmetic* mean.

10.10.4   Geometric Correction

The primary limitation on the geometric integrity of a single SAR image is the reference surface used to make the ground range projection. Earlier in this chapter we looked at the distortions associated with the use of a flat projection surface, as shown in Figure 10.7. Note that the reference surface need not be flat, like it is in the figure. Ideally it should be the ground topography itself so that a particular pixel is then projected onto its true x, y and z coordinate. The greater the difference between the reference surface and the actual surface, the greater the geometric distortion in the projected image. A perfect distortion-free product can therefore only be produced when sufficiently accurate topographic information is available to be used as the reference surface. An image projected onto the actual topography is known as a "geocoded" product and is equivalent to an orthorectified image in optical remote sensing (i.e., topographic distortions have been removed and each pixel lies in its proper geographic location). For predominately flat areas (open water or countries like The Netherlands being good examples) the locational errors due to the topography are relatively insignificant but are progressively worse as the topography becomes more variable. The difference between the geographic location of a point and its projected location on the reference surface can be estimated from:

$$\Delta R_g \simeq \frac{h}{\tan \theta_i} \qquad \text{[m]}$$

where h is the height of the point above the reference surface used in the translation from slant to ground range distance and θ_i is the incidence angle in this part of the image. For areas with varied topography, even a very basic digital elevation model (DEM) has the potential to provide greatly improved results.

Ground control points   The effect of topography on a radar image can be very localised. The oblique measurements made by an imaging radar mean that large variations in h over a small region results in large distortions over that area. Optical systems, especially spaceborne systems that fly at such high altitudes compared to the size of h and the area of the image coverage, are affected by topography much less. In such a case, the use of a collection of many tens of ground control points will be sufficient to geocorrect an optical image to a reasonable degree of accuracy using some polynomial function to warp the image.

With variable topography in a radar image, however, it is not necessarily appropriate to use a collection of ground control points with known ground and image locations, followed by some polynomial warping of the image. This standard optical approach will work reasonably well for radar data when the topographic variation is minimal, but in the general case the geometric distortion in a radar image is so variable (and possibly too localised within a small area) that such an approach would not be sufficient to correct for all the topographically induced distortions.

This incomplete process may be applied in some data formats and the image may be described as "geocorrected" but this might only mean that it has been rotated, rescaled and realigned to match the slant range pixels projected onto the reference surface to some geographic projection — it does not mean the data has been corrected for topographic distortion. An image with complete geometric correction using topography would normally be referred to as *geocoded* or *orthorectified*.

When topographic data is not available prior to the image acquisition, it may be possible to use other techniques to derive the topography from the SAR data itself — two techniques for doing this are briefly described in the next section, whereas the third technique of SAR interferometry is discussed in Chapter 11.

Although they have limited use for geometric correction, ground control points in SAR imagery are important for matching separate images in,
for instance, multitemporal datasets or mosaics. These points should be
known in three dimensions — they do no necessarily allow correction for
the topographic distortion, but they provide tie points to allow image pairs
to be co-registered or to allow some geographical reference for the data.
In order to guarantee the integrity of the ground control point and its visibility in the image, it is often the case that metallic corner reflectors will
be deployed in convenient locations during an data acquisition campaign.
"Convenient" means both accessibility on the ground as well as sensible
deployment in a region with very low backscatter (such as a bare field) to
increase the contrast (and therefore visibility) of the corner reflector in the
final image.

10.10.5   Limitations of Geometric Correction

There are a number of points to note following the above discussion. The
first is that areas of layover and shadow remain areas of layover or shadow
regardless of the type of geometric correction that has been carried out.
Radar shadow has no data and so there is no way to reinstate that information. Geometric correction merely re-projects the areas of shadow so that
they cover their proper geographic area.

Layover and shadow

Layover is a problem because you have a combination of different signals from different parts of the ground. Not even the clever technique of
interferometry (see Chapter 11) is always able to disentangle these signals
properly so you are left with the frustrating, but unavoidable, situation of
having only a small number of slant range pixels which represent a much
larger area on the surface. These pixels must therefore be "stretched out"
so that a large number of pixels in the final corrected image will have the
same pixel value.

For both layover and shadow, the only way to get around the problem
is to use a combination of many images acquired from different look directions, the simplest case being to take two images from opposite directions.
A better alternative would be to take two images from perpendicular directions, but this is a more complicated scenario for planning flight-lines.
Satellite coverage is constrained by the orbital paths, so that almost opposite view directions are achieved near the equator and perpendicular views
from mid-latitudes, but only similar view directions are possible near the
poles.

One important aspect of geometric correction is that it involves changing the properties of the image. By moving the location of pixels and by
projecting from slant to ground range, requires re-sampling of the data
points. In addition to the stretching out of foreshortened and layover pixels on upslopes, geocoding will also squash-up pixels on downslopes to
provide a weighted average value for the equivalent corrected pixel. Such
resampling is inevitably variable across any particular image, so that the

Image statistics

local statistics of the image become variable across the imaged scene —
i.e. the probability distribution functions given earlier in this chapter may
no longer apply generally across an entire image. The final statistics will
depend on the size of the area being considered and the degree of distortion
it contains. When dealing with such images it is appropriate to determine
the equivalent number of looks for the particular part of the image you are
studying.

10.11   SAR Data Formats

The complete processing chain from the I and Q measurements to the generation of an image is rather more elaborate than is appropriate for the
scope of this book. However, it is appropriate to consider some of the
different formats that SAR data is made available.

SLC

In using SAR imagery (rather than doing the data processing) the nearest a "user" is likely to come to dealing with raw SAR data is the single look complex (SLC) format. "Single look" means that the azimuth
compression has been carried out using the full azimuth bandwidth of the
sensor and therefore gives the highest azimuth spatial resolution but as a
consequence it suffers from maximum speckle. It is "complex" because
each pixel is a complex number (i.e., one real and one imaginary component) that represents the amplitude and phase of the wave corresponding
to that resolution cell.

A full resolution SLC image can be displayed for viewing by representing the grey level of the image with the amplitude of the signal for each
sample bin. Such an image will look extremely elongated because the dimensions of the resolution cell on the ground are usually rectangular —
the range compression defines the range dimension; the azimuth compression defines the azimuth dimension. For SLC data the azimuth dimension
is usually smaller, so when displayed with square image pixels it is much
longer in the azimuth direction.

SLC data has not been projected onto any reference surface and remains in slant range coordinates. SLC data format is a requirement for
interferometry as it contains the phase information.

Multi-look detected

For many applications that rely on measuring RCS, the SLC data format is not very useful because of the high level of speckle. An alternative
is to have the image processed using the multilooking procedure described
in Section 10.1, which processes sub-apertures. This is *multi-look detected*
(MLD) format. Commonly the image may be produced with somewhere
between 3 and 6 looks which is considered to provide a good trade off between spatial resolution (which degrades in proportion to the number of
looks) and radiometric variation due to speckle. Since multilooking is an
incoherent procedure, each pixel in an MLD image is a real (digital) number that represents the average amplitude of the signal within the pixel. A
relationship such as the one in Equation (10.28) will be provided with the
data so that the digital number can be converted to a calibrated σ^0 or, more

likely, β^0.

A MLD image will also specify whether it is in slant range or ground range coordinates. It will not have been fully geocorrected but will have used some reference surface for the ground-projection. For airborne data the reference surface is likely to be the (flat) average terrain height. For spaceborne systems it will be a smooth (curved) ellipsoid such as WGS84, or equivalent.

Precision images (PRI) are MLD images that have been resampled into square pixels, rotated to account for the view direction of the instrument and warped by some predefined operation so that the projected image pixels are properly georeferenced onto some geographical coordinate system. This is not geocoded and so still exhibits topographic distortion. It would only be truly geographically correct if the actual ground surface matched the reference surface. The effective number of looks may vary across the scene. *PRI*

To provide a proper geocoded product (i.e., orthorectified) requires a DEM from some source (which may be from the radar system itself through stereo processing or interferometry). The MLD image is then resampled so that image pixels lie in their planimetric (map-like) position on the new grid. Again, the resampling means that the statistics and the effective number of looks will vary across the scene. *Geocoded or orthorectified*

Since this data product is a true planimetric image it is the most useful type of SAR product when carrying out mapping or area analysis.

Imagery from polarimetric sensors are often simply given as one of the above formats for each of the HH, VV, HV, VH channels. The role of fully polarimetric data was discussed in more detail in Chapter 4 where it was stressed that the relative phase information of each of these channels must be retained in order to analyse the full polarimetric data. MLD data can therefore only be used to study multi-polarimetric responses based on the powers and not fully polarimetric analysis. *Polarimetric data formats*

As alternatives the data may be stored as a covariance matrix or a coherency matrix whereby the correlations between the different channels are recorded. These matrices have the advantage of being suitable for simple averaging to increase the equivalent number of looks and hence reduce speckle while maintaining the polarimetric phase information. The details of these matrices are discussed in Chapter 4.

10.12   Extracting Topography from SAR Images

Many readers may have come across the technique of SAR interferometry for determining high resolution ground topography, and this will be described in more detail in the following chapter. However, that is not the only way in which topography can be extracted from radar imagery. Two other methods are briefly described below and are described here as they do not require any additional data processing other than the images described above.

10.12.1   Stereo SAR Radargrammetry

The limitation of a single radar image is that it only measures the *distance* to the surface features — there is an inherent directional ambiguity. Photography has a similar problem, but in this case the direction is known well but the range is ambiguous. The problem is solved in aerial photography by acquiring a second image from a different location so that the *differences* in look angle between the two images (parallax) is used to infer the range. With direction *and* range, the surface topography can be determined.

The same approach can be applied to radar imagery but now it is not differences in look angle, but differences in slant range that are measured. The topographic distortion in the slant range image is in proportion to both the surface height and the look angle θ_l, as shown in Figure 10.12, which also illustrates how the different viewing positions result in different ground range projected positions of the target point, P (P_1' and P_2'). By measuring the differences in slant range between the two images we can derive the look angle and hence estimate the surface height.

The two images are acquired with a separation known as the *baseline*, B. From the figure we can apply the cosine rule such that

$$R_1^2 = R_2^2 + B^2 - 2R_2B \cos\left(90 - \theta_l + \alpha\right)$$

and using the relationship $\cos(-x) = \cos(x)$

$$R_1^2 = R_2^2 + B^2 + 2R_2B \cos\left(\theta - \alpha\right) \qquad (10.34)$$

$$\cos\left(\theta - \alpha\right) = \frac{R_1^2 - R_2^2 - B^2}{2R_2B} \qquad (10.35)$$

$$\theta = \arccos\left(\frac{R_1^2 - R_2^2 - B^2}{2R_2B}\right) + \alpha. \qquad (10.36)$$

To determine topographic height, we assume that the baseline angle, α, is known and that

$$h = H - R_1 \cos\theta.$$

R_1 is known directly from the measurements (it is one dimension of the image) and H is a flight parameter. To be able to determine the height effectively, large baselines are required so that the differences in slant range distance are equivalent to many pixels in the final image, otherwise the differences in distortion cannot be estimated accurately enough.

10.12.2   SAR Clinometry

One further method of topographic inference from SAR imagery is clinometry, whereby the variation in imaged backscatter values is assumed to be only a result of variations in local slope. The method has its most prolific use in the context of areas of slowing varying topography and constant land cover — ideally, bare soil or rock, so that the assumption that the mean scattering properties remain the same over a large area. For this rea-

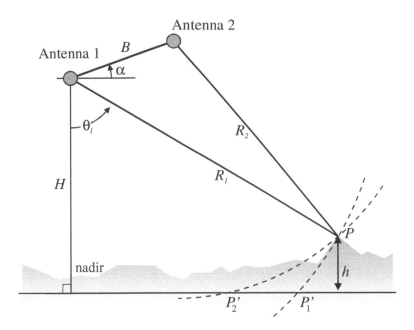

FIGURE 10.12 The geometry of SAR stereo radargrammetry. When the baseline distance, B, is sufficiently large between two image acquisitions, the point P will be projected onto two distinct locations on the reference surface: i.e. points P_1' and P_2'. The difference between these points in the two images is proportional to the height, h, of P above the reference surface.

son it has been used in the context of planetary surfaces (specifically the Magellan SAR imagery of Venus) where surface properties are expected to remain relatively unchanged over entire imaged scenes.

10.13 Further Reading

You will find some straightforward theory of SAR in Elachi (1987). For more detailed information on the processing of SAR (and this is very detailed) you should head to Curlander and McDonough (1991). The best summary of how to handle SAR images, including image statistics, speckle filtering and segmentation can be found in Oliver and Quegan (1998) . For details on the geometry and geometric correction, the best source is Leberl (1990).

11
INTERFEROMETRY

Forcing yourself to use restricted means is the sort of re-
straint that liberates invention. It obliges you to make a
kind of progress that you can't even imagine in advance.
— Pablo Picasso (1881–1973), *Life With Picasso*[94].

The term *interferometry* refers to techniques that utilise the coherent
properties of electromagnetic radiation to compare two or more waves. In
remote sensing, such techniques can be used to address two problems that
have been discussed in previous chapters: the poor spatial resolution of
passive microwave sensors and the directional ambiguity in radar images
that results in significant spatial distortions.

In radar imaging we find that interferometric techniques are actually
much more powerful than simply resolving topographic influences, but
can also be used to measure physical surfaces changes in the order of centimetres.

11.1   The Need for Interferometric Measurements

The need for interferometric measurements stems initially from practical
issues of trying to improve the spatial resolution of passive imagers and
altimeters, and finding some means of correcting the topographic distortion in imaging radar. All the previously discussed roles of microwave
sensors therefore still apply, but with potentially better quality measurements. This is particularly true for passive sensors from space whereby
the move to longer wavelengths (L-band and longer) to allow better penetration through vegetation for increased sensitivity to soil moisture and

---

[94]   Cited in "Picasso: In His Words", edited by Hiro Clark, Pavilion Books,
1993.

improved sensitivity to ocean salinity, brings the requirement for larger antennas.

Additionally, though, we will also find in this chapter that interferometric techniques applied to imaging radar provide an entirely new set of measurements — high resolution digital elevation models (DEMs) of the surface (or near surface) and differential DEMs (DifDEMs) characterising surface movements. These DEMs and DifDEMs are achieved with extremely high vertical precision (~centimetres), even from space. The DEMs can be made over surfaces that have traditionally been challenging for stereo-photography (due to lack of contrast, such as snow-covered glaciers) and DifDEMs, with their centimetre precisions over many kilometres, now provide large scale information on glacial movement, tectonic displacements and subsidence that are simply impossible by any other means.

DEMs

The importance of global DEMs is primarily a military one, and it is no surprise that the US military had an input to the Shuttle Radar Topography Mission (SRTM) — this was a mission carried out by the Space Shuttle Endeavour in 2000. In its 11-day mission it collected data over most of the land surfaces between 60° North and 54° degrees South — about 80% of all the land on Earth. To date, this data set is unrivalled in its coverage and quality, although it is severely limited over mountainous areas and contains a vegetation bias (since the C and X-band SAR could only see the top of vegetation canopies).

DifDEMS

Difference DEMs provided by the technique of *differential* interferometric SAR are now proving to be unique data sets. The first demonstration of the technique demonstrated the value of DifDEMs to earthquake monitoring. The Landers earthquake in California in 1992 was measured by successive passes of the European ERS satellite and allowed a widescale map of the ground movements associated with this seismic event. It is the stability of the satellite platforms and the ability to measure the phase that allows SAR to achieve these amazing feats. The same technique has since been applied successfully to mapping subsidence in many cities (as a consequence of changes in hydrology, tunnelling or changes in mining activity), the swelling of active volcanoes and the crustal deformations due to changes in ocean tides.

In terms of precision these measurements are rivalled only by ground-based GPS, but these only provide a sparse grid of point measurements, whereas radar interferometry provides the data over an entire swath.

11.2 Principles of Interferometry

11.2.1 Phase Measurements

At this stage you should consider re-reading Section 3.1 to review the concepts of phase, interference and coherence. The key things to take from those sections are that:

- The phase of a wave can be a sensitive measure of path length. A single measure of phase is not usually very useful as the exact number of complete wave cycles is rarely known, but a comparison between two or more phase measurements can provide a measure of path length *difference* to within a fraction of a wavelength (even over many hundreds of kilometres). If the exact number of cycles is not known, then there is a phase ambiguity such that the phase difference is known only to within a complete cycle (2π).

- For two receivers (or transmitters) in close proximity to each other, the path length differences will change as a function of look angle to the source (or transmission direction).

- Coherence is a comparative measure of the similarity of two waves, or collections of waves. Low coherence means that two wave patterns are not well-correlated; i.e., you could not use knowledge of one to predict the other. Similarly, a high coherence means the wave patterns are highly correlated, so that if you knew one you would be able to predict the other with a high degree of certainty.

The phase of a detected wave is directly related to the complete path that the wave has travelled, from original source to detector. In remote sensing, this may be a *single* path length if the radiation is emitted from the surface and detected by an instrument, or a *double* path length when the instrument transmits the original wave as well as detecting it when it returns. But how can we use the phase information to tell us something about this path length? Answer: we can't.

In the passive system, we don't know anything about the distance to the source – we cannot infer distance from simply detecting emitted radiation. With a radar system we can measure distance, but not to the accuracy of a fraction of a wavelength — that is practically not feasible.

If the exact number of whole wavelengths that have been travelled in the path length is not known, then the measurement of phase will not tell us very much. If we consider a C-band radar system operating at 5.6cm, then an uncertainty in the distance would have to be much less than this in order for the phase measurement to make sense. The phase difference between two different measurements, however, *does* tell us something useful. The exploitation of such measurements is the technique known as interferometry.

11.2.2   Application of Dual Systems

The general principle of interferometry is that two or more measurements are combined in some way to give more information than was available from each measurement on its own. The measurements may be separated in space or in time, or both, and they may be made by the same antenna or different ones, as long as the recorded data is compatible. This usually means the instruments must be identical, although it is possible to use different instruments if each one can be characterised with sufficient accuracy that you can account for the differences.

In earlier chapters we have already seen examples where a number of measurements have been combined to produce a collective performance much better than the sum of the single measurements. The aperture synthesis described in Section 5 is a perfect example of such an approach. In the case of SAR it is the same instrument making very many measurements from different locations along a flight path.

It is worth noting here that the distinction between aperture synthesis and interferometry is really only a matter of how many different measurements you are dealing with. In aperture synthesis the measurements are separated by no more than a fraction of a wavelength whereas in interferometry they can be separated by many wavelengths. In SAR, for instance, the PRF must always be high enough so that the antenna has moved by less than half a wavelength between pulses — it is the high density of measurements along the flight path that allows the technique to approximate one measurement from a very long antenna.

More Venus imaging   In interferometry, we are usually only dealing with a handful of sparse measurements. The distinction is similar to the distinction between Figure 3.8 which shows the pattern from only two sources, and Figure 3.12 which has a number of sources in a line (and separated by less than a wavelength). We can use the two source case to consider a first example of the principle of interferometry using two antennas to tackle the ambiguity problem we had in Section 4.1 when describing radar imaging of Venus. That discussion was stopped short on reaching the problem of the North-South ambiguity. What was required was some means of distinguishing between echoes from the Northern and Southern hemispheres — despite being from different hemispheres, these echoes had the same time delay and Doppler shift and so were indistinguishable.

N-S ambiguity solution   The solution comes from combining Figure 3.8 with the Venus imaging geometry so that we have the case shown in Figure 11.1. The difference is that we now use two antennas separated by some distance d parallel to the rotational axis of Venus (i.e., in the Venusian North-South direction, rather than relative to Earth). The North and South hemisphere echoes can now be differentiated by their *phase difference*. Remember that this interference pattern can represent the phase difference measured from different directions. In the direction toward P_1 (the Northern hemisphere) echoes arriving at S_1 and S_2 are generally in phase, whereas they are mostly out of phase for the equivalent southern latitudes (direction of P_2). The exact phase difference varies as we move from North to South, but in this case all we need to do is distinguish between two very broad directions — each antenna still does range-Doppler processing, but the phase difference at the two antennas can now be used to differentiate signals from the two hemispheres.

Note that there is still the potential for some ambiguity here: echoes

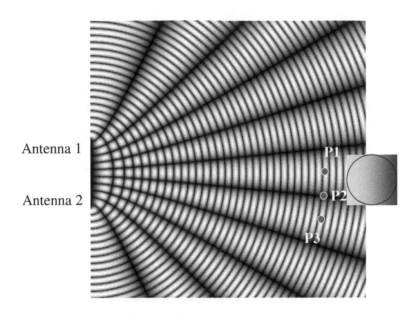

FIGURE 11.1 The North-South ambiguity can be overcome by measuring phase differences from two appropriately spaced antennas. With reference to Figure 3.8, note that the phase differences of the echoes from the Northern hemisphere are different from those arriving from the South.

from the direction P_3 also give echoes that are in phase. For the Venus case we avoid this problem by making sure it matches the geometry of Figure 11.1 — in this case there are no echoes from P_3. The distance between the two antennas is the key thing here as it defines the narrowness of the beams within the interference pattern. This distance d, is called the interferometric baseline, or simply "the baseline" (when loosely used in an interferometric context). For Venus imaging the baseline would be chosen to minimise ambiguity in the phase difference returns.

11.2.3 Interferometry for Resolving Direction

Resolving the N-S ambiguity for imaging Venus offers a good illustration of how interferometry can be used as a means of determining direction. In the active case of radar imaging, we are resolving an ambiguity by differentiating the echo direction using measurements of phase differences. Throughout this book it has been stressed that microwave systems are not good at measuring direction. Now we find that interferometry allows us to partly get around this problem by providing some information on the direction of a signal. In passive systems it should therefore be possible to use interferometric techniques to improve angular resolution beyond the aperture limit of Equation (9.25). A passive interferometer using a number of antennas would still generate a 2-D image, but at potentially higher spatial resolution than each antenna is capable of attaining on its own.

Interferometry might also be used to remove the ambiguity associated with the topographic distortion of radar imagery since this was due to an ambiguity of angle — it is the inability of a radar imager to differentiate the look directions of echoes that results in foreshortening and layover. If we can infer both the range *and* the look angle to the target then it should be possible to map the full 3-D surface without distortion.

In fact humans use the principle of interferometry all the time to determine the direction of low frequency sounds. In a similar way to the use of two eyes in stereo to detect the small changes in look angle (parallax) that can be used to infer the distance to an object, we use two ears to determine directions. Stereo vision resolves the ambiguity of distance associated with a single eye, whereas stereo hearing ("binaural audition") allows resolving the ambiguity of angle resulting from the use of one ear. One ear of an echolocating bat, for instance will allow discrimination of distance to a target with high accuracy[95], but to resolve azimuthal direction, bats, and humans, need to use binaural sensing.

The Interaural Phase Difference (IPD), proposed by Lord Rayleigh, is one of the vital indicators of sound source direction (Middlebrooks and Green, 1991) and relies on a similar approach to the Venus imaging described above . Instead of two microwave antennas, we have two ears. The IPD is related to the pathlength difference between a waveform arriving at the ear nearest the sound source and arriving at the ear furthest away. This pathlength difference results in a shift of the waveform of a fraction of a cycle which is detected by the ear. Such a phase shift between sound arriving at each ear can provide the information needed to find the corresponding azimuth angle of the sound source. Despite a diffraction limit giving each ear an angular resolution of more than 90° for low frequency sounds, the IPD can provide an azimuthal resolution of 5–10° for humans, and as little as 1.5° for echolocating bats (Masters et al, 1985).

Microwave interferometry performs a similar function to the IPD by measuring the phase difference between the two signals measured at different locations.

As one might imagine, the practical details of interferometry are somewhat more subtle than this introductory section might suggest. The rest of this chapter will focus on some of the details. To introduce some of the key issues we will begin by considering interferometry in the context of passive microwave imaging.

11.3 Passive Imaging Interferometry

The Venus example in the previous section gives a very general impression of the principle behind interferometry. But let us now generalise to the case

---

[95] A single ear can also provide some degree of information on the elevation angle of a sound source as a result of the elevation dependent frequency response of the pinna that provides cues to the elevation of the location of a sound.

of passive microwaves instead of just radar. Can we use the interferometric principle to achieve a *passive* microwave image of Venus — i.e., based entirely on the phase difference measurements and not the range-Doppler processing as used by the radar? In fact the technique of passive interferometry has been well-established for many decades within the field of radio astronomy for achieving angular resolutions much greater than the individual radio telescopes can achieve on their own. The famous Very Large Array, for instance, is probably the world's most recognisable astronomical radio observatory. It consists of 27 radio antennas in a Y-shaped configuration on the Plains of San Agustin, New Mexico. Each antenna is 25 meters in diameter but by combining the data interferometrically it can achieve a resolution equivalent to an antenna 36km across!

The first problem we have with passive microwaves is that the signal is almost completely incoherent. In radar we have control over the signal we transmit, so we can make sure the signal maintains a constant frequency, with a consistent phase. We also control the timing, so we can identify different echoes. Emitted thermal radiation from a surface, on the other hand, covers a broadband of frequencies and has amplitude and phase patterns that are virtually completely random. Differentiating between the signals coming back from different parts of the surface is not possible as they appear effectively like noise.

The way to get around this is two-fold. The first step is to make sure the signal being measured has as narrow a bandwidth as is practical. This can be achieved with a well-designed optimised radiometer that filters out all but a narrow range of frequencies. The problem of incoherence is dealt with by recognising that a signal is always completely coherent with *itself*. Coherence is a measure of correlation and so any signal, no matter how random it may appear, will be totally correlated with itself. To describe it in the context of predictability, you just have to note that you can always predict a signal pattern if you already have a copy of that very same signal!

As long as we are dealing with a relatively small area of ground, over a relatively short period of time, and over a small difference in angle between antennas 1 and 2, then even thermal emission has a sufficient degree of correlation between the two signals to be able to measure it above the noise. The processing step that is required is therefore to perform a correlation between the two complex signals being measured at the two antennas, so that

$$\Gamma = \langle E(t_1)E^*(t_2) \rangle \tag{11.1}$$

where the correlation is made over some finite time interval. Note that this is the same form as the coherence defined in Equation (3.17) but that the correlation is over time t, rather than space. This complex correlation matches those signals that originate from the same area on the target since the signals will only correlate with themselves and nothing else.

In the passive imaging case, interferometry therefore gives a higher resolution 2-D image— it does not give any more information than you would get from a larger antenna. When using interferometry with many antennas

(> 2) it starts to become clearer how this configuration can replicate a large antenna. For this reason the technique may often be interchangeably referred to as "interferometric radiometry" or "synthetic aperture radiometry", although strictly speaking the distinction is based on the distance between each receiver.

SMOS

The first spaceborne interferometric radiometer will be the The Soil Moisture and Ocean Salinity Mission (SMOS), which is an L-band (1.4GHz) 2-D interferometric radiometer. It has three long and thin antennas that form a "Y" when deployed and will achieve a ground spatial resolution of 30–90km. The interferometry here is in two dimensions so as to emulate a large circular antenna and so achieve a resolution equivalent to a 9m circular antenna, while actually using three 4×0.25m antennas. The importance of this configuration is the dramatic savings in size and weight — a 9m diameter circular antenna is not yet a viable option for a spaceborne system.

The result is a long wavelength passive imager that can provide a ground resolution sufficient for making important global measurements on a regular basis.

11.4   Radar Interferometry

Interferometry in the context of imaging radar is an extremely powerful tool, but requires some elaborate explanation.

In the first instance, we will consider how the principle of using interferometry to solve Venus's N-S ambiguity can be extended to tackle the problem of poor resolution in altimetry. This will then be expanded to address the topographic distortion in imaging radar, and finally, the technique of differential interferometry will be described.

11.4.1   Interferometric Altimetry

The description of altimetry in Section 3 simply left the angular resolution limit of an altimeter as a practical constraint on microwave altimetry, but in this chapter the use of interferometry has allowed for an improvement on the angular resolution of a single antenna by combining phase and amplitude information from two or more antennas that are nearby to each other. This same approach can be applied to radar altimeters to improve the resolution in the cross-track direction.

We saw in Section 4.2 that the resolution can also be improved in the along-track direction by using the motion of the sensor and performing aperture synthesis. By employing a coherent altimeter signal, a larger virtual antenna can be synthesised and the along-track resolution can be reduced to hundreds of metres instead of the usual tens of kilometres. However, that is aperture synthesis, not interferometry as the sampling will be made at a high PRF, thus leaving only a small distance between each echo

reception. Interferometry in altimetry is performed by having two antennas aligned in the cross-track direction. The resulting geometry in the x-z
plane, is identical to Figure 11.1 but rotated by 90°. The phase difference
therefore contains information about the direction of the scattered signal
and the problems of a typical pulse-limited altimeter are minimised in the
cross-track direction (to compliment the improved resolution of the aperture synthesis).

The Cryosat mission is an ESA Earth Explorer Opportunity altime-   *Cryosat*
try mission aimed at determining variations in the thickness of the Earth's
continental ice sheets and marine ice cover. Its primary objective is to test
and quantify the prediction of thinning polar ice due to global warming
and to do so requires both centimetre accuracies in range but also high
spatial resolutions in order to provide altimetric coverage of the steep margins of the ice sheets, and to systematically monitor changes in the thickness of Arctic sea ice. It therefore employs both aperture synthesis in the
along-track direction and interferometric processing in the across-track direction by using two antennas 1m apart using an instrument called SIRAL
(SAR Interferometric Radar Altimeter). SIRAL is a high-resolution Ku-
band (2.2cm) radar altimeter based on the heritage of the French Poseidon
instrument from TOPEX/Poseidon and Jason altimeter missions.

When the topography is relatively flat (i.e., over open ocean and the
interior of ice sheets) the altimeter will operate in a conventional pulse-
limited mode known as Low Resolution Mode (LRM). In "SAR" mode
Cryosat will achieve along track spatial resolutions of 250m using aperture
synthesis processing of pulses from a single antenna. Additionally, it can
fly in "SARIn" mode of operation whereby a second antenna is added to
form an interferometer (with 1m baseline) across the satellite track. Echoes
from pulses transmitted by one of the antennas are received by both antennas simultaneously and analysis of the phase difference between the two
isolates the cross-track location of the echo. This mode is intended to provide improved elevation estimates over the kind of variable topography that
would cause problems for conventional altimeters (refer back to Chapter
9). Around the complex topography of the sloping margins of ice sheets,
it would not otherwise be possible to assign a position (and therefore elevation) to a given range measurement. And it is just these areas around
the edges of the ice-sheets that are the most dynamic and therefore most
sensitive to climate change.

Note that the Cryosat team chose the term SARIn so as to avoid confusion with InSAR, the interferometric technique for radar imagery.

11.4.2 Interferometric SAR

In imaging radar interferometry is not used for improving the resolution —
the spatial resolution of SAR is already impressively good due to chirped
pulses and aperture synthesis. The value of interferometry for SAR is, in
the first instance, to resolve the ambiguity of direction that results in topo-

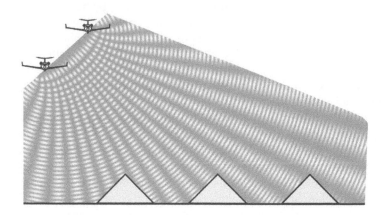

FIGURE 11.2 This diagram makes the link between radar interferometry and the interference patterns described in Section 4. In particular, take a look back at Figure 3.8 — the radial pattern corresponding to lines of equal phase difference which is equal to lines of equal look angle. Projected onto the ground surface, these would be a set of lines, or "fringes". Note that many fringes map onto the up-slope of the nearest pyramid whereas the down-slopes corresponds to only a single fringe.

graphic distortion within the image. Take another look at Figure 10.7. This side-looking geometry is required for radar imaging so that time delay is approximately equivalent to distance along some reference surface. However, if the topography is variable, then the time delay is not equivalent to distance along the reference surface and the resulting image exhibits the topographic distortions of foreshortening and layover. This was attributed to the fact that the radar could not differentiate signals that arrive back after similar echo delay times but from different look directions.

If we can determine the difference in angle between these two echoes, then we would be able to account for the distortion due to topography and in so doing determine the nature of that topography. This is effectively the same problem that we had with the N-S ambiguity in the radar imaging of Venus and so we should be able to apply the interferometric technique to imaging radar and resolve this ambiguity.

For SAR, the interferometric approach requires taking two measurements displaced in the across-track direction. The associated interference pattern is illustrated in Figure 11.2. In order to introduce some terminology, the various viewing geometries employed for radar interferometry will first be described, followed by further discussion on how the technique provides the information required, and the practical and theoretical limitations of the method.

11.4.3 InSAR Viewing Geometries

We limit the current discussion to what is known as "across-track interferometry" (XTI), which means that the two antennas are displaced in the across-track direction as shown schematically in Figure 11.2. This is directly analogous to the Venus imaging interferometry[96].

Since interferometry in imaging radar is almost exclusively done using SAR systems, the rest of this chapter will refer to InSAR[97]. The term InSAR usually refers to across-track interferometry only.

The directional ambiguity in imaging radar lies in the range plane — i.e., the across-track direction, perpendicular to the flight direction. To resolve this ambiguity in a similar manner to Section 3 above, we therefore require two antennas, separated across-track by a distance B. We can now redraw Figure 11.2 to simplify the geometry of the imaged scene as shown in Figure 11.3 (not to scale). We can see from this figure that the distance B, which is the interferometric baseline, can be described in different ways, as summarised in Figure 11.4. Simply giving its length is not sufficient, since that does not tell us about the geometry of the path length differences to each antenna. We therefore also need to specify a baseline angle, α, measured between the baseline, B, and the horizontal. Alternatively, we could give the horizontal, B_y, and vertical, B_z, components, which can make the geometry simpler to deal with, or we could give the perpendicular (or effective) baseline, $B_\perp$, and the parallel baseline, $B_\parallel$, which are more directly related to path length differences. If the term "baseline" is used without qualification, it is usually either the baseline length, B, or the effective baseline, $B_\perp$, that is being discussed.

There are a number of ways in which the dual-system geometry can be achieved. The most effective is to have a single platform that carries two antennas. This is *single-pass* interferometry as it requires only a single *Single-pass* flight over the scene. It can be operated in two different modes. The first mode transmits from one antenna and measures returned echoes at both. The second mode is known as ping-pong mode which alternates between each antenna so that each transmits (and receive) in succession.

Carrying two antennas on a single platform is not always possible. Size and weight restrictions on both aircraft and satellites mean that two antennas are much more impractical or costly, for instance, and larger baselines beyond the width of the platform are not possible. An alternative is to use a *Repeat-pass* *repeat-pass* geometry whereby the two measurements at A_1 and A_2 are acquired during different overpasses of the instrument. A key difference with single-pass interferometry is that now each measurement can only be made with transmission and reception by one antenna at a time. For more elabo-

---

[96] In a later section the concept of "along-track" interferometry will be described, which, although having many basic principles in common, is a quite different measurement approach and is used for entirely different applications.

[97] The term InSAR is now in common usage amongst the geoscience community, but it is still occasionally referred to as IfSAR. The latter term appears to be more popular within the field of surveying and geomatics.

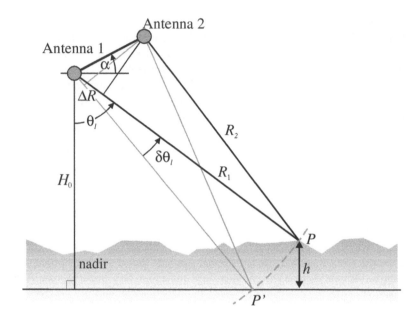

FIGURE 11.3 The detailed geometry of across-track interferometry highlighting the various angles of interest. The ultimate aim is to determine the topography height, h , which is found by inferring the look angle, θ_l, for a given range (or more precisely, the look angle difference, $\delta\theta_l$, between $P\prime$ and P. .

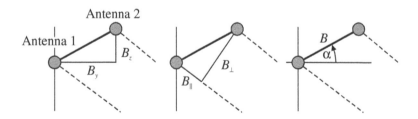

FIGURE 11.4 Since the two antennas need not lie exactly in the horizontal plane, we need two numbers to describe the separation. The three common ways to do this are by using the horizontal and vertical component; the perpendicular and parallel component; or the baseline angle and baseline length.

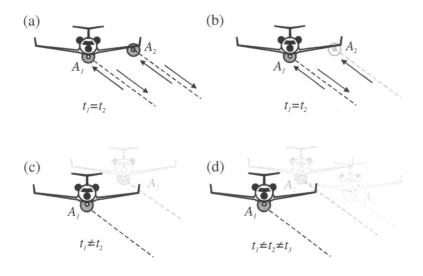

FIGURE 11.5 A summary of across-track InSAR configurations: (a) single-pass with two active antennas; (b) single-pass with only one active antennas but two receiving antennas; (c) dual-pass using a single antenna; (d) multi-pass using a single antenna.

rate applications it is necessary to use *multi-pass* interferometry, whereby many parallel flight lines are flown.

These different configurations are summarised in Figure 11.5.

The time difference (or temporal baseline) between the passes may be as short as a half an hour, the time it takes an aircraft to turn and re-fly the scene, or as long as 40 days for a satellite system with a long repeat period. It could even be many years later. The problems associated with the temporal baseline and the impact it can have on the interferometric coherence will be discussed later in this chapter.

In the passive case we most certainly could not do repeat pass interferometry since emitted thermal radiation is coherent only over extremely small time scales. A temporal baseline of even a fraction of a second would mean that the coherence would be lost, and so there would be no meaningful phase difference values between the measurement pairs.

Radar systems have an added advantage in this regard as they utilise scattered radiation rather than emitted radiation and in a SAR system the transmitted signals must remain coherent over many pulses for the aperture synthesis to be effective. Inter-pulse coherency was a requirement for aperture synthesis for the same reason that it is a necessary for interferometry: that we need to be able to combine the echoes in a coherent manner. In the same way, a SAR will be able to maintain coherency of the transmitted signals between different overflights of a scene. In fact a spaceborne SAR system can maintain coherence over many years (as long as the imaged surface remains unchanged).

For now, in order to consider the idealised case we are going to assume that the scattered waves remain coherent between passes. What is exactly meant by this statement will become more apparent in Section 4.4 when we discuss what happens when it does not remain coherent. For now we consider only the (unrealistic) theoretical ideal of the surface remaining completely unchanged across the temporal baseline, no changes in the atmosphere between the scene and the instrument, and no instrument noise.

Geometry   The phase difference that would be measured between A_1 and A_2 for a given look angle, θ_l, is related to the path length difference, which is given by

$$\Delta R = R_2 - R_1 \qquad [\text{m}]$$

for single pass interferometry, while for repeat-pass or single pass ping-pong it is given by

$$\Delta R = 2\left(R_2 - R_1\right).$$

The extra factor of 2 arises from the there-and-back path length. In single pass geometry the "there" path is the same for both and so does not contribute to the phase difference.

The phase difference $\delta\phi$ is then found by determining the fraction of whole wavelengths that will fit into ΔR so that:

$$\delta\phi = 2\pi\frac{\Delta R}{\lambda} = k\Delta R \qquad [\text{rad}]. \tag{11.2}$$

Where k is the wavenumber. If you wanted the angle in degrees instead of radians you would multiply through by 360 instead of 2π.

The geometry of SAR interferometry differs from stereo SAR in that the baseline is very much smaller than the range distance, so that the path length difference is virtually equal to the parallel baseline. The absolute phase difference can then be expressed as a function of the average look angle so that

$$\begin{aligned}
\Delta R &\approx B_\parallel = B\sin\left(\alpha - \theta_l\right), &&\text{for } B \ll R \tag{11.3} \\
\delta\phi &= kB\sin\left(\alpha - \theta_l\right).
\end{aligned}$$

Note that the for any given pair of measurements, B, k and α are constant, leaving the absolute phase difference as a function of the look angle, θ_l. For imaging radar the look angle will never approach zero (nadir) and is unlikely to get below 15° or above about 70°.

Equation (11.3) now mathematically describes what has been discussed qualitatively already — that the interferometric phase difference tells us something about the direction, or look angle, of a signal. Now, rather than distinguishing between North and South hemisphere signals, the aim is to consider the signal for each image pixel and determine the direction (the look angle) to that pixel by measuring the phase difference between the two measured signals. The pixel's slant range is already determined by the radar imaging process, and so additional knowledge of the pixel's look angle allows the pixel to be assigned a true ground range and a height (altitude). There is then no need to project the radar image onto a simple reference surface, as is done for a single measurement — instead, the signal can be directly assigned its true location.

Now, if only it were that simple. In practice there are a number of practical problems that need to be overcome.

The first problem is apparent in Figure 11.2. Unlike the Venus case, the *Problems* echoes from across the ground surface go through many cycles of being in-phase and out of phase. Since it is only the *relative* phase difference that is measured directly, and not the absolute phase difference, the echoes from many different patches on the terrain will have the same relative phase difference.

One way to observe what is happening is to simulate an InSAR measurement and over-plot the target area using, in the first instance, grey levels to represent the relative phase difference measurements. This kind of image is known as an interferogram. A hypothetical interferogram over a *Interferogram* pyramid feature lying on a region of flat Earth is shown in Figure 11.6. The parameters used to generate this image were chosen to simulate a typical pattern you might expect from a satellite InSAR geometry. Note that this image shows the phase values for each patch on the surface in true ground range coordinates — i.e., not how it would appear in an interferogram created immediately from the two input images. The foreshortening in such an image would make Figure 11.6 harder to see what is going on.

Rather than using grey levels, interferograms are often coloured by assigning the range of hues to the full 2π phase difference cycle — an example of such a colour interferogram is shown in Figure 11.7. The resulting lines of equal colour (hue) are known as *fringes*, and in this kind of interferogram they represent lines of equal look direction. As it stands, there is an ambiguity as to exactly which look direction the fringes correspond to, so it is not a simple case of equating a phase difference to a unique look angle.

Notice that the fringes become squeezed together on an up-slope and stretched apart on a down-slope (from the perspective of the antenna). The reason for this should be apparent from reference to Figure 11.2. From the perspective of the instrument(s) the up-slope covers a wide range of look angles, while the downslope equates to only a small range of look angles. Across the surrounding flat Earth, the fringes appear regular but in fact they will have increasing width as the range distance increases. This is due to a further aspect of the geometry: for an equivalent interval of look angles, the corresponding ground distance covered increases with distance from the nadir (due to the oblique geometry). This variation is only usually apparent in airborne data, rather than spaceborne data, due to the much larger range of look angles[98].

The interferogram is the first step in interferometry. Ultimately, however, what is of interest is to convert the phase differences into a look angle and thereafter a measure of height above a reference surface. As before

---

[98] Just for completeness, it should also be mentioned that the change of look angle across the swath is a consequence of the effective baseline changing. The effect on the fringes will depend upon the relative position of the two antennas.

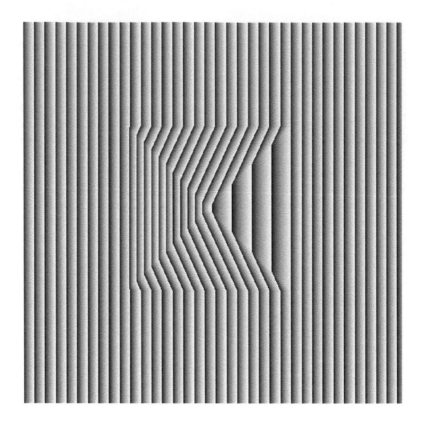

FIGURE 11.6 An simulated (noise-free) interferogram using a satellites geometry for an area that is flat except for a single pyramid in the centre of the image. The image has been geocoded so that it is a planimetric representation of the fringes, rather than a true interferogram (i.e. there are no spatial distortions in the image). The range direction is towards the right. Note that the fringes are squashed together on the near-side slope and stretched apart on the back-slope.

our reference surface can be an average terrain height, or a standard terrestrial reference surface (a datum). The latter would mean the final products would be in a standard geographical coordinate system. The trick here is to derive a new interferogram that only shows the phase differences that are different from what we would expect from the reference surface. The geometry of the radar with respect to the reference surface is known so it is possible to simulate the interferogram we would expect for the reference surface alone — the resulting interferogram for our pyramid example is shown in Figure 11.8.

This is often referred to as the "flat-Earth" interferogram, although the reference surface need not be flat — it may be a local or global ellipsoid, for example (such as WGS84).

Flat-Earth correction   By subtracting the flat-Earth interferogram from the measured inter-

FIGURE 11.7 A colour interferogram using hue (phase difference), saturation (coherence magnitude) and intensity, instead of RGB. In this way, the hue cycle corresponds to the cyclic phase angle and the saturation represents the "meaningfulness" of the phase, such that the phase angle becomes less visible when the saturation gets low (since the phase will be noisy).

ferogram, we create a new interferogram where the fringes now refer to a look angle *difference* between the actual surface and the reference surface. For the pyramid example, the *flat-Earth corrected* interferogram is shown in Figure 11.9. Since on this example the ground around the pyramid is our reference surface, there is no difference between the look angle for the reference surface and the look angle for the ground — the difference is consistently zero across the area surrounding the pyramid. Most interesting, however, is that the new fringes on the pyramid seem to follow the pattern of what you would expect from a set of contour lines. The fringes now represent lines of equal look angle separation between the terrain and the reference surface. These fringes are therefore very nearly (but not exactly) equivalent to contour lines — so the terrain surface is encoded within the phase difference measurements, which is exactly as required.

A key difference between the fringes in Figure 11.9 and contour lines is that we still do not know the *absolute* phase difference, which is what is required to calculate the topography of the terrain — the fringes are still

Phase unwrapping

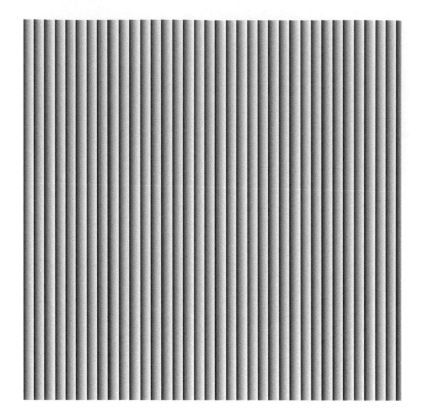

FIGURE 11.8 A simulated "flat-Earth" interferogram for the case in Figure 11.6. Where the topography matches the flat-Earth reference surface (in this case, the area around the pyramid) the fringes are identical to the interferogram in Figure 11.6.

in cycles of 2π. It is as if we have a set of unlabelled contour lines. Such an interferogram is referred to as being "wrapped" because it continuously wraps around to zero again, whenever it reaches 2π. What is required is a system of "unwrapping" to convert these cyclic phase differences to absolute phase differences. Figure 11.10 (a) illustrates the problem, while 11.10 (b) shows the solution, using a cross-section through an arbitrary interferogram (in *approximately* the range direction, as the fringes can vary like this in any direction). In (a) the phase resets to zero each time it reaches 2π because the original measurements cannot tell us the absolute number of cycles that each echo has undergone — only the relative phase difference of the two waves.

The principle behind phase unwrapping, is that we assume the terrain surface is continuous, by which we mean there are no discrete, instantaneous jumps in the terrain. It can then be assumed that if we start from one location and count up the phase values, when we get to 2π, the next little phase increment $\partial\phi$ is now re-assigned as $2\pi + \partial\phi$, and not simply $\partial\phi$.

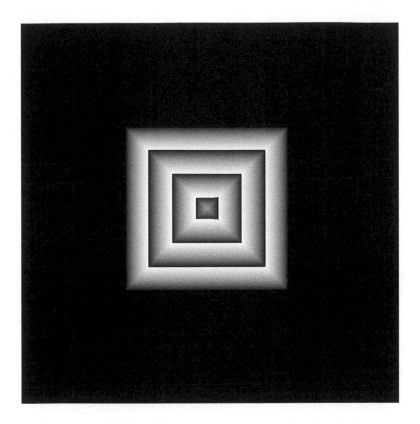

FIGURE 11.9 The flat-Earth corrected interferogram is obtained by subtracting Figure 11.8 from Figure 11.6 (the original interferogram). The result is an interferogram that very nearly matches the contours of the surface topography.

All the subsequent phases have 2π added until we reach the next cycle, 4π, when all through the next cycle the phases are reassigned as $4\pi + \partial\phi$, and so on, until the entire region has been converted to an absolute phase value, as shown in 11.10(b).

There are a number of practical problems to consider with phase unwrapping. Firstly, the description above assumes we know the absolute phase of the starting point. This is not always the case as it is not always possible to get sufficiently accurate knowledge of the location of both antennas relative to the reference surface to determine the absolute phase difference, and hence 3-D location, of one point in the image. An alternative is to use ground control to accurately determine the location of at least one known point within the image. If the position is known accurately then the phase differences, and therefore relative heights, are determined relative to this point. The Shuttle Radar Topography Mission, for instance, required laser range finding systems to measure the 3-dimensional baseline to within a few millimetres over the 60m length of the mast. The plan

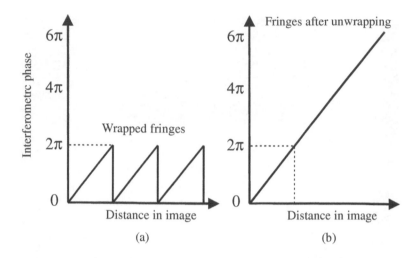

FIGURE 11.10 If we consider a slice through some fringes on an interferogram, as in (a), we would see that they "wrap" around 2π. To determine the absolute phase it is necessary to "unwrap" the fringes, as in (b). This is possible only if you have an uninterupted succession of fringes so that you can add-up each new cycle of 2π.

was to measure the baseline with such high accuracy that ground control campaigns could be minimised. Orbital stresses, oscillations, astronaut activity and thermal expansion/cooling due to the extreme conditions in orbit meant that the antenna boom actually moved by as much as 15cm during the flight. For this instrument an error in the 3-D baseline of 3mm would result in an error in the retrieved surface elevation of about 9m.

Fringe ambiguity   The second problem for phase unwrapping is related to the geometry. In order to successfully phase unwrap it is necessary to be able to distinguish the individual fringe cycles. The pixel size is therefore one restricting factor — in the interferogram each fringe cycle must be larger than a few pixels so that it is discernible in the data. If there is the equivalent of a whole cycle *within* a single pixel we lose the continuity in the phase unwrapping since we will miss one (or more) of the 2π steps. This has to *Sensitivity* be traded-off with the need for sensitivity. The closer together the fringes, the more sensitive the interferometer is to topographic height and so the higher the quality of the DEM. The sensitivity to height is primarily governed by the baseline, such that a longer baseline gives narrower fringes and so greater sensitivity.

This is an important issue since we saw above that the local terrain slope, or more generally, the local incidence angle has an impact on the width of the fringes. The smaller the incidence angle, the narrower the fringes. This is in contrast to the ground range dimension of an individual pixel which increases in size as the local incidence angle gets smaller (which is why we get foreshortening). This puts a severe restriction on the interferometric configuration. In a practical interferometric system, it is therefore necessary to try to maintain sufficient fringe width given a known range resolution of the system and an estimate of the local terrain variation. The controllable factor is then the baseline, which can be adjusted to optimise the visibility of the fringes.

When the foreshortening becomes so extreme that layover occurs, even interferometry cannot help. Layover will sufficiently distort the fringes in the interferogram that phase unwrapping algorithms have difficulty in making sense of them. One approach could be to use a radar configuration that operates at a much higher look angle (larger incidence angles). This would minimise the foreshortening and layover, but would increase the areas of radar shadow. In shadow there is no ground signal, only noise, and so there are no meaningful fringes. It is sensible therefore to simply mask out shadowed regions in the interferograms. Phase unwrapping algorithms generally prefer areas of shadow as they can be ignored completely, whereas areas of layover are harder to identify and so add an undeterminable error to the unwrapped interferogram.

Now, layover is determined by the local slope, down at the level of a *Layover* few pixels, and need not always be the on the scale of mountains. Landscapes can have many small discontinuities such that the local slope is very steep: the edge of a cliff, a motorway embankment, the side of building or even a tree, are all examples of sudden slope changes. How does an interferometer deal with this? From the point of view of phase unwrapping, at least, the key criteria is that the phase difference across this discontinuity does not change by more than once phase cycle. Because of the change in height, there is an associated change in the flat-Earth corrected phase. If this jump simply goes from 1.2π to 1.7π then the phase unwrapping can accommodate that without a hitch. If the jump in absolute phase across a single pixel is actually from 1.2π to 3.7π ($= 2\pi+1.7\pi$) then the phase unwrapping will assume that it was only 1.2π to 1.7π and one whole cycle will have been missed. The height associated with one full cycle (for a particular viewing geometry) is the *height ambiguity*.

The final problem for interferometry is the interferometric coherence itself and this is such a big topic that it deserves a whole new section to discuss it.

11.4.4   Interferometric Coherence Magnitude

Up to this point, it has been assumed that a completely meaningful measurement of phase difference can be made. In fact this is often not the case and it is necessary to consider in some more detail how the interferometric phase difference is actually determined. It is tempting to think that we could simply take two single look complex SAR images taken from slightly different, but parallel, flight paths and simply subtract the phase value of one from the phase value of the other to produce an interferogram. This is feasible, but along with all the sensible phase differences, you would also be making phase differences of, for example, areas of radar shadow that are simply noise, with no way to discriminate the good measurements from the bad. Two noise values will still give a phase difference even though it has no physical meaning.

What is required is a method that emphasises the phase differences that are "meaningful". By meaningful we mean that the phase differences are telling you something about the geometry rather than just random fluctuation due to spurious phase signals such as noise, or any of the factors discussed below that interrupt the consistency of the geometry. It has no physical meaning to measure a phase difference between two noisy (random) signals — this is simply more noise. The phase difference must be measured between effectively the same signal. Note, however, that this is the same problem that we had with interferometric radiometry in Section 3 — trying to find meaningful phase differences of signals that originate from the same target area. In radar interferometry we therefore use the same approach of measuring a complex correlation, which is the same as the complex coherence, and we use an expression such as:

$$\gamma = \frac{\langle E(t_1)E^*(t_2) \rangle}{\sqrt{\langle |E(t_1)|^2 \rangle \langle |E(t_2)|^2 \rangle}}. \tag{11.4}$$

For SAR, the $\langle ... \rangle$ brackets cannot refer to averaging over time since a SAR does not sufficiently record the temporal response of the signal. Instead, it is necessary to invoke the assumption of ergodicity that was introduced when discussing speckle in Section 10.1. Ergodicity means the temporal average can be replaced by a spatial average, where we are assuming that across a small window the image surface properties remain constant. The coherence is then found from an N-sample window such that,

$$\gamma = \frac{\sum_N p_1 p_2^*}{\sqrt{\sum_N |p_1|^2 \sum_N |p_2|^2}}, \tag{11.5}$$

where p_1 and p_2 are the SLC complex pixel values and N is the number of pixels. As before, γ is therefore a complex number.

The phase angle of γ gives us the phase difference, as required for the interferogram, but the amplitude, or magnitude, $|\gamma|$, gives us a measure of the "meaningfulness" of that measurement. The denominator normalises the magnitude of γ so that it ranges from 0 to 1. The technique of SAR interferometry is therefore a type of *correlation* interferometry (as opposed to additive interferometry).

Degree of coherence   Since $|\gamma|$ gives us a measure of the quality of the correlation it is called the *degree of coherence*. Note that this is really the coherence of the interference *pattern*, not the individual waves that make up the echoes (since these must be coherent in order to conduct aperture synthesis). If the pattern of complex values over the N pixels are completely correlated then they are fully coherent and $|\gamma| = 1$. At the other extreme, if $|\gamma| = 0$ then the patterns are uncorrelated and are said to be incoherent. For the most general case when $0 < |\gamma| < 1$, the patterns are said to have *partial* coherence.

Rather than referring to "coherence magnitude" all the time, it is common practice to simply refer to this value as "the coherence". This coherence magnitude is a very powerful tool and some applications of this parameter are described in the next section. It can also be used as a penalty function to optimise the co-registration of two (or more) SLC SAR images. For interferometry to be effective, the images must be co-registered to within a fraction of a pixel (one tenth of a pixel is a good rule of thumb) otherwise the phase differences are not being measured for the same target area. The coherence magnitude helps here since it will reach a local maximum when the pixels are most closely registered — i.e., the complex signals in the two images will be most similar.

In the ideal case, the coherence magnitude would always equal unity, meaning that the phase differences contain 100% useful information — this is what was assumed in the examples above. But in practice this is never quite the case as there are a number of factors that will reduce the magnitude of the coherence. Since coherence is a measure of correlation, these factors are said to "decorrelate" the interferometric signal.

11.4.5 Decorrelation

The first source of decorrelation has already been mentioned — system noise. Every radar measurement contains some element of noise, as was discussed in detail in Chapter 6, and this random signal will not be correlated between different images (because the noise originates within the system not from the target). Noise decorrelation has an impact in proportion to the SNR — the noise will only really impact on interferometric coherence when the return echo is very low. In areas of radar shadow, there is no echo so noise dominates and, although we could still measure a phase difference, the magnitude of the coherence will be zero and so we know the phase difference is meaningless. *Noise decorrelation*

Likewise, smooth surfaces that exhibit specular reflection, such as calm water surfaces, tarmac roads and some ice surfaces, will give such a low echo signal that the noise can dominate the signal and the coherence magnitude will be very low. Noise in the instrument adds unpredictable random fluctuations in the amplitude and phase of the measurement. This noise is additive, so is less of a problem for bright targets.

Other instrument factors will also introduce some decorrelation: the co-registration will not always be perfect, the generation of the individual images (from repeat-pass) may be subject to motion errors, and the impulse response function may also mean that the complex coherence is not exactly measuring the correlation between the same two signals.

When the concept of coherence was introduced in Section 3.1 it was *Space and time* noted that coherence tends to be measured over either space or time. The ergodicity assumption was also invoked above, implying some correspondence between temporal and spatial properties. Similarly, waves tend to decorrelate as a consequence of separation in *either* space or time. In optical theory they tend to refer to coherence *length* (space) or coherence *interval* (time). In microwave interferometry separation in either space or time will likewise decorrelate the signal. There are two sources of decorrelation that are geometrical (space) factors: *baseline* decorrelation and *volume* decorrelation. Decorrelation over time is referred to as *temporal* decorrelation. These three factors are described below.

Baseline decorrelation   In the description of speckle in Section 10.1 the interference pattern of the scattered wave was compared to the pattern shown in Figure 3.13. This is what gave us the speckle pattern. In radar interferometry we are effectively correlating speckle patterns, since we make our complex over a spatial area. However, interferometry also requires a spatial baseline (since there is no phase difference information if $B = 0$) and so there must be change in the speckle pattern associated with this shift in the location of the instrument. If we increase the separation between measurements, the scattered field pattern will progressively become less and less similar, and so the signals will decorrelate. The degree of decorrelation will depend upon the separation distance (the baseline, B) and the regularity of the scattered field. For a hypothetically flat surface the degree of baseline decorrelation can be estimated theoretically and so some adjustment can be applied to the data.

Volume decorrelation   When the surface is not flat, we have an additionally problem — in particular when there is a range of heights within the region being used to estimate coherence. This effect is most significant when there is a large vertical spread of the scatterers within the target area, such as you would find in a forest. This is known as volume decorrelation and is of particular importance when making interferometric measurements over vegetated land surfaces (although note that ice and snow may also act as a volume scatterer).

In general, therefore, decorrelation increases (coherence goes down) as either the detectors or the scatterers are increasingly separated in space. Indeed, some might argue that these effects are actually the same phenomenon. The next thing to consider is how coherence changes with time.

Temporal decorrelation   In the passive interferometry case, the signal source was a natural thermal source with the amplitude and phase varying randomly in time so that even within a tiny fraction of a second the signal was no longer coherent. In radar, however, the signal is under the control of the instrument, so that consistent amplitudes and phases can be generated even over very long time periods. Temporal decorrelation is only a problem for repeat-pass interferometry where there is a time delay between the image acquisitions. However, if the target area remains identical, then even a repeat-pass interval of many years will still result in the same scattered field and so the data will remain coherent.

For remote sensing of the Earth's surface very few targets will remain unchanged over time. Water, sand and vegetation are blown by the wind. Vegetation grows or is burnt or harvested. Soil gets wet and dries out, and it might also freeze and thaw. Snow falls and eventually melts. Agricultural land is sown, harvested and ploughed. Each of these kinds of change in the surface can have a significant impact on the measured coherence since the physical changes are larger than the size of the wavelengths used.

The time-scales of relevance are different for each type of surface. Water can decorrelate in about 0.1s because the surface moves. Vegetation can also decorrelate quickly due to wind-induced movement — on a windy day, forests may decorrelate within seconds, the degree of decorrelation being dependent upon the wind speed and the radar wavelength being used. Shorter wavelengths will mostly detect the smaller leaves and branches so will decorrelate in even a light breeze, whereas longer wavelengths that penetrate to the larger branches and trunks will not decorrelate so easily. *Timescales*

At the other extreme are the few surfaces that are known to remain stable over long periods of time. Artificial structures, bare rock, stony deserts (where there is little in the way of erosional processes) and the surface of glaciers and ice-sheets do not change significantly over time scales of months to years. The slow creep of glaciers and the displacement of ground surfaces due to tectonic activity or subsidence, are a special exception. In these cases the surface scatterers will approximately maintain their *relative* position to each other and so the scattered field (the speckle pattern) may still remain relatively consistent. In such cases the overall average phase across an area may change slightly, but the pattern will remain approximately the same[99].

11.4.6   Summary of Decorrelation

The issue of decorrelation can be both advantageous and problematic.

Decorrelation is a severe problem for the generation of an unambiguous interferogram and puts a limit on the effectiveness of phase unwrapping. If the coherence drops below about 0.3 then the integrity of the phase difference measurement is severely compromised and the phase information becomes virtually useless. There is a very neat way of visualising the interferogram that clearly represents this process by incorporating not just the phase information but the average intensity of the two radar images as well as the coherence. Rather than using an RGB colour composite, this is done using the hue, saturation and intensity channels. The average backscatter intensity is allocated to the intensity channel and the phase is allocated to the hue as was done in Figure 11.7. The clever thing is that the coherence is allocated to the saturation so that in areas of low coherence, the saturation of the hues drops so that they are progressively less visible.

---

[99]   "Approximately" is used here because it all depends on the extent of the surface displacement and any associated distortions across the imaged area — all of these considerations must be taken in proportion to the size of the wavelength being used.

The fringes are therefore only apparent where the saturation (coherence) is high.

Total coherence

To generalise the total coherence for a given measurement we can model it as a function of three independent coherences, such that the total coherence magnitude is given by

$$|\gamma| = |\gamma_{\mathrm{noise}}| \cdot |\gamma_h| \cdot |\gamma_t| \,. \tag{11.6}$$

where $|\gamma_{\mathrm{noise}}|$ is the decorrelation due to noise, γ_h the decorrelation due to the vertical spread of the scatterers and γ_t is the temporal decorrelation. Interferometric processing will try to account for baseline decorrelation as part of its aim of maximising the total coherence. Additionally, for narrow scattering layers, γ_h can be assumed close to unity. The others terms can be maximised by applying phase noise filtering (Stebler et al 2002, Zebker and Villasenor 1992). The noise term refers to additive thermal noise and measurement noise but for simplicity it can be considered to include all SNR terms including the SAR processing (which may be of low quality) and baseline decorrelation (if it has not been accounted for).

Advantage

The advantage of decorrelation is that it can tell you something about the target area that is being imaged and so can be used for inferring surface properties or image classification. Increasing quantities of vegetation, for instance, will approximately equate to decreasing coherence (due to a reduction in γ_h) and the sensitivity of coherence to wavelength-scale changes in the surface also offer a staggering potential for change detection.

Landcover classification

Since the temporal coherence is related to physical changes on the surface, the coherence over time can be used to differentiate different surface types. This relationship is also dependent upon the wavelength as different wavelengths will be more/less sensitive to different parts of the target. In the context of biomass mapping, for instance, there is some evidence to show a sensitivity of repeat-pass coherence to above-ground biomass such that increasing biomass gives progressively lower coherence. However, this is not very robust, since the key factor is not just the vertical distribution of the scatterers but the relative contribution from each. For instance, if we consider, say, L-band data for low, medium and high biomass values, we might observe high coherence at low biomass since the signal is mostly scattered from the flat ground beneath the vegetation. At medium biomass levels, the microwave response may be fairly equally distributed through the height of the canopy — there may still be a ground signal, but it is not dominant. However, at high biomass levels, it may be that the forest is dense enough that most of the signal again comes from relatively narrow range of height, albeit this time from high in the canopy, so that the coherence is higher than it is for medium biomass!

Using coherence magnitude alone for estimating biomass is therefore not a straightforward business. It is made even more difficult by the temporal decorrelation caused by the wind-induced movement of the tree canopy. However, it is just this sensitivity to movement that can also be exploited

FIGURE 11.11 An interferometric coherence image from two ERS-2 scene acquired in 1997 and 1999. Despite the two year separation in the data acquisitions, the Nasca pampa remains a bright region of high coherence in the centre-right of the image. This is a stony desert surface that is eroded only by occasional run-off from the mountains (the dark fluvial features cutting across the plateau) and humans (the dark lines are dirt tracks). This is in contrast to the highly decorrelated (dark) river valleys that are used for agriculture.

by interferometric measurements. Since coherence can be disrupted by *Surface Change* wavelength-scale changes in the location of surface scatterers repeat pass InSAR can map very small surface changes on an otherwise stable surface.

Random changes in the physical location of the surface scatterers between data acquisitions of the order of a wavelength (5cm for C-band) RMS are sufficient to decorrelate the interferometric signal and are therefore visible in the coherence data (Zebker and Villasenor, 1992). Since vegetation and soil moisture changes may also reduce the coherence, surface change studies have focused on desert areas, where even the process of a vehicle driving across an area of desert can be seen in the coherence data. The coherence image in Figure 11.11 clearly shows tracks formed by the random motion of small stones on dirt tracks across the otherwise very stable Nasca plateau in Peru. Such changes would not necessarily produce a noticeable effect between the corresponding intensity images since the average radar cross-section would not change.

11.5 Practical DEM Generation

High resolution DEM generation from airborne and spaceborne interferometric SAR (InSAR) is now a well established technique and commercial products derived from (airborne) interferometric mapping are now becoming increasingly available. For mapping purposes, short wavelength systems (particularly X and C bands) have been the preferred option since small antennas with a short baseline can be used so that single pass InSAR is feasible with a single aircraft. It can therefore provide a precision of a few centimetres and the degrading influence of the presence of vegetation in repeat-pass InSAR is minimised.

SRTM   The use of X-band repeat-pass interferometry from space has been demonstrated on Shuttle Imaging Radar campaigns during the 90's, and more recently the NASA's Shuttle Radar Topography Mission (SRTM) has demonstrated the wide systematic coverage that X-band single-pass InSAR can achieve. SRTM carried both an X and C-band single-pass interferometer, although the X-band data only provided intermittent coverage. The X-band system was operated by the German Aerospace Establishment (DLR) and they estimate their DEM products to have a vertical accuracy of 4 m, albeit with limited coverage.

Although the resulting DEMs may be of high quality, however, the use of short wavelength microwaves will mean the DEM has a vegetation bias — it represents the near-top-of-the-canopy and not the ground surface, as the bulk of the energy of these short wavelengths do not penetrate very far into vegetation. The alternative of moving to P-band and longer wavelengths brings the challenge of having to use a repeat-pass configuration, as well as the other practical limits of flying long wavelength sensors (i.e., obtaining a licence to transmit P-band signals for airborne campaigns, and Faraday rotation for spaceborne missions).

One alternative approach is polarimetric InSAR (Cloude & Papathanassiou, 1998) which has the potential to derive both true ground DEMs and tree height directly, and is described in a little more detail in Section 6.

11.5.1   InSAR Processing Chain

Processing SAR data in an interferometric manner requires an elaborate but methodical processing chain to provide useful and reliable results. A complete description of this processing chain is beyond the scope of this book, but for completeness, here is a brief summary of some of the major steps.

Pre-processing   Interferometry can only be performed on single look complex (SLC) data or raw, unprocessed data. Imaged data will have lost the crucial phase component of the data. Unprocessed data is preferable since the data can be processed to optimise the interferometry, rather than optimising image properties. Another key piece of information required is an estimate of the baseline, and how it might change across the image (from converging orbits, or from perturbed flight paths).

Co-Registration   Accurate co-registration of the two (or more) images is vital for interferometry since the estimate of coherence requires the measurements to overlap by 90% or more. Poor co-registration results in poor coherence, and therefore a less reliable phase measurement.

In InSAR terminology the terms *master* and *slave* refer to the separate images being used — the latter is the image that is adjusted to co-register with the former, so that the master image always remains unchanged.

The two co-registered complex images are then cross-correlated to gen- *Interferogram*
erate an interferogram followed by the flat-earth correction to convert the raw interferogram to an interferogram relative to the chosen reference surface.

The flat-Earth corrected interferogram is wrapped — it contains phase *Phase unwrapping*
difference values only from $0 - 2\pi$. The next step is therefore to unwrap this interferogram so that the result is the absolute phase differences between the two image acquisitions. These phase values can then be used to infer the pixel height through the relationships given in Equation (11.3).

There are some other subtle features that I have avoided here — the last stage requires iteration, for instance — but this list should at least give some idea of the basis steps involved in generating a DEM from a pair of radar images.

11.6   Vegetation Height Estimation

In Section 5 it was mentioned that one limitation of InSAR DEMs is that they have a vegetation bias, i.e., they represent a ground-plus-canopy surface and not the ground surface alone. This can also be used to our advantage, as it means that InSAR may provide information on the vegetation height.

In order to elaborate on this issue, it is convenient at this point to in- *Scattering phase cen-*
troduce the concept of the *scattering phase centre*. The InSAR geometry *tre*
in the previous section was limited to considering surfaces, with the notion that the phase corresponds to an average phase value for that pixel. When considering the phase difference geometry, it was assumed that the phase difference corresponds in some way to the geometric "centre" of the pixel. This is an appropriate model when considering surfaces, but when we introduce the volume scattering effect of vegetation, we have a different question — to what physical height, h_v, will the average phase difference geometry correspond?

To ask this in another way: what is the retrieved height when you produce a DEM with a vegetation-bias and how does this compare with the actual vegetation height?

The answer will depend upon the radar frequency and the nature of the scattering within the range cell. Even if the actual geometric phase differences are distributed over a wide range due to the vertical spread of scatterers, the InSAR processing will treat it as one single scatterer within a pixel with an equivalent scattering phase centre. This scattering phase centre will normally lie somewhere between the top of the canopy and the underlying ground surface — a longer wavelength means a scattering phase centre closer to the ground. In some circumstances, the scattering phase centre may even be located where there is physically no vegetation. One such example has been observed in emergent trees in tropical forests, where the high emergent tree falls in the same range bin as regular canopy some distance away. The scattering phase centre then falls some way between the emergent crown and the canopy, giving a virtual tree crown horizontally displaced and much shorter than the actual tree.

With this knowledge in mind, we can consider four InSAR methods of determining vegetation height: single frequency InSAR, dual-frequency InSAR, multi-baseline interferometry and polarimetric interferometry.

11.6.1   Single Frequency

The first two methods use a short wavelength SAR system, such as X-band, as these will not penetrate very far into vegetation. Realistically, the system will be single-pass to avoid the temporal decorrelation across the vegetation that would easily occur at short wavelengths. The retrieved In-SAR DEM will approximate a top-of-canopy surface. Strictly speaking, even X-band systems do not represent top-of-canopy since even at centimetre wavelengths there is some penetration into the vegetation layer, as well as the spatial averaging associated with a finite pixel size. An underestimation of the canopy height in the order of many wavelengths would therefore always be expected, unless a suitable backscatter model has been used to adjust for this penetration.

If the top surface of the vegetation can be estimated from the X-band data, then all that is needed next is an estimate of the ground surface, which can then be used to subtract a ground DEM from the vegetation DEM and hence determine the canopy height. The difference between the two methods is how the ground surface DEM is determined. The first method only requires the single pass data, but requires the horizontal dimension of the vegetation to be small (i.e. in the region of tens of metres) and the underlying surface topography to vary only slowly. The ground DEM can then be interpolated across the vegetation using unvegetated surface heights in the surrounding area. This method has the advantage of only requiring data from a single radar system, but falters when the vegetation is broader than 100m or so, or if the ground surface varies significantly over this distance so that the interpolation is no longer accurate. A better approach, if the data is available, is to use another ground DEM source, such as a national mapping product, in which case the accuracy of the height retrieval is dependent upon the accuracy of this DEM.

11.6.2   Dual-Frequency

The second method of canopy height estimation is to use a second In-SAR system using longer wavelengths such as P-band or VHF. Since these wavelengths are long enough penetrate the vegetation and reach the ground they can provide topographic information about the ground surface. This ground DEM is then subtracted from the short-wavelength DEM. This technique requires using two different InSAR systems, and since the P-band (or longer) requires a much wider baseline, it would also normally require repeat pass configuration.

11.6.3   Polarimetric Interferometry and Multibaseline Interferometry

Two further techniques offer alternative methods to deriving canopy heights and vertical density distributions. These are polarimetric interferometry and multi-baseline interferometry, both of which rely on models that describes the polarimetric and/or interferometric response from a vegetation canopy of a certain size and geometric properties.

Polarimetric interferometry (PI) uses a single frequency InSAR but exploits the polarimetric information to distinguish between canopy and ground. Forest canopies tend to act as random volumes, with no preferred orientation apparent in the polarimetric response, whereas the ground surface or trunk-ground interactions have distinctive polarimetric responses. By measuring how the coherence varies with different polarisations, and comparing this with a modelled response, an estimate of canopy height can be made (assuming the extinction through the canopy is known).

The key constraint on this method is the ratio of the scattering contribution from the canopy to that of the ground surface. In ideal conditions there is a approximately equal contributions from each so that the scattering phase centre is located somewhere deep within the canopy. For most forests L-band seems to be the most appropriate for this technique since the signal can penetrate to the ground but still has a large contribution from the canopy.

If the extinction is not known, a further measurement must be made, either at a different wavelength (such as X-band, which would help constrain the canopy top height) or a further baseline measurement.

The use of many baselines is the basis for multi-baseline interferometry which incorporates data from many (>4) passes. This need not be a polarimetric system. Instead the information comes from how the complex coherence measurements across the different baselines, in much the same way as the passive interferometric radiometry uses a number of antennas together. Unlike the passive case, the antennas can be separated in time since the SAR remains coherent, but in this case the temporal decorrelation will be a limitation.

11.6.4   SAR Tomography

SAR tomography is an extension of the last two methods, but at the time of writing remains an interesting but understudied technique whereby a dozen or so data acquisitions are made using a polarimetric system. This technique has the potential to retrieve fully volumetric data, with spatial variations in backscatter properties in the vertical as well as the horizontal, effectively creating a hologram of the entire target area.

11.7   Differential SAR Interferometry

SAR interferometry is principally used for determining geometrical heights (topographic or vegetation canopy) as described in the previous sections. However, there is a very exciting extension to InSAR called *differential* interferometry, or DifSAR, which has now become common place in the field of geophysics, vulcanology and glaciology for the measurement of small ground and ice movements.

The principle behind DifSAR is best described by considering a repeat pass InSAR configuration but with zero baseline — i.e., $B = 0$, so that the instrument flies exactly the same flight path. If there has been no change of the surface characteristics within fractions of a wavelength RMS, then the SAR images should be identical (including the speckle patterns). As long as the local slope is not too steep the resulting coherence will be equal to one and the phase differences between each image will be zero. The effect of any random motion of the scattering elements on the surface will be to decorrelate the signals and add noise to the phase difference measurements, but even then the average phase difference would be zero.

However, a small (less than 10 wavelengths or so) *systematic* movement of the surface between the two image acquisitions has a completely different effect. "Systematic" implies that the relative positions of the surface scattering elements within a pixel remain constant, but that the absolute range distance to that surface area from the instrument is changed. This small change in range will result in a phase difference between the before and after images. This situation is illustrated in Figure 11.12. Note that the same geometry arises if we move the sensor by a small distance rather than the ground surface.

The important dimension in DifSAR is the component of the motion in the slant range direction, not the absolute surface motion. The resulting range displacement, D_R, is therefore given by

$$D_R = |\mathbf{m}| \sin(\theta_i - \zeta) \tag{11.7}$$

where $\mathbf{m}$ is the motion vector describing the direction and magnitude of the surface movement so that $|\mathbf{m}|$ is the magnitude and ζ is the direction from the horizontal (in the ground range direction).

The coherence will remain high because the signal pattern remains unchanged except for a consistent phase change across the region of surface that has undergone movement. This assumes that the area of ground movement covers more image pixels than are used to calculate the coherence in Equation (11.5). The phase change associated with component D_R is then

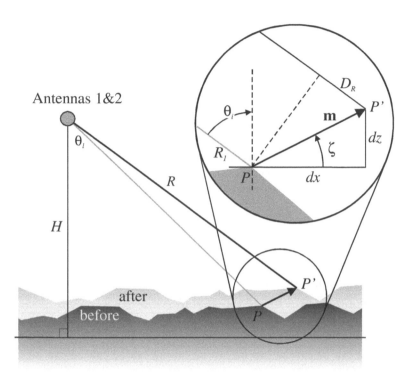

FIGURE 11.12 The geometry of differential InSAR. The diagram assumes that the second flight line is indentical to the first, so that both antennas are geometrically in the same location. During the interval between the two pasess the ground moves location, even though the fine scale features on the surface remain the same. At any location we can define this ground displacement with the vector, **m**.

given by

$$\phi_D = 2\pi \frac{D_R}{\lambda}, \tag{11.8}$$

$$= \frac{2\pi}{\lambda} |\mathbf{m}| \sin(\theta_i - \zeta),$$

so that

$$|\mathbf{m}| = \frac{\lambda \phi_D}{2\pi \sin(\theta_i - \zeta)}, \tag{11.9}$$

meaning that the phase difference can be used to measure ground displacements of a fraction of a wavelength.

One limitation of this technique is that it is the full 3-D vector **m** that is ultimately required in most applications, but InSAR only provides the component of **m** in the range direction. One common approach when using spaceborne InSAR is to determine two components of **m** by making measurements from both the ascending and descending orbits.

In geophysics, where ground displacements caused by seismic activity or as a result of activity within volcanoes, can now be mapped to within centimetres over areas of 100×100km. Crustal deformation measurements with such accuracies are possible with Global Positioning Systems, but only as point-measurements over a network that can only provide limited coverage (and must be placed *before* the event).

A second example of ground deformation comes in the form of environmental impact assessment. Throughout the 20th century there have been significant changes in ground water (including sea level rise), subterranean mining activities and urban construction (including underground transportation). Such changes can impact on urban areas in the form of subsidence, the slow sinking of land. Since this movement is of most relevance in urban areas, and the movements is on the scale of centimetres over a few years, InSAR is ideally suited to map such changes. Urban areas are good targets for short to medium-term InSAR as they provide bright backscatter returns and do not change significantly over time, so they remain coherent over many months to years.

Glaciology   The other discipline where small movements of the surface are of interest is glaciology. The slow creep of glaciers down a valley or the progression of an ice-sheet can be mapped with great precision using InSAR, thus aiding the modellers in their understanding of ice flow. Ground-based survey techniques and GPS provide some ground control, but are unable to offer the frequent and wide coverage available from spaceborne SAR systems.

11.7.1   Considerations and Limitations

One important practical consideration in DifSAR is the timing. If the movement is a discrete jump, like an earthquake, it is necessary to make measurements before and after the event. For continuous motion, such as glaciers, the important thing is that the temporal separation is such that the movement is not too large (compared to a pixel) that the coherence is lost. Over ice, coherence may also be reduced if the surface of the ice changes too much, through snow accumulation, freeze-thaw, or structural deformation.

For ground displacement studies, the requirement for high coherence tends to limit the study to areas of low vegetation cover. There is some possibility, however, that this limitation may eventually be circumvented to some extent by low frequency spaceborne sensors, or the use of polarimetric interferometry, discussed in Section 6. Note also that DifSAR must be done with repeat-pass configuration (since there must be time for the movement to occur) so that all the problems associated with this mode apply (such as temporal decorrelation).

Practicalities   In practice, it is often difficult or inappropriate to fly a zero baseline, especially with satellite systems, where orbit stability requirements do not always allow such accurately predetermined baselines. Differential interferometry is therefore commonly employed with non-zero baselines so that

D_R is actually an extra phase shift in addition to the phase difference resulting from the finite baseline (and the surface topography).

In this case, two images are not sufficient, even for the case of a discrete movement as it is not possible to differentiate the phase difference due to topography and that due to the movement — we need a method to distinguish the two different contributions. One method is to use an accurate DEM of the imaged area, so that the topographically-induced phase shift can be modelled and subtracted from the differential interferogram (instead of a flat Earth correction, this can be considered a "true Earth" correction). An alternative is to use at least three scenes. Qualitatively, these can be thought of as having two scenes to provide the topographic phase correction, and one further scene to determine the displacement phase.

Once the differential interferogram has been generated, the problem of phase unwrapping still applies, so that there is an implicit assumption when doing DifSAR processing that the displacements are continuous (no discrete jumps) and vary only slowly with ground distance (e.g., centimetre variations in the displacements over tens or hundreds of metres). This may seem a restrictive condition, but it actually makes it perfectly suited to applications in geophysics and glaciology. The ability to measure centimetre-scale surface displacements has had a significant impact on these areas of study.

A further limitation on DifSAR is the extent of the ground displacement. On the one hand it must be large enough to be measured across a collection of pixels, but on the other it cannot be too extensive across the imaged scene. If it is, the problem becomes one of knowing the baseline to high precision. If there was a constant displacement of half a wavelength across the entire scene (say, 3cm at C-band), this would be indistinguishable from a shift in the baseline in the range direction of 3cm!

However, for most applications, the area coverage of the displacement is a proportion of the image area — the differential fringes can then be considered as relative ground movements. It is therefore usual to have some kind of ground control for these studies whereby the ground movement is known to high accuracy but in only a small number of discrete locations. The control points are then used to calibrate the measured fringes in the interferogram.

11.7.2 Atmospheric Water Vapour

In Chapter 9 we saw that one of the limitations on high accuracy altimetry was the influence of atmospheric water vapour on the velocity of the microwaves. In interferometry this slowing down of the waves has an effect on the measured phase of the signal. This is not a problem for single-pass interferometry since the phase shift is the same for both signals so that the interferometric phase difference remains the same regardless of the influence of water vapour. It is a problem for repeat-pass because the water vapour concentration may have changed during the time elapsed between passes. Fortunately, water vapour does not change significantly over short distances, so for most topographic retrieval using InSAR the effect is not a significant problem. However, for differential InSAR these small variations across a scene can have a large impact on the assessment of ground displacements. One precaution that can be taken is simply to make many repeat passes so that real trends in the ground can be seen above the more random fluctuations of the atmospheric effects.

11.8   Permanent Scatterer Interferometry

At the time of writing, one of the most interesting new developments in SAR interferometry has been the increasingly widespread application of the technique known as Permanent (or Persistent) Scattering Interferometry (PSI). This is a form of differential SAR interferometry. It also measures small ground displacements, but it does not require a continuous spatial area of high coherence magnitude to do the phase unwrapping, but instead uses a long time series to identify small differences in relative location across the scene. The key requirement is that the dominant scattering being observed originates from permanent scatterers — i.e. scatterers that persist throughout the time interval of the image acquisitions. In urban areas there may be many hundreds of permanent scatterers within a square kilometre, often including very effective corner reflectors at the right-angled boundaries between walls and the ground. Actual radar corner reflectors can also be deployed to provide detailed information for specific features, such as bridges, dams and pipelines, this technique often being referred to as CRInSAR.

The key limitation is that the phase difference between scene acquisitions must be less than 2π, so that there is no phase ambiguity. One can then imagine a similar scenario to figure 11.10, except instead of "distance in image" the x-axis would be labelled "time". The small phase differences observed in the time sequence for a single pixel are then unwrapped into a steady change over time, which can be converted to a component of the displacement, as discussed for differential interferometry.

Note that the phase difference observed for each individual pixel cannot be taken in isolation since many different factors change this absolute phase — the viewing geometry and the changes in atmospheric conditions, being the two most significant ones. Some reference locations must be used where the displacements are known (or there is confidence that there is no displacement). Figure 11.13 demonstrates the staggering level of sensitivity in this technique, whereby 30+ SAR scenes over central London have been combined over a 5 year period to measure relative ground displacements on a scale of millimetres per year! The image shows the total displacement over this period, but the actual data can show the temporal pattern in displacement for each individual pixel over the time period. The line of subsidence is associated with an extension to an underground rail tunnel, and is well within limits predicted by the engineers. The rising

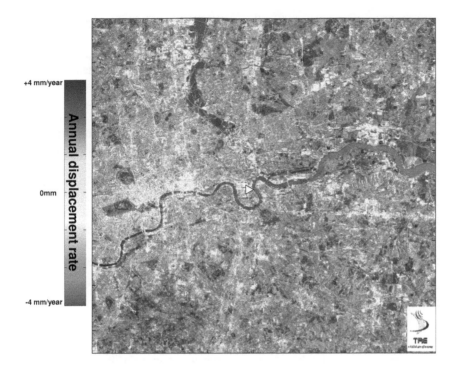

+4 mm/year

Annual displacement rate

0mm

-4 mm/year

FIGURE 11.13 This remarkable image shows the results of PSInSAR for central London over a period of 1995-2000. Note the red line to the left of the image, just below the river. This is subsidence due to the tunnelling of a new extension to the London underground. (Image is courtesy of TRE. ESA-ERS data were used for PS processing).

ground to the North of the river is thought to be the result of recent redevelopment that has seen a number of industries move out of this area to be replaced by commercial and residential buildings. When they were operational the industries extracted ground water, whereas now this water is left to saturate the soils. The result is a gradual swelling of the soils as the ground water returns to more natural levels.

There has been some success in applying this same technique in non-urban areas, but the character of a natural landscape is such that the surface is rarely very permanent so that you might expect tens rather than hundreds of permanent scatterers per square kilometre. Natural permanent scatterers are most likely to be associated with bare rock, so that you might expect most success in areas of minimal vegetation cover and large exposed areas of rock (rather than soil or sand as this is not permanent).

11.9 Along-Track Interferometry

Finally, it is worthwhile saying something about along track interferometry (ATI). The approach of ATI is that the antennas are separated in the along-track direction — i.e. the baseline is parallel to the flight path. This is an important tool in the analysis of water surface currents and the identification of moving targets. It was discussed earlier that repeat-pass interferometry cannot be used over water because water decorrelates so quickly. We have also seen, though, that repeat-pass interferometry can detect very small surface changes (Figure 11.12). A fast moving target, or the surface of flowing water, has the same geometry as Figure 11.12 but on a very short timescale, so we could imagine getting useful interferometric information if our temporal baseline was in the order of a few milliseconds — within this timescale the surface would not have moved very far, so that differential interferometric techniques could be applied to retrieve the rate of change of the surface. This is effectively the nature of ATI: a repeat pass configuration using two antennas that are following the same flight path (zero spatial baseline) but taking images within milliseconds of each other. It is not referred to as repeat-pass because its actually a single-pass system, but conceptually it is acting like a repeat pass interferometry with an extremely short temporal baseline.

11.10   Further Reading

Interferometry as a topic does not usually get very good coverage in introductory text books. You are better to attempt the introductory parts of books such as Hanssen (2001). Another recommendation is Henderson (1998) or for a detailed, but concise, summary, try Bamler and Hartl (1998).

APPENDIX A
Summary of Useful Mathematics

A.1 Angles

You will come across two units of measurement for linear angles: degrees and radians. We also want to consider angles in two dimensions, and for that we use *solid angles*.

A.1.1 Degrees

By definition there are 360 degrees in one full revolution (i.e. within a circle). You might well ask why 360? Well, why not — you need to define an angle somehow, and 360 is a useful number because it divides easily by 2, 3, 4, 5, 6, 8, 9, 10, 12, etc. Before pocket calculators such considerations were very important. Degrees can be given in decimal degrees or as degrees/minutes/seconds (common in latitude/longitude definitions).

A.1.2 Radians

Radians derive from an alternative definition of a unit of angle. Whereas degrees are very useful in a practical sense (applied mathematics) for calculating actual values, radians are generally more useful in abstract mathematics such as algebra. The reason for this is that we now define the angular unit not as some arbitrary fraction of a full circle, but as a proportional measure based on the radius of the circle. If you draw a circle with radius r, then measure out a distance r along the circumference to form a sector of the circle, then one radian is defined as the angle of that sector at the centre of the circle. Since the equation for the total circumference of a circle is $C = 2\pi r$, there must be 2π "r" along the circumference. The angle of one full revolution (circle) is therefore equal to 2π radians.

   Note that radians are unitless in the sense that they are a ratio of two distances.

   Radians are commonly the default we use when using most programming languages — if you wish to use degrees you must convert between radians and degrees.

A.1.3   Steradian (solid angle)

Solid angles can be thought of as "2-D angles". They are carved out volumes in space, rather than areas and so form cones, rather than triangles. We can define the solid angle equivalent of a radian by defining the *steradian* (sr) as a unit of solid angle subtended by an area of r^2 at the centre of a sphere of radius, r. To put it another way, if you paint a circle of area r^2 on the surface of a sphere of radius, r, then the cone this area forms with the centre of the circle corresponds to a solid angle of one steradian. Since the surface area of such a sphere is $4\pi r^2$, then the solid angle of a entire sphere is 4π and for a hemisphere it must be 2π.

A.2   Some Useful Trigonometric Relations

$$
\begin{aligned}
\sin^2 a + \cos^2 a &= 1 \\
\sin 2a &= 2\sin a \cos a \\
\cos 2a &= \cos^2 a - \sin^2 a = 2\cos^2 a - 1 \\
\sin \tfrac{1}{2}a &= \sqrt{\frac{1 - \cos a}{2}} \\
\cos \tfrac{1}{2}a &= \sqrt{\frac{1 + \cos a}{2}} \\
\sin(a \pm b) &= \sin a \cos b \pm \cos a \sin b \\
\cos(a \pm b) &= \cos a \cos b \mp \sin a \sin b \\
\sin a + \sin b &= 2\sin \tfrac{1}{2}(a+b)\cos \tfrac{1}{2}(a-b) \\
\cos a + \cos b &= 2\cos \tfrac{1}{2}(a+b)\cos \tfrac{1}{2}(a-b) \\
\sin(-a) &= -\sin a \\
\cos(-a) &= \cos a \\
\sin(a \pm \tfrac{\pi}{2}) &= \pm\cos a \\
\cos(a \pm \tfrac{\pi}{2}) &= \mp\sin a
\end{aligned}
$$

A.3   Logs and Exponentials

Logarithms are a convenient way to express relationships to do with power terms. The "logarithm to the base 10", say of a number x, is the number you have to put 10 "to the power of", in order to get x. For example, $\log_{10}(100)$ is 2, since to get 100 you must put 10 to the power of 2. Similarly, $\log_{10}(1000) = 3$, $\log_{10}(10000) = 4$, and so on. Note that the logged terms are going up as powers of ten (100, 1000, 10000,..) but the logarithms are going up in linear steps (1, 2, 3, 4...) — that is the value of using logarithms to characterise values that range over many orders of magnitude.

A.3.4   Some Fundamental Properties

$$\ln(ab) = \ln a + \ln b, \qquad (-\pi < \arg a + \arg b \leq \pi)$$

$$\ln\left(\frac{a}{b}\right) = \ln a - \ln b, \qquad (-\pi < \arg a + \arg b \leq \pi)$$

$$\ln(a^n) = n \ln a, \qquad (n \text{ integer}, -\pi < n \arg a \leq \pi)$$

$$
\begin{aligned}
10^x &= (e.10/e)^x \\
\log_{10} LHS &= x \\
\ln(RHS) &= x(\ln(e) + \ln(10) - \ln(e)) \\
&= x \ln(10)
\end{aligned}
$$

$$\ln(\exp z) = z$$

$$\exp(\ln z) = z$$

$$\frac{d}{dx} e^x = e^x$$

$$e^a e^b = e^{(a+b)}$$

$$(e^a)^b = e^{ab}$$

A.3.5   Special values

$e = 2.71828...$ $\ln 1 = 0$
$\ln 0 = -\infty$ $\ln(-1) = \pi i$
$\ln(\pm i) = \pm\frac{1}{2}\pi i$ $\ln e = 1$
$e^0 = 1$ $e^\infty = \infty$
$e^{-\infty} = 0$ $e^{\pm \pi i} - 1$
$e^{\pm\frac{\pi i}{2}} = \pm i$ $\int_1^e \frac{dt}{t} = 1$

$$e^{2\pi ki} = 1, \qquad (k \text{ any integer})$$

A.3.6   Series Expansions

$$\ln(1+x) = x - \frac{1}{2}x^2 + \frac{1}{3}x^3 - \ldots, \qquad (|x| \leq 1 \text{ and } z \neq -1)$$

$$\ln(x) = \left(\frac{x-1}{x}\right) - \frac{1}{2}\left(\frac{x-1}{x}\right)^2 + \frac{1}{3}\left(\frac{x-1}{x}\right)^3 - \ldots, \qquad (Rx \geq \frac{1}{2})$$

$$e^z = \exp z = 1 + \frac{z}{1!} + \frac{z^2}{2!} + \frac{z^3}{3!} + \ldots, \qquad (z = x + iy)$$

A.4   Complex Numbers

The complex-number representation of waves offers an alternative description which is mathematically simpler to work with. In fact, the complex exponential form of the wave equation is used extensively in microwave remote sensing, as well as in optics, and both classical and quantum mechanics. Additionally, many of the unique features of SAR images are far easier to understand once you get the hang of the complex notation.

The complex number z has the form

$$z = x + iy,$$

where $i = \sqrt{-1}$. Note that sometimes the symbol j is used instead of i, often with the former being used by electrical engineers and the latter by physicists and mathematicians. Since all these disciplines are involved with microwave remote sensing you are likely to come across both. In conventional mathematics, the value of i is not defined — you simply have to imagine what it is like, and hence the term "imaginary" number. All imaginary numbers can then be represented as a real number multiplied by i. When we combine both real and imaginary numbers, then we call it a "complex" number.

The real and imaginary part of z, above, are respectively x and y, where both x and y are themselves real numbers. We can illustrate this graphically using what is known as an *Argand* diagram, an example of which is shown in Figure A.1. The axes of an Argand diagram are the real and imaginary components. In fact, if you need justification for using complex numbers, you can simply imagine (!) them as being a convenient way to deal with two orthogonal coordinates. The x and y terms of a Cartesian coordinates system are entirely independent, and so are the real and imaginary components of a complex number. We can then use a single complex number to represent two coordinates.

An even more convenient method of representing these values can be found if we consider the z in terms of polar coordinates (r, θ), with r being the length of the line in Figure A.1 and θ is the angle it makes with the real (x)-axis. We then have the two components,

$$x = r\cos\theta, \qquad y = r\sin\theta$$

and

$$z = x + iy = r(\cos\theta + i\sin\theta).$$

The connection to Euler's constant, e, comes via "Euler's formula" giving the special relationship that,

$$e^{i\theta} = \cos\theta + i\sin\theta.$$

For those of you not mathematically inclined, you will have to take this expression as "given". Otherwise, you need a little knowledge of the calculus, so that you can take the differential, with respect to θ, of $z = \cos\theta + i\sin\theta$, where $r = 1$. This yields $dz/d\theta = iz$, which after integration (which should give you what you started with) gives $z = e^{i\theta}$.

The value of such a relationship is that we can now write

$$z = re^{i\theta} = r\cos\theta + ir\sin\theta,$$

where r is said to be the *magnitude* of z, and θ is the *phase angle* of z, in radians. The magnitude is usually denoted by $|z|$ and referred to as the *modulus* or *absolute value* of the complex number — in wave terminology it is called the *amplitude*.

The *complex conjugate*, indicated by an asterisk, is found by replacing i wherever it appears, with $-i$, so that

$$\begin{aligned} z^* &= (x + iy)^* = (x - iy) \\ z^* &= r(\cos\theta - i\sin\theta) \end{aligned}$$

and

$$z^* = re^{-i\theta}.$$

The operations of addition and subtraction of complex numbers is just like real numbers, except that you never mix real and imaginary numbers — they never combine together:

$$z_1 \pm z_2 = (x_1 + iy_1) \pm (x_2 + iy_2)$$

and therefore

$$z_1 \pm z_2 = (x_1 \pm x_2) + i(y_1 \pm y_2).$$

Notice that this process is very much like the component addition of vectors.

Multiplication and division are more easily expressed in polar form, so that

$$z_1 z_2 = r_1 r_2 e^{i(\theta_1 + \theta_2)}$$

and

$$\frac{z_1}{z_2} = \frac{r_1}{r_2} e^{i(\theta_1 - \theta_2)}.$$

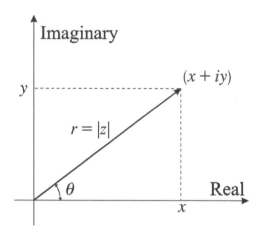

FIGURE A.1 Argand diagram showing the complex notation geometrically. The real and imaginary components are treated separately, in the same manner as vector components.

The important thing to remember when doing multiplication of complex number is that $i \times i = -1$, by definition.

It should be apparent here that the use of the r and θ provide a much neater way of dealing with complex number operations.

Now, there are some useful relationships in complex mathematics that are of value in wave calculations. From the ordinary trigonometric addition formulas, for instance, we find that

$$e^{i(\theta_1 + \theta_2)} = e^{i\theta_1} e^{i\theta_2}.$$

The modulus of a complex quantity is given by

$$\begin{aligned} |z| &= \sqrt{zz^*} \\ &= \sqrt{x^2 + y^2}. \end{aligned}$$

By remembering that $\cos 2\pi = 1$ and $\sin 2\pi = 0$, we have

$$e^{i2\pi} = 1.$$

Similarly

$$e^{i\pi} = e^{-i\pi} = -1, \quad e^{\theta + i2\pi} = e^{\theta} e^{i2\pi} = e^{\theta} \quad \text{and} \quad e^{\pm i\pi/2} = \pm i.$$

Any complex number can be represented as the sum of a real part $\mathrm{Re}(z)$ and an imaginary part $\mathrm{Im}(z)$

$$z = \mathrm{Re}(z) + i\,\mathrm{Im}(z).$$

A.5 Vectors

Complex numbers might be helpful for dealing with two independent variables at the same time, but vectors allow you to handle 2 or more parameters at the same time. The most obvious example is when we want to deal with sets of 3-D or even 4-D coordinates (such as x, y, z and t). With vectors we have a convenient way of expressing a set of such coordinates by putting them in a column, such that:

$$\mathbf{p} = \begin{pmatrix} x \\ y \\ z \\ t \end{pmatrix}.$$

The convention is to use lower case letters for vectors, but they should be underlined or in bold to distinguish them from scalar variables.

Colloquially, we consider vector quantities to have direction *as well as* a magnitude, which is what separates them from a scalar quantity, but this notion can be difficult to visualise if we go to 4 or more dimensions and besides, vectors are much more powerful than simply incorporating direction. The power of vectors is that they can be dealt with in special ways that allows very time consuming algebra to be done very quickly. It should also be clear that there is a direct relationship to "arrays" in the context of computer programming.

The basis of using vectors is the special vector algebra that defines addition and multiplication. In the following summary, we consider 3-element vectors defined by

$$\mathbf{a} = a_1\mathbf{i} + a_2\mathbf{j} + a_3\mathbf{k}$$

where a_1, a_2 and a_3 correspond to the scalar lengths along the orthogonal set of axes defined by unit-length vectors, $\mathbf{i}$, $\mathbf{j}$ and $\mathbf{k}$.

A.5.7 Law of Vector Algebra

Commutative law for addition: $\mathbf{a} + \mathbf{b} = \mathbf{b} + \mathbf{a}$
  Associative law for addition: $\mathbf{a} + (\mathbf{b} + \mathbf{c}) = \mathbf{a} + \mathbf{b} + \mathbf{c}$
  Associative law for scalar multiplication: $m(n\mathbf{a}) = (mn)\mathbf{a} = n(m)\mathbf{a}$
  Distributive law: $(m + n)\mathbf{a} = m\mathbf{a} + n\mathbf{a}$
  Distributive law: $m(\mathbf{a} + \mathbf{b}) = m\mathbf{a} + m\mathbf{b}$

A.5.8 Cross or Vector Product

$$\mathbf{a} \times \mathbf{b} = ab\sin\theta\,\mathbf{u}, \qquad 0 \leqq \theta \leqq \pi$$

where θ is the angle between $\mathbf{a}$ and $\mathbf{b}$, and $\mathbf{u}$ is a unit vector perpendicular to the plane of $\mathbf{a}$ and $\mathbf{b}$ (such that $\mathbf{a}$, $\mathbf{b}$, $\mathbf{u}$ form a *right-handed system*).

We can determine the vector elements of $\mathbf{a} \times \mathbf{b}$ through the following expression:

$$\mathbf{a} \times \mathbf{b} = \begin{vmatrix} \mathbf{i} & \mathbf{j} & \mathbf{k} \\ a_1 & a_2 & a_3 \\ b_1 & b_2 & b_3 \end{vmatrix} = (a_2 b_3 - a_3 b_2)\mathbf{i} + (a_3 b_1 - a_1 b_3)\mathbf{j} + (a_1 b_2 - a_2 b_1)\mathbf{k}.$$

Note also that:

$$\mathbf{a} \times \mathbf{b} = -\mathbf{b} \times \mathbf{a}$$

and

$$\mathbf{a} \times (\mathbf{b} + \mathbf{c}) = \mathbf{a} \times \mathbf{b} + \mathbf{a} \times \mathbf{c}.$$

A.6   Matrices

Matrices can be considered as multidimensional vectors. More exactly it should be thought of as the other way around: vectors are a special case of a matrix, in that they are a matrix with a single dimension (a "column matrix"). In this book we deal almost exclusively with 2-D matrices, a general description of which is given by

$$\mathbf{M} = \begin{pmatrix} m_{11} & m_{12} & \cdots & m_{1j} \\ m_{21} & m_{22} & \cdots & m_{2j} \\ \vdots & \vdots & & \vdots \\ m_{i1} & m_{i2} & \cdots & m_{ij} \end{pmatrix}.$$

The subscripts ij refer to the row and column identifier for each element. Note that the convention for matrices is to use upper case letters in bold (or double underlined).

The elements with $i = j$ are the diagonal elements, a term appropriate only for square matrices. The sum of the diagonal elements of a square matrix is the *trace*, such that:

$$\text{trace}(\mathbf{M}) = \sum_{i=0}^{n} a_{ii}.$$

Note that sometimes you might find this incorrectly referred to as the *span* of the matrix. If these elements equal unity and all the other elements equal zero, we have the special matrix known as the *identity* matrix, denoted by $\mathbf{I}_n$ (the subscript n representing the number of elements), for example:

$$\mathbf{I}_4 = \begin{pmatrix} 1 & 0 & 0 & 0 \\ 0 & 1 & 0 & 0 \\ 0 & 0 & 1 & 0 \\ 0 & 0 & 0 & 1 \end{pmatrix}.$$

In dealing with matrices, it is sometimes advantageous to devise several operations that generate new matrices from a given starting matrix, M. In the algebra of real numbers we can transform a variable such as x with operations such as x^{-1} to generate a new number. In complex algebra, we can perform a complex conjugate z^* of a complex number z, and again generating one number from another. In developing operations for matrices, we should first realise that matrices can also have elements that are complex numbers — such matrices are particularly useful in the analysis of polarimetric data.

Three common ways of generating new matrices from a starting matrix M are:

Definition 1 *Complex conjugate matrix:*

$$\mathbf{M}^* \qquad [\mathbf{M}^*]_{ij} = M_{ij}^*$$

So that each complex element in M is transformed into its complex conjugate.

Definition 2 *Transpose matrix:*

$$\mathbf{M}^T \qquad [\mathbf{M}^T]_{ij} = M_{ji}$$

In this case, the rows and columns are interchanged, so that if M is a column matrix (i.e. a vector),

$$\begin{pmatrix} a \\ b \\ c \\ \vdots \end{pmatrix}^T = \begin{pmatrix} a & b & c & \cdots \end{pmatrix}$$

or for a 2-D matrix,

$$\begin{pmatrix} a & b & c \\ d & e & f \\ g & h & i \end{pmatrix}^T = \begin{pmatrix} a & d & g \\ b & e & h \\ c & f & i \end{pmatrix}.$$

Definition 3 *Hermitian conjugate matrix:*

$$\mathbf{M}^\dagger \qquad [\mathbf{M}^\dagger]_{ij} = M_{ji}^* = [\mathbf{M}^T]_{ij}^*$$

The asterix denotes complex conjugation.

The Hermitian conjugate M equals $\mathbf{M}^T$ if M has only real elements.

In some special cases, it may be that one or more of the matrices derived above are identical to the original matrix M, or some simple version of it ($-M$ for instance). This possibility leads to the following definitions of special kinds of matrix (noting that all but the first must be a square matrix):

Definition 4 *Real matrix:*

$$\mathbf{M}^* = \mathbf{M}$$

Definition 5 *Symmetric matrix:*

$$\mathbf{M}^T = \mathbf{M}$$

Definition 6   *Hermitian matrix*
:

$$\mathbf{M}^\dagger = \mathbf{M}$$

Definition 7   *Antisymmetric matrix:*

$$\mathbf{M}^T = -\mathbf{M}$$

Definition 8   *Unitary matrix:*

$$\mathbf{M}^\dagger = \mathbf{M}^{-1} \qquad (\text{i.e., } \mathbf{M}\mathbf{M}^\dagger = \mathbf{M}^\dagger\mathbf{M} = \mathbf{I})$$

A.6.9   Matrix Algebra

There are also some important things to realise in matrix algebra, the most important thing to remember is that the order of matrix multiplication is important. The term $\mathbf{AB}$ means that matrix $\mathbf{B}$ is pre-multiplied by $\mathbf{A}$, or $\mathbf{A}$ is post-multiplied by $\mathbf{B}$. Note, however, that $\mathbf{AB} \neq \mathbf{BA}$. so matrices are not commutative in multiplication. Matrix multiplication is associative, though, so that $\mathbf{A}(\mathbf{BC}) = (\mathbf{AB})\mathbf{C}$.

Some other important rules in matrix algebra are as follows:

$$\begin{aligned}
(\mathbf{AB})^{-1} &= \mathbf{B}^{-1}\mathbf{A}^{-1}, \\
(\mathbf{AB})^{T} &= \mathbf{B}^{T}\mathbf{A}^{T}, \\
(\mathbf{AB})^{\dagger} &= \mathbf{B}^{\dagger}\mathbf{A}^{\dagger}.
\end{aligned}$$

bibliography

Allen, J. and Neely, S.: 1992, Micromechanical models of the cochlea, *Physics Today* .

Bakewell, S.: 2000, It's alive, *Fortean Times* (139), 34–39.

Bamler, R. and Hartl, P.: 1998, Synthetic aperture radar interferometry, *Inverse Problems* **14**, R1–R54.

Barath, F. et al.: 1993, The Upper Atmosphere Research Satellite Microwave Limb Sounder Instrument, *Journal of Geophysical Research* **98**(D6), 10751–10762.

Cloude, S. and Pottier, E.: 1996, A review of target decomposition theorems in radar polarimetry, *IEEE Transactions on Geoscience and Remote Sensing* **34**(2), 498–518.

Cloude, S. and Pottier, E.: 1997, An entropy based classification scheme for land applications of polarimetric SAR, *IEEE Transactions on Geoscience and Remote Sensing* **35**(1), 68–78.

Curlander, J. and McDonough, R.: 1991, *Synthetic Aperture Radar: Systems and Signal Processing*, Wiley – Interscience, New York.

Elachi, C.: 1987, *Introduction to the physics and techniques of remote sensing*, John Wiley and Sons.

Elachi, C.: 1988, *Spaceborne radar remote sensing: applications and techniques*, IEEE Press.

Feynman, R. P., Leighton, R. B. and Sands, M.: 1963, *The Feynman Lectures on Physics*, Vol. I: Mainly Mechanics, Radiation and Heat, Addison-Wesley, Reading, MA.

Freeman, A.: 1992, SAR calibration: An overview, *IEEE Transactions on Geoscience and Remote Sensing* **30**(6), 1107–1121.

Freeman, A. and Durden, S.: 1998, A three-component scattering model for polarimetric SAR data, *IEEE Transactions on Geoscience and Remote Sensing* **36**(3), 963–973.

Fung, A. K.: 1994, *Microwave Scattering and Emission Models and Their Applications*, Artech House, Norwood, MA.

Gernsback, H.: 2000, *Ralph 124C 41+: a romance in the year 2660*, University of Nebraska Press, Lincoln, NE.

Griffin, D.: 1957, More about bat radar, *Scientific American* (199), 40–44.

Gurney, R., Foster, J. and Parkinson, C. (eds): 1993, *Atlas of satellite observations related to global change*, Cambridge University Press, Cambridge.

Hanssen, R.: 2001, *Radar interferometry : data interpretation and error analysis*, Kluwer Academic, Dordrecht; London.

Hecht, E.: 2001, *Optics*, Addison-Wesley.

Henderson, F. and Lewis, A. (eds): 1998, *Principles and Applications of Imaging Radar*, Wiley, New York.

Houghton, J., Taylor, F. and Rodgers, C.: 1984, *Remote Sounding of Atmospheres*, Cambridge University Press, Cambridge.

Janssen, M. (ed.): 1993, *Atmospheric Remote Sensing by Microwave Radiometry*, Wiley and Sons, New York.

Leberl, F.: 1990, *Radargrammetric Image Processing*, Artech House, Norwood, MA.

Masters, W., Moffat, A. and Simmons, J.: 1985, Sonar tracking of horizontally moving targets by the big brown bat eptesicus fuscus, *Science* (228), 1331–1333.

Metzner, W.: 1991, Echolocation behaviour in bats, *Science Progress* (75), 453–465.

Middlebrooks, J. and Green, D.: 1991, Sound localization by human listeners, *Annual Reviews in Psychology* **42**, 135–159.

Oliver, C. and Quegan, S.: 1998, *Understanding Synthetic Aperture Radar Images*, Artech House, Boston and London.

Peckham, G.: 1991, Instrumentation and measurement in atmospheric remote sensing, *Reports on Progress in Physics* **54**, 531–577.

Pope, K., Rey-Benayas, J. and Paris, J.: 1994, Radar remote sensing of forest and wetland ecosystems in the Central American tropics, *Remote Sensing of Environment* **48**, 205–219.

Preissner, J.: 1978, The influence of the atmosphere on passive radiometric measurements, *Proceedings of the Symposium on Millimeter and Submillimeter Wave Propagation Circuits: AGARD Conference*, Vol. 245, pp. 1Ű–13.

Rodgers, C. D.: 1976, Retrieval of atmospheric temperature and composition from remote measurements of thermal radiation, *Reviews of Geophysics and Space Physics* **14**(4), 609–624.

Sagan, C.: 1997, *The Demon-Haunted World*, Headline Book Publishing, London.

Schanda, E.: 1986, *Physical Fundamentals of Remote Sensing*, Springer-Verlag, Berlin.

Skolnik, M. I.: 2002, *Introduction to Radar Systems*, McGraw-Hill.

Skou, N.: 1989, *Microwave Radiometer Systems: Design and Analysis*, Artech House, Inc.

Svendsen, E. et al.: 1983, Norwegian remote sensing experiment: evaluation of the nimbus-7 scanning multichannel microwave radiometer for sea ice research, *Journal of Geophysical Research* (88), 2781–2791.

Twomey, S.: 1977, *Introduction to the mathematics of inversion in remote sensing and indirect measurements*, Developments in Geomathematics 3, Elsevier Scientific Publishing Company, Amsterdam.

Ulaby, F. and Elachi, C. (eds): 1990, *Radar polarimetry for geoscience applications*, Artech House, Norwood, MA.

Ulaby, F., Held, D., Dobson, M., McDonald, K. and Senior, T.: 1987, Relating polarization phase difference of SAR signals to scene properties, *IEEE Transactions on Geoscience and Remote Sensing* **25**(1), 83–92.

Ulaby, F., R.K.Moore and Fung, A.: 1981, *Microwave Remote Sensing: Active and Passive*, Vol. I: Fundamentals and radiometry, Addison-Wesley.

Ulaby, F., R.K.Moore and Fung, A.: 1982, *Microwave Remote Sensing: Active and Passive*, Vol. II: Radar remote sensing and surface scattering and emission theory, Addison-Wesley.

Ulaby, F., R.K.Moore and Fung, A.: 1986, *Microwave Remote Sensing: Active and Passive*, Vol. III: From theory to applications, Artech House.

van der Sanden, J.: 1997, *Radar remote sensing to support tropical rain forest management*, PhD thesis, Wageningen Agricultural University, Wageningen, The Netherlands.

van Oevelen, P.: 2000, *Estimation of areal soil water content*, PhD thesis, Wageningen Agricultural University, Wageningen, The Netherlands.

van Zyl, J.: 1989, Unsupervised classification of scattering behaviour using radar polarimetry data, *IEEE Transactions on Geoscience and Remote Sensing* **27**(1), 36–45.

van Zyl, J., Zebker, H. and Elachi, C.: 1987, Imaging radar polarization signatures: Theory and observation, *Radio Science* **22**(4), 529–543.

von Bekesy, G.: 1957, The ear, *Scientific American* (197), 66–78.

Wadhams, P.: 1994, Remote sensing of snow and ice and its relevance to climate change processes, *in* R. Vaughan and A. Cracknell (eds), *Remote Sensing and Global Climate Change*, Vol. 24 of *NATO ASI Series*, Springer, pp. 215–221.

Waters, J.: 1993, Microwave limb sounding, *in* M. Janssen (ed.), *Atmospheric Remote Sensing by Microwave Radiometry*, Wiley and Sons, New York, chapter 8.

Zebker, H., van Zyl, J. and Held, D.: 1987, Imaging radar polarimetry from wave synthesis, *Journal of Geophysical Research* **92**, 683–701.

Zebker, H. and Villasenor, J.: 1992, Decorrelation in interferometric radar echoes, *IEEE Transactions on Geoscience and Remote Sensing* **30**, 95–959.

index

measurement noise
    in forward model, 187
noise equivalent sigma
    nought, 168
noise equivalent temperature,
    168
phase noise filtering, 330
receiver temperature, 175
source, 167
speckle as noise, 289
thermal, 330
vector
    in forward model, 187
non-random surfaces, 127
normalised radar cross-section,
    108, 109, 118, 129, 131,
    168, 227
    from wind ripples, 131
    in image data, 288
    of a volume, 108, 118
    of mixed targets, 148
    of vegetation, 263
NRCS, *see* normalised radar
    cross-section

oceans, 129, 209
    altimetry, 233
    carbon dioxide uptake, 210
    gravity waves, 129
    gravity-capillary waves, 129,
        261
    imaging radar, 260
    microwave properties, 129
    salinity, 212
    sea foam, 130, 211
    sea surface temperature, *see*
        sea surface temperature
    thermal expansion, 210
    two-scale roughness model,
        130
    wave spectrum, 129
    wind speed, 213
Oersted, Hans Christian, 11
opacity, 95, 97, 114–116, 118,
    189, 208, 218, 263
    in the inverse model, 190
    of cloud ice layer, 202
optical depth, *see* opacity
optical scattering, *see* scattering
    regimes

orientation angle, 69, 85
oxygen, 182, 185
ozone, 99, 100, 183
    as a greenhouse gas, 183
    hole, 183
    ultravoilet absorption, 182

parallax, 302, 310
partial coherence, *see* coherence
partially polarimetric, 80
passive imaging
    polarimetry, 207
    viewing geometries, 207
Pauli basis, *see* basis
PDF, *see* probability distribution
    function
pedestal height, 88
penetration depth, 56
    definition, 55
period, of a wave
    definition, 26
periodic surfaces, 127
permeability
    magnetic, *see* magnetic per-
        meability
    of free space, 53
    relative, 54
permittivity
    of free space, 54
    relative, 54
phase change
    at a boundary, 123
    in corner reflectors, 146
photons, 24
Planck function, 58, 60, 95, 96,
    188, 212
Planck's constant, 58
Planck, Max, 57
Poincare sphere, 73, 77, 83–85
Poincare, Jules Henri, 73
pointing angle, *see* boresight an-
    gle
polarimeter
    definition, 82
    radar, 82
polarimetric basis, *see* basis
polarimetric decomposition, 90
polarimetric entropy, 90
polarimetric phase difference,
    90